Yuli K. Godovsky

Thermophysical Properties of Polymers

With 172 Figures and 25 Tables

Springer-Verlag
Berlin Heidelberg New York
London Paris Tokyo
Hong Kong Barcelona Budapest

Prof. Dr. Yuli K. Godovsky
Karpov Institute of Physical Chemistry
Ul. Obukha 10
103064 Moscow, USSR

ISBN 3-540-54160-8 Springer-Verlag Berlin Heidelberg New York
ISBN 0-387-54160-8 Springer-Verlag New York Berlin Heidelberg

© Springer-Verlag 1992
Printed in the United States of America

Typesetting: Springer-TEX-Inhouse-System
02/3020-543210 Printed on acid-free paper

Preface

Among various branches of polymer physics an important position is occupied by that vast area, which deals with the thermal behavior and thermal properties of polymers and which is normally called the thermal physics of polymers. Historically it began when the unusual thermo-mechanical behavior of natural rubber under stretching, which had been discovered by Gough at the very beginning of the last century, was studied 50 years later experimentally by Joule and theoretically by Lord Kelvin. This made it possible even at that time to distinguish polymers from other subjects of physical investigations. These investigation laid down the basic principles of solving the key problem of polymer physics – rubberlike elasticity – which was solved in the middle of our century by means of the statistical thermodynamics applied to chain molecules. At approximately the same time it was demonstrated, by using the methods of solid state physics, that the low temperature dependence of heat capacity and thermal expansivity of linear polymers should follow dependencies different from that characteristic of nonpolymeric solids. Finally, new ideas about the structure and morphology of polymers arised at the end of the 1950s stimulated the development of new thermal methods (differential scanning calorimetry, deformation calorimetry), which have become very powerful instruments for studying the nature of various states of polymers and the structural heterogeneity. Hence, the development of polymer physics is closely related to the development of ideas about the thermal behavior and the thermophysical properties of polymers.

This book summarizes the theoretical ideas and the main results of experimental investigations in the thermal physics of polymers. It consists of two parts. In the first part the main thermophysical properties of polymers, such as heat capacity, thermal conductivity and diffusivity, and thermal expansivity are considered. The second part deals with the thermal behavior of polymers under deformation and fracture. Both parts also include a brief consideration of the relevant experimental methods.

Consideration of the thermal properties and thermal behavior of polymers is based on general ideas of solid state physics and molecular physics. Special attention is paid both to the common features in the thermophysical properties of polymeric and nonpolymeric systems and to the difference in their thermal behavior resulting from the particular molecular and supermolecular structure of polymers. The main feature of molecular structure of polymers consists of large local anisotropy of the force field due to a sharp difference in the inter- and intramolecular interactions. Normally the interatomic forces in macromolecules are about two orders of magnitude larger than intermolecular ones and this difference is the source of the large anisotropy of the physical properties of polymers. Due to this difference, macromolecules keep their individuality in any polymer system. Under the action of orientation fields, the local anisotropy and individuality of macromolecules transform to the macroscopic anisotropy, which is especially characteristic of oriented polymers. Moreover, the local anisotropy and individuality of macromolecules lead to some specific temperature dependencies of heat capacity of solid polymers, the neagative thermal expansivity along the macromolecules in polymer lattices and some unusual thermophysical features.

Normally, mechanical deformation of polymers are accompanied by heat effects. The character of the heat effects resulting from the elastic (reversible) deformation of solid polymers can be established using the methods of thermoelasticity. However, remaining in the frames of phenomenological thermodynamics, its is impossible to display the molecular origin of the entropy and internal energy changes during deformation and establish the relationships between them. The answer to these questions may be obtained with the help of equations of state. Using some simple equations of state for solid polymers and molecular networks in this book, a generalized thermo-mechanical treatment of rubberlike elastic deformations of polymer networks and reversible deformations of solid polymers is given. This approach is used consistently for the description of the reversible thermo-mechanical behavior of various polymers and polymer materials. An attempt was made to connect this thermo-mechanical behavior with molecular and structural changes resulting from the deformations.

One of the striking features of solid polymers consists of their ability to be transformed into the oriented (drawn) state. In contrast to the reversible deformations the plastic deformation, cold drawing and orientation of solid polymers is always accompanied with the dissipation of energy. The main problem here is the relationship between the energy dissipated during the plastic deformation and cold drawing and the energy absorbed during this process, as well as the connection of the latent energy with the molecular and structural changes resulting from the cold drawing and orientation. Finally, fracture of polymers is inevitably accompanied by thermal effects due to local plastic deformation and rupture of macromolecules. Both these sources are localized in cracks and this location may lead to enormous local overheating in the vicinity of the growing cracks. All these problems are considered in the second part of this book.

This book is a revised and extented version of the original Russian book "Thermal Physics of Polymers" containing updated information and new material.

Yuli K. Godovsky

Table of Contents

Preface

List of Abbreviations

Part One: Thermal Properties of Polymers

1   Heat Capacity 3
1.1   Basic Concepts of the Heat Capacity of Solids 3
1.2   Temperature Dependence of Heat Capacity of Polymers ... 6
   1.2.1   Theoretical Frameworks 7
   1.2.2   Experimental Results and Comparison with
          Theoretical Concepts 17
1.3   Temperature Transitions and Heat Capacity 28
   1.3.1   Amorphous Polymers 28
   1.3.2   Crystalline Polymers 36
   References 40

2   Thermal Conductivity 43
2.1   Basic Concepts of Thermal Conductivity of Solids 43
2.2   Temperature Dependence of Thermal Conductivity of
     Polymers 45
   2.2.1   Amorphous Polymers 46
   2.2.2   Crystalline Polymers 49
2.3   Effect of Molecular Parameters 56
2.4   Anisotropy of Thermal Conductivity – Effect of
     Orientation 59
2.5   Effect of Radiation 65
2.6   Effect of Pressure 66
2.7   Thermal Conductivity of Filled Polymer Materials 67
2.8   Thermal (Temperature) Diffusivity 70
   References 72

3   Thermal Expansion 75
3.1   Basic Concepts of Thermal Expansion of Solids 75
3.2   Equation of State of Polymers 76
3.3   Gruneisen Parameters of Polymers 79
3.4   Thermal Expansion of Polymeric Crystals 84

3.5   Thermal Expansion of Drawn Polymers 94
      3.5.1   Negative Thermal Expansion of Drawn Crystalline
             Polymers 94
      3.5.2   Anisotropy of Thermal Expansion 98
3.6   Thermal Expansion of Filled Polymers and Polymer–
      Matrix Composites 101
      References 105

4   Experimental Methods and Instrumentation 107
4.1   Heat Capacity 107
      4.1.1   Adiabatic Calorimetry 107
      4.1.2   Differential Scanning Calorimetry (DSC) 109
4.2   Thermal Conductivity and Diffusivity 110
      4.2.1   Steady-State Methods 112
      4.2.2   Unsteady-State Methods 114
4.3   Thermal Expansion 121
      References 122

**Part Two: Thermal Behavior of Polymers
           under Mechanical Deformation and Fracture**

5   Thermomechanics of Glassy and Crystalline Polymers .. 127
5.1   Phenomenological Aspects of Thermomechanics of Elastic
      Materials 128
5.2   Linear Thermomechanics of Quasi-Isotropic Hookean
      Solids 130
      5.2.1   Uniform (Volume) Dilatation and Compression 130
      5.2.2   Simple Elongation and Compression 132
      5.2.3   Shear Torsion 134
5.3   The Thermoelastic Effect in Glassy and Crystalline
      Polymers 134
5.4   Thermomechanics of the Undrawn Glassy and Crystalline
      Polymers 137
5.5   Thermomechanics of Drawn Polymers 141
      5.5.1   Drawn Amorphous Polymers 141
      5.5.2   Drawn Crystalline Polymers 143
5.6   Microphase-Separated Block Copolymers with a Solid
      Matrix 158
5.7   Filled Solid Polymers 160
5.8   Biopolymers 160
      References 161

**6   Thermomechanics of Molecular Networks and
    Rubberlike Materials 163**
6.1   Thermomechanics of Molecular Networks (Theory) 163
      6.1.1   Thermomechanics of Gaussian Networks 163

6.1.2   Thermomechanics of Non-Gaussian Networks 170
6.1.3   Phenomenological Equations of State 171
6.1.4   Thermoelasticity of Liquid Crystalline Networks .. 174
6.2   Thermomechanical Behavior of Molecular Networks 175
   6.2.1   Entropy and Energy Effects at Small and
          Moderate Deformations 175
   6.2.2   Thermomechanics at Large Deformations 183
   6.2.3   The Thermoelasticity of Mesophase Networks 186
6.3   Thermomechanical Behavior of Rubberlike Materials 189
   6.3.1   Stress Softening: Thermomechanics and
          Mechanism 189
   6.3.2   Energy Contribution 192
   6.3.3   Mesophase Block Copolymers 198
   6.3.4   Random Copolymers 200
   6.3.5   Elastomer Blends 201
   6.3.6   Bioelastomer Materials 201

**Appendix Thermomechanics of the New Models
        of Rubber Elasticity** 203
References 206

**7   Thermodynamic Behavior of Solid Polymers in Plastic
    Deformation and Cold Drawing** 211
7.1   Temperature Effects During Plastic Deformation and
     Cold Drawing of Glassy and Crystalline Polymers 212
   7.1.1   Uniform Neck Propagation 214
   7.1.2   Self-Oscillated Neck Propagation 221
7.2   Thermodynamics of Plastic Deformation and Cold
     Drawing of Glassy and Crystalline Polymers 225
   7.2.1   Plastic Deformation at Uniaxial Extension (Cold
          Drawing) 225
   7.2.2   Plastic Deformation at Simple Compression 231
7.3   Thermal Behavior of Cold Drawn and Plastically
     Deformed Glassy and Crystalline Polymers and the
     Nature of the Stored Energy 233
   7.3.1   Amorphous Polymers 236
   7.3.2   Crystalline Polymers 241
7.4   Thermomechanical Behavior of Hard Elastic Fibers and
     Films 244
   References 246

8   Thermal Behavior of Solid Polymers Under Fracture 249
8.1   Thermal and Temperature Effects Resulting from the
     Formation and Growth of Cracks in Solid Polymers 249
   8.1.1   Estimation of Temperature Rise at the tip of
          Propagating Cracks 252

8.1.2   The Local Thermal Processes During the Crack
Growth in Solid Polymers 254
8.2   Energetics of Chain Rupture in Stressed Polymers 259
8.2.1   Energetics of Elementary Scissions of Stressed
Macromolecules 259
8.2.2   Experimental Data on the Energetic Effects
Resulting from the Degradation of Macromolecules
in Stressed Polymers 263
8.3   Self-Heating During Cyclic Deformation and Thermal
Fatigue Failure 268
References 270

9   **Experimental Methods and Instrumentation** 271
9.1   Measurements of the Temperature Changes During
Deformation 271
9.1.1   Elastic Deformation (Isoentropic Measurements) .. 272
9.1.2   Temperature Rise in the Necking of Polymers 276
9.1.3   Temperature Rise in the Zone of Rupture of
Polymers 278
9.2   Temperature Dependence of Stresses (Isometric
Measurements) 282
9.3   Deformation Calorimetry (Isothermal Measurements) 284
9.3.1   Gas Calorimeters 284
9.3.2   Heat-Conducted Deformation Calorimeters
(Tian–Calvet Type) 288
9.4   Calorimetric Methods for Investigating the Energy State
of Deformed Polymers 291
9.4.1   Dissolution Calorimetry 292
9.4.2   DSC Measurements 292
References 295

Subject Index 297

List of Abbreviations

Polymer Names

EPR	Ethylene-propylene rubber
NBR	Acrylonitrile-butadiene rubber
NR	Natural rubber
PA	Polyamide
PAr	Polyarylate
PBA	Poly(buthyl acrylate)
PBR	Polybutadiene rubber
PBT	Poly(buthylene terephtalate)
PC	Polycarbonate
PCR = PCP	Polychloroprene rubber
PCTFE	Poly(chlorotrifluoroethylene)
PDES	Poly(diethylsiloxane)
PDMS	Poly(dimethylsiloxane)
PE	Polyethylene
LDPE	Low density polyethylene
HDPE	High density polyethylene
LLDPE	Linear low density polyethylene
UHMWPE	Ultra high molecular weight polyethylene
PEMA	Poly(ethyl methacrylate)
PEO = POE	Polyethylene oxide
PET	Poly(ethylene terephtalate)
PIB	Polyisobuthylene
PMMA	Poly(methyl methacrylate)
P4MP1 = PMP	Poly(4-methylpentene-1)
α-PMS	Poly(α-methylstyrene)
POM	Polyoximethylene = Polymethylene oxide
PP = i-PP	Polypropylene, isotactic polypropylene
PPO	Poly(phenylene oxide), Poly(propylene oxide)
PS	Polystyrene
PSu	Polysulfone
PTFE	Poly(tetrafluoroethylene)
PU	Polyurethane
PVA	Poly(vinylacetate)
PVAl	Poly(vinyl alcohol)

PVC	Poly(vinyl chloride)
PVF_2	Poly(vinylidene fluoride)
SBR	Styrene-butadiene rubber
SBS	Styrene-butadiene block copolymers

Abbreviations not Referring to Polymer Names

DSC	Differential scanning calorimetry
DTA	Differential thermal analysis
HB model	Hosemann–Bonart model
HSH model	Hess–Statton–Hearl model
IR	Infrared
PP model	Peterlin–Prevorsek model
SAXS	Small angle X-ray scattering
WAXS	Wide angle X-ray scattering

List of Main Symbols

a	Thermal diffusivity
A_0	Radius of unstrained sample
a, a_1, a_2	Parameters of van der Waals equation of state of network
c	Sound velocity
c	Rate of crack propagation
d	Fractal dimension
e	Uniform (volume) strain
f	Strength of bonds
f_{th}	Theoretical strength of bonds
f	Retractive force
f_s	Entropy component of retractive force
f_u	Energy component of retractive force
f	Fraction of the flexible bonds
f_c	Function of orientation
h	Planck's constant
k_B	Boltzmann's constant
k	Wave vector
l	Length of cracks
$\bar{l}$	Phonon mean free path
m	Continuum dimension
r	End-to-end distance of network chains
r_{max}	Maximum extension of network chains
$\langle r^2 \rangle$	Mean square end-to-end distance of network chains in undeformed sample
$\langle r^2 \rangle_0$	Mean square end-to-end distance of unperturbed chains
$\langle r^2 \rangle_i$	Mean square end-to-end distance of network chains in the reference state
w	degree of crystallinity
x_i	generalized coordinate
A_0	Cross-sectional area of sample
C, C_v, C_p	Heat capacity
C	Constant in the Gaussian equation of state for rubber elasticity
C_1, C_2	Constants in the Mooney–Rivlin equation
D, D_m, $\bar{D}_m$	Strain functions in van der Waals equation of state

D	Debye function
D	Bond dissociation energy
E	Einstein function
E	Modulus of elasticity
E_{th}	Theoretical modulus of elasticity
E_c	Conformational energy of chains
E_h	Mole energy of holes
E_{AB}	Modulus of elasticity of amorphous intrafibrillar regions
E_{AI}	Modulus of elasticity of amorphous interfibrillar regions
F	Free energy
F^*	Free energy of deformation of solids
G	Free enthalpy
G^*	Free enthalpy of deformation of solids
G_0	Shear modulus
H	Enthalpy
H^*	Enthalpy of deformation of solids
H_{st}	Stored enthalpy
K	Modulus of elasticity (volume)
K_c	Stress intensity factor
L_0	Length of undeformed sample
L	Length of deformed sample
L_i	Length of samples in the reference state
M	Molecular weight
M	Twisting couple
N	Avogadro number
P	Pressure
P_i	Internal pressure
P_T	Thermal pressure
Q	Heat
R	Molar gas constant
S	Entropy
T	Absolute temperature
T_g	Glass transition temperature
T_m	Melting point
V	Volume
V	Rate of deformation
U	Internal energy
U^*	Internal energy of deformation of solids
U_{st}	Stored internal energy
W	Mechanical work
X	Portion of interfibrillar amorphous regions
α	Volume thermal expansion
β	Linear thermal expansivity

γ	Gruneisen parameter
γ_i	Mode Gruneisen parameter
γ_T	Thermodynamic Gruneisen parameter
γ_L	Lattice thermodynmaic parameter
γ	$= (\delta \ln \langle r^2 \rangle_0 / \delta \ln V)_{T,L}$
γ_s	Shear deformation
δ	Solubility parameter
δ	Thickness of plastic zone
ε	Strain (uniaxial)
η	Heat to work ratio
θ	Characteristic temperature
θ_E	Einstein temperature
θ_D	Debye temperature
κ	Volume compressibility
χ	Thermal conductivity
κ_L	Linear compressibility
λ	Elongation (or compression) ratio
λ_Q	Elongation corresponding to inversion of heat
λ_u	Deformation corresponding to inversion of internal energy
λ_m	Limiting chain extensibility
λ_f	Elongation corresponding to inversion of force
μ	Poisson's ratio
ν	Number of elastically active network chains
ρ	Density of vibrational state
ρ	Density
ξ, ξ_i	Generalized force
σ	stress
τ	Shear stress
τ	Time constant
ψ	Volume fraction of filler
ψ	Angle with the draw direction
ω	Frequency
ω_i	Mode frequency
ω	Internal energy to work ratio
ω_E	Einstein frequency
ω_D	Debye frequency
Γ	Surface energy of solids
Γ_{eff}	Effective surface energy

Part One:
Thermal Properties of Polymers

1 Heat Capacity

1.1 Basic Concepts of the Heat Capacity of Solids

The heat capacity of any solid body is determined by the normal modes of vibrations available to its structure. If the spectrum of vibrational states, $\rho(\omega)d\omega$, which gives the number of modes of vibration whose frequencies lie between ω and $\omega + d\omega$, is established for a solid body the heat capacity C of a unit volume of the body can be immediately determined as

$$C = k_B \int \rho(\omega) \left(\frac{\hbar\omega}{k_B T} \right)^2 \frac{\exp(\hbar\omega/k_B T)}{[\exp(\hbar\omega/k_B T) - 1]^2} . \tag{1.1}$$

Hence, the key assumption in the treatment is that $\rho(\omega)$ is known. The simplest assumption concerning $\rho(\omega)$ was made by Einstein [1] who has suggested that the vibrational spectra consist of only a single frequency ω_E (Fig. 1.1). It means that the thermal motion of N structural elements of the lattice is equivalent to the thermal motion of 3N independent linear harmonic vibrators with the same frequency. Hence, in this case

$$\rho(\omega) = 3N\,\delta(\omega - \omega_E) \tag{1.2}$$

and the heat capacity for such a solid is

$$C = 3Nk_B \left(\frac{\hbar\omega_E}{k_B T} \right)^2 \frac{\exp(\hbar\omega_E/k_B T)}{[\exp(\hbar\omega_E/k_B T) - 1]^2} \tag{1.3}$$

or

$$C = 3N_B E \left(\frac{\theta_E}{T} \right) \tag{1.4}$$

where

$$E \left(\frac{\theta_E}{T} \right) = \left(\frac{\theta_E}{T} \right) \frac{\exp(\theta_E/T)}{[\exp(\theta_E/T) - 1]^2} \tag{1.5}$$

is the Einstein function and $\theta_E = \hbar\omega_E/k_B$ is the Einstein temperature.

Although such a simple vibrational spectrum does not correspond to any real solid, nevertheless, using this simple model Einstein could, in principle, explain the low temperature behavior of solids. According to Eq. (1.4) at low temperature, namely at $T \ll \theta_E$, the heat capacity decreases approximately exponentially with decreasing temperature, while at high temperatures $T \gg \theta_E$ it reaches the constant value of 3R (Dulong-Petit law). The

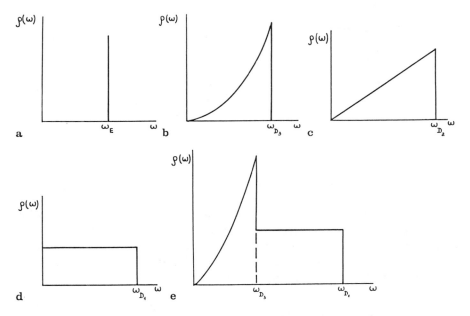

Fig. 1.1. Spectral density of various models of solids: **a** – according to the Einstein model; **b** .. according to the Debye model; **c** – according to the Tarasov model for the two-dimensional non-interacting layers; **d** – according to the Tarasov model for the one-dimensional non-interacting chains; **e** – according to the Tarasov model for the one-dimensional interacting chains assuming 30% three-dimensional modes

Einstein functions are widely used for calculations of the contributions to the heat capacity of some narrow parts of spectra or local vibrations.

There are two theoretical approaches to the evaluation of the vibrational spectrum of solids: the continuum approach and the lattice dynamic approach [2–4]. The simplest treatment of the continuum approach is that due to Debye [5] in which one ignores the atomic structure of any solid body as well as dispersion and relates frequency and wave vector κ by $\omega = (c/2\pi)\kappa$, where c is the sound velocity, for all frequencies. The distribution of the acoustic modes in the Debye model is given by (Fig. 1.1)

$$\rho(\omega) = \frac{3N\omega^2}{\omega_D^3} \qquad \omega \leq \omega_D$$
$$\rho(\omega) = 0 \qquad \omega > \omega_D \tag{1.6}$$

for each mode, namely one longitudinal and two transverse. The cutoff frequency ω_D is introduced to limit the total number of degrees of freedom to the total number N of particles in the system. By introducing Eq. (1.6) to Eq. (1.1) one can arrive at the following contribution of the acoustic vibrational modes to the heat capacity

$$C = 3Nk_B \frac{3}{\omega_D^3} \int_0^{\omega_D} \left(\frac{\hbar\omega}{k_B T}\right)^2 \frac{\exp(\hbar\omega/k_B T)}{[\exp(\hbar\omega/k_B T) - 1]^2} \omega^2 d\omega \tag{1.7}$$

or

$$C = 3k_B D \left(\frac{\theta_D}{T} \right) \tag{1.8}$$

where

$$D \left(\frac{\theta_D}{T} \right) = 3 \left(\frac{T}{\theta_D} \right) \int_0^{\omega_D/T} \frac{(\theta/T)^4 \exp(\theta/T)}{[\exp(\theta/T) - 1]^2} \, d \left(\frac{\theta}{T} \right) \tag{1.9}$$

is the Debye function (for three-dimensional continuum) and $\theta_D = \hbar\omega_D/k_B$ is the Debye temperature. At high temperatures $T \gg \theta_D$ we again arrive at the Dulong-Petit law $C = 3R$, while at low temperatures ($T \leq \theta_D/12$) the heat capacity follows the T^3-Debye law

$$C = 3Nk_B \frac{4\pi^4}{5} \left(\frac{T}{\theta_D} \right)^3 . \tag{1.10}$$

The Debye model allows us to understand many of the general features of the heat capacity of ordinary isotropic solids. It leads to correct results at low temperature where the atomic structure can be ignored, because in this temperature range only long-wave vibrations are excited, which is considerably larger than the lattice parameters. Because in the long-wavelength limit the details of the atomic structure are unimportant the Debye model is also suitable for the amorphous solids, since all that is required for the existence of a normal mode spectrum is the existence of stable equilibrium positions for each oscillating unit. However, the disordered dielectric solids (inorganic glasses or glassy amorphous polymers) demonstrate the anomalous thermal behavior at very low temperatures: besides the Debye contribution to the specific heat an additional linear with temperature contribution exists which is the consequence of disordered structure of glasses.

It has been established experimentally that many elastically isotropic crystalline solids follow T^3-law at low temperatures, although the temperature region where T^3 takes place can be considerably more narrow than that predicted by the Debye theory [4]. The continuum Debye model was developed for the elastically isotropic solids, therefore, it is not completely suitable for the application to highly anisotropic solids. In this case the velocity of the elastic waves in various directions will be different and there will be no definite polarization of the vibrations. Of course, the limited temperature laws of the heat capacity will take place in this case as well both for high temperature ($C = 3R$) and low temperature ($C \simeq T^3$ in the vicinity of $0\,K$), however, in the intermediate temperature interval the heat capacity behavior can be significantly different from that predicted by the Debye formula. Solids with chain-like (one-dimensional) or plane-like (two-dimensional) structures are as a rule highly anisotropic due to a large difference in the forces along the chain or in the plane and between them. Most polymers belong to this class of solids and their heat capacity behavior normally does not follow completely the Debye formula.

The second approach to the calculation of the vibrational spectra of solids consists in a detailed analysis of the vibrational modes using the techniques of lattice dynamics (Born–von Karman method) [2, 3]. The calculations normally start from the choice of force constants corresponding to a particular crystalline structure of the solid under consideration with a subsequent analysis of the possible vibrational branches both the acoustic (whose frequency goes to zero) and optical nature. As a rule the detailed nature of the vibrational spectra of lattice vibrations for any solid is very complex and can be found only after time-consuming numerical computations. After the spectral density $\rho(\omega)d\omega$ of a solid is established due to such calculations the heat capacity can be estimated according to Eq. (1.1).

Both these approaches have been used for analysis of the temperature dependence of the heat capacity of crystalline and amorphous polymers.

1.2 Temperature Dependence of Heat Capacity of Polymers

Similar to other solids the heat capacity of polymers is determined by their vibrational spectra. A key feature of the polymer chains is its one-dimensional nature. Due to the chain-like nature of macromolecules the heat capacity of polymers can be considered from the point of view of two contributions: lattice vibrations and characteristic vibrations resulting from internal motions of the repeating unit or its rather independent hindered rotation or conformational isomerisation. The lattice (skeleton) vibrations are low frequency acoustic vibrations and give the main contribution to the heat capacity of solid polymers at low temperatures. The skeleton vibration is a universal feature of the vibration spectrum of all polymers while the existence and the nature of the optical modes depend upon the exact nature of the repeating unit. The analysis of the lattice vibrations can be peformed on the basis of the skeleton approximation when any macromolecules can be considered as a structureless chain of point masses equal to the mass of the repeating unit. Because for many carbochain polymers the geometry of the repeating unit and force constants are similar the only parameter which changes in transition from one polymer to another is the mass of the repeating unit.

The characteristic vibrations of the side groups are the optical vibrations, therefore, their frequencies lie considerably higher relative to the skeleton vibrations. Their contribution to the heat capacity at low temperature is negligible, it becomes visible at temperatures above approximately 100 K. Finally, due to the existence in macromolecules of various isomers with different energies a contribution to the heat capacity resulting from the conformational transitions is also possible. In this section we will consider the theoretical predictions of the temperature dependence of the heat capacity following from various approaches. Similar to other solids the most interesting features of the temperature dependence of the heat capacity occur at low temperatures.

1.2.1 Theoretical Frameworks

Continuum Models

The first simple continuum model in which the chain-like structure of macro-molecules was attempted to take into account was the Tarasov model [6–9]. Tarasov suggested one should consider linear macromolecules as elastic bars and planar structures as elastic plates and subsequently Debye introduced the one- and two-dimensional (low-dimensional) continuum for the description of their vibrational spectra. The generalized distribution of vibrational states is then given by

$$\rho(\omega) = 3mN\omega_m^{-m}\omega^{m-1} \tag{1.11}$$

where $m = 1, 2$ or 3 for one-, two-, and three-dimensional continuum correspondingly. The shapes of the spectra are shown in Fig. 1.1. By introducing the distribution (1.11) into Eq. (1.1) one gets

$$C = 3mR \left(\frac{T}{\theta_m}\right)^m \int_0^{\theta_m/T} \frac{(\theta_m/T)^{m+1}\exp(\theta_m/T)}{[\exp(\theta_m/T) - 1]^2}\, d(\theta_m/T)$$

$$= 3mR\, D_m\left(\frac{\theta_m}{T}\right) \tag{1.12}$$

where $D_m\left(\frac{\theta_m}{T}\right)$ is the m-dimensional Debye function and the $\theta_m = \hbar\omega_m/k_B$ are equivalent temperatures. Equation (1.12) is valid for non-interacting chain-like and plate-like structures. For high temperatures it gives $C = 3R$ similar to the Debye model and for low temperatures it gives

$$C_1 = \pi^2 R(T/\theta_1)^1 \tag{1.13}$$

$$C_2 = 43.27\, R(T/\theta_2)^2 \tag{1.14}$$

$$C_3 = \frac{12}{5}\pi^2 R(T/\theta_3)^3\ . \tag{1.15}$$

It is seen that for the one-dimensional continuum, C_1 is a linear function of temperature, while for the two-dimensional continuum, C_2 is a parabolic function of temperature. Hence, the low temperature behavior of the heat capacity depends drastically upon the dimensionality of the continuum.

In the above consideration it has been assumed that there is no intermolecular interaction between structural elements of the continuums. In real solids such interactions between the structural elements exist and the low temperature heat capacity behavior depends drastically on this interaction. To take into account the intermolecular interaction, Tarasov suggested combining the three-dimensional and one- or two-dimensional continuums. At low temperatures when long-wavelength excitations occur, the structural elements will interact with one another through the weak interchain forces and the vibrations will be essentially three-dimensional, therefore, the three-dimensional Debye continuum will be satisfactory. At higher temperatures, shorter wavelength and higher frequencies will exist and these normal modes will be essentially vibrations along the bars or in the planes and will, therefore, be essentially one- or two-dimensional in nature. Hence, in his model, Tarasov assumed that all of the low-frequency modes are three-dimensional and their spectrum can be calculated as in the Debye model up to the same cut-off frequency, ω_3. The remaining higher frequency vibrational modes are considered as a one-dimensional or a two-dimensional continuum with some final cut-off frequency, ω_1 or ω_2, which is determined by the requirement that the total number of frequencies be equal to the number of repeating units (N). The resulting spectral density will then be

$$\rho_m(\omega) = \begin{cases} \frac{9N\omega^2}{\omega_3^2 \omega_m^m}, & 0 \leq \omega \leq \omega_3 \\ \frac{3mN\omega^{m-1}}{\omega_m^m}, & \omega_3 \leq \omega \leq \omega_m \end{cases} . \tag{1.16}$$

In this treatment it is assumed that the fraction ω_3/ω_m of the total number of vibrational states is of a three-dimensional character and the remaining vibrations are of one-dimensional or two-dimensional character. The spectral density corresponding to the Tarasov model is shown in Fig. 1.1.

Inserting the Tarasov density into Eq. (1.1) leads to the following expressions for the total heat capacity of two transverse and one longitudinal mode and for one mole of repeating units

$$C_{m,3}/R = D_m\left(\frac{\theta_m}{T}\right) - \left(\frac{\theta_3}{\theta_m}\right)^m \left[D_m\left(\frac{\theta_3}{T}\right) - D_3\left(\frac{\theta_3}{T}\right)\right] . \tag{1.17}$$

In contrast to the Debye model the Tarasov model has two parameters θ_3 and θ_m. At low temperatures the three-dimensional modes dominate the calculation and one can arrive at the following equation

$$C_{m,3}/R = \left(\frac{\theta_3}{\theta_m}\right)^m \frac{12\pi^4}{5} \left(\frac{T}{\theta_3}\right)^3 . \tag{1.18}$$

According to estimations given by Tarasov the T^3-dependence of the interacting structures should exist up to $T \simeq 0.1\,\theta_3$. At the intermediate temperature, $T \geq \theta_3 \ll \theta_m$, suggesting that $(\theta_3/\theta_m) \ll 1$, all the three-dimensional vibrations are excited and the one-dimensional or two-dimensional modes dominate the temperature dependence of the heat capacity. In this temperature interval a weaker temperature dependence occurs

$$C \sim \left(\frac{T}{\theta_m}\right)^m , \quad T \leq 0.1\,\theta_m . \tag{1.19}$$

Thus, the Tarasov model predicts the expected cubic dependences of the heat capacity at low temperatures for structures of all dimensionality and a linear dependence for the one-dimensional continuum and a quadratic dependence for the two-dimensional continuum at intermediate temperatures. At higher temperatures, Eq. (1.17) again leads to the Dulong-Petit law.

The Tarasov model has some physical shortcomings. The model suggested that both the longitudinal and transverse modes have the similar dispersion relationship: $\omega \sim \kappa$. In contrast to isotropic solids, in which there are no vibrational modes which do not obey this relationship, in real chain-like or plate-like structures, as emphasized for the first time by Lifshitz [10], the main contribution to the low temperature heat capacity gives vibrational modes, which have a quite different dispersion relationship, namley $\omega \sim \kappa^2$. These modes correspond to the acoustical low frequency bending vibrational modes (the bending waves). The existence of these bending waves leads to a principally different temperature dependence of the heat capacity in that temperature range where one can neglect the interaction between the low-dimensional structures, i.e. $C \sim T^{1/2}$ for chain-like structures and $C \sim T^1$ for plate-like structures. Using the continuum approach language, one can conclude that in the Tarasov model a macromolecule is modeled by a stiff linear string in which only the longitudinal vibrations are possible, while real macromolecules must be considered as thin flexible needles or bars, in which the bending waves can be excited. Similarly, the two-dimensional plate-like structures behave like flexible membranes rather than like stiff plates. Hence, as a matter of fact, in the Tarasov model the main contribution to the low temperature heat capacity of the low-dimensional continuums was not taken into account. Therefore, Eq. (1.19) cannot predict, in principle, the low temperature behavior of the heat capacity of chain-like and plate-like structures. Even if such dependencies occur in experimental studies it cannot be considered as a proof of the chain-like or plate-like structure of the solid under study and no useful information regarding the nature of vibrational spectrum can be achieved.

A further shortcoming concerns the artificial method by which in the isotropic Debye continuum the anisotropy was introduced. The degree of anisotropy in the Tarasov model is characterized by the ratio (θ_3/θ_1), which can change from 1 (isotropic system) to 0 (highly anistropic system). A critical consideration of the method of describing the anisotropic properties and the physical assumptions of the Tarasov model was given by Lifshitz [10, 11] and Stepanov [12].

Lifshitz [10] (see also [13]) studied theoretically the heat capacity of thin needles and films and paid attention to the fact that at low temperature the most important role is played by the bending waves with the dispersion relationship $\omega \sim \kappa^2$. Their existence leads to the unusual low temperature behavior of thermal properties, namely the heat capacity and the thermal expansion. This very important result was taken into account in his rigorous theory of the thermal properties of anisotropic crystals, consisting of the chain-like molecules and atomic (molecular) layers. The chains and layers were considered as one-dimensional and two-dimensional continuums. The dispersion relationships for thin needles are

$$
\begin{aligned}
\omega_1 &= c\kappa_z && \text{– longitudinal wave} \\
\omega_{2,3} &= \gamma\kappa_z^2 && \text{– transverse bending waves}
\end{aligned}
\tag{1.20}
$$

where κ_z is the projection of the wave vector on the axis of the needle, c is the velocity of the elastic waves, $\gamma = ca\nu/\pi$ (a is the distance between the atoms, ν is the "transverse stiffness" of the needle, dimensionless parameter). The resulting distributions are

$$\rho_1(\omega) = \frac{N}{\omega_{m1}}$$

$$\rho_{2,3}(\omega) = \frac{N}{2\omega^{1/2}}\omega^{-1/2} \tag{1.21}$$

and the heat capacity of the thin needles is

$$C_1 = Nk_B\gamma_1 \left(\frac{T}{\theta_1}\right)^{1/2}\left[1 + \gamma_2\left(\frac{T}{\theta_1}\right)^{1/2}\right] \tag{1.22}$$

where θ_1 is the characteristic temperature of the needle, $\gamma_1 \simeq 2.3\nu^{-1/2}$ $\gamma_2 \simeq \nu^{-1/2}$. The first term in Eq. (1.22) corresponds to the bending waves and the second one – to the longitudinal waves. At low temperatures one can neglect the term (T/θ_1) in comparison with $(T/\theta)^{1/2}$ and one can arrive at

$$C_1 = Nk_B\gamma_1\left(\frac{T}{\theta}\right)^{1/2}. \tag{1.23}$$

Similarly for the noninteracting layers at low temperatures

$$C_2 \sim Nk_B\left(\frac{T}{\theta_2}\right) \tag{1.24}$$

where θ_2 is the characteristic temperature of the layers. Hence, the low temperature heat capacity of the noninteracting needles and layers is almost totally determined by the bending waves while the contribution of the longitudinal waves to the heat capacity can be neglected. This conclusion is quite different from that of the Tarasov theory for the noninteracting one-dimensional and two-dimensional structures.

The interaction between the chains and layers were made on the model of the highly anisotropic hexagonal crystalline lattice. The introduction of the interaction leads to some characteristic temperature intervals of the heat capacity. For the interacting chains these intervals are as follows

$$C = A\left(\frac{T}{\theta}\right)^3 \qquad \text{at} \quad T \ll \eta^2\theta \tag{1.25}$$

$$C = B\left(\frac{T}{\theta}\right)^{5/2} \qquad \text{at} \quad \eta^2\theta \ll T \ll \eta\theta \tag{1.26}$$

where A and B are the coefficients which include the elastic modulus of the chains. In the temperature interval $\eta\theta \ll T \ll \theta$ the influence of the inter-chain interaction can be neglected, therefore, the temperature dependence of the heat capacity follows Eq. (1.23).

Correspondingly for the interacting layers

$$C = A \left(\frac{T}{\theta}\right)^3 \quad \text{at} \quad T < \eta^2\theta \tag{1.27}$$

$$C = B \left(\frac{T}{\theta}\right)^2 \quad \text{at} \quad \eta^2\theta \ll T \ll \eta\theta \tag{1.28}$$

where A and B are the coefficients which include the elastic modulus of the layers. In the temperature interval $\eta\theta \ll T \ll \theta$ where the influence of the interaction between the layers can be neglected the temperature dependence of the heat capacity follows Eq. (1.24).

Thus, Lifshitz demonstrated that for highly anisotropic crystals in which one can identify some structural elements like chains or layers the temperature dependence of the heat capacity, in those temperature regions where the interaction between these elements can be neglected, is determined by the bending vibrations of the chains and layers.

An elegant theoretical consideration of the vibrational spectra of highly anisotropic low-dimensional crystals and their low temperature heat capacity has been performed by Kosevich [14]. This consideration supports existence of the temperature regions where the heat capacity of the chain-like and layers structures follows Eqs. (1.23)–(1.28). The Kharkov physical school is very active now in studying the lattice dynamics of the highly anisotropic crystals and thin films and their low temperature heat capacity [15–17].

Stepanov [12] also considered rigorously the influence of the high elastic anisotropy of the hexagonal crystalline lattice on the vibrational spectrum and heat capacity without any direct identification of the low-dimensional structures in such crystals. He has demonstrated that existence of the high elastic anisotropy leads to a special temperature dependence of the heat capacity, which occurs after the usual $C \sim T^3$ dependence. The behavior of the one- and two-dimensional structures are treated as particular cases. It has been especially emphasized that, in contrast to the Tarasov model, any calculation of the vibrational spectra must be performed in the three-dimensional wave vector space. Finally, he has also pointed out that the characteristic dependencies of the heat capacity occur only then when a very high elastic anisotropy in solids really exists. A similar approach and similar conclusions were also published later in [19].

Normal Mode Spectrum of Polymer Chains

It is well known now that any continuum model cannot be expected to give a completely accurate picture of the heat capacity of any solid. Since the heat capacity represents an average over many vibrational modes and is insensitive to the fine details of any normal mode spectrum, very little of the detail of $\rho(\omega)$ can be deduced from the heat capacity measurements; only the general features and the limiting behavior at low temperatures can be obtained with any certainty. If a complete picture is to be obtained, more detailed and realistic calculations are needed, which, in contrast to the continuum approach, make use of the actual structure of the solid and the interatomic forces. Such calculations were carried out for many macromolecules using the lattice dynamic approach. Below we will consider the main principles of such calculations and some typical results.

The normal mode spectrum of a single isolated macromolecule can be obtained using the lattice dynamic calculations [2–4]. In a real macromolecule, in contrast to the usually considered mathematical model, both the longitudinal and transverse modes are possible. The normal mode spectrum of the macromolecules with fixed ends, consisting of only one vibrating element and having the rotational symmetry, includes one longitudinal and two degenerate transverse modes. It can be shown [4], that for the long waves the dispersion relationships for these modes are quite different: for longitudinal $\omega \sim \kappa$, while for transverse $\omega \sim \kappa^2$. Therefore, at low temperatures $(T \ll \theta = \hbar\omega/k_B)$ the longitudinal modes contribute to the heat capacity according to T^1, while transverse modes – according to $T^{1/2}$. The total heat capacity of the isolated macromolecules at low temperatures is

$$C = aT^1 + a_{1/2}T^{1/2} \; . \tag{1.29}$$

At high temperatures it is constant $C \sim 3R$. Hence, we see that the lattice dynamic calculation gives the low temperature heat capacity of the isolated macromolecules in full accord with the Lifshitz model, rather than the Tarasov model.

In the macromolecules considered there is only one vibrating element in the unit cell. If the unit cell contains n vibrating elements the normal mode spectrum includes also the optic vibration branches. According to the theory [2–4] the total number of the vibrational branches has to be 3n. Thus, even a simple macromolecule like $(-CH_2-)_p$ should have 9 branches. Miyazawa and coworkers [20] carried out extensive calculations of the vibrational modes of the isolated PE chain. The dispersion curves and spectral density are shown in Fig. 1.2. It is seen that the acoustic and optical modes do not overlap and there is a considerable gap between them. All optical modes have relatively high frequencies. It means that only the vibrations which are derived from the acoustical modes of the isolated chain will contribute importantly to the low temperature heat capacity of PE. The corresponding estimations showed that the low frequency optical modes $(750–1000 \, cm^{-1})$ contribute approximately 2% of the total heat capacity at $120 \, K$ and about 40% of the heat capacity at $400 \, K$, while the high frequency optical modes have practically no contribution to the heat capacity at room temperature.

The picture of the dispersion curves shown for PE in Fig. 1.2 is generally typical for all isolated polymer chains. A change of the chemical structure of the chains is accompanied by a change of the number of branches and the interval of their frequencies. The general feature of the normal mode spectrum of the isolated chains is the existence of 2 acoustic and $n-2$ optical branches.

How will the normal mode spectrum of the isolated chain change, if the intermolecular interaction is "switched"? In contrast to the optical modes,

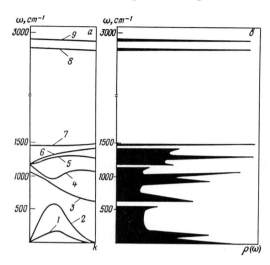

Fig. 1.2. Dispersion relations (**a**) and spectral density (**b**) for the isolated PE chain [20]. *1* – C-C internal rotations; *2* – C-C bending; *3* – CH$_2$ rocking; *4* – C-C stretching; *5* – CH$_2$ twisting; *6* – CH$_2$ wagging; *7* – CH$_2$ scissoring; *8,9* – CH$_2$ stretching

for which the intermolecular interaction is only a slight perturbation from the isolated chain, the acoustic branches change drastically due to the interchain interaction, in spite of their smallness in comparison with the intramolecular interaction. Figure 1.3 shows the normal mode spectrum of PE, calculated with taking into account the interchain interaction according additive scheme [4, 20]. The elastic constants of the interchain interaction were chosen in such a way in order to reach a good coincidence with the low temperature heat capacity. Comparison of Fig. 1.2 and Fig. 1.3 shows that the main perturbations in the spectrum after switching of the interchain interaction have occurred in the low frequency part of the spectrum, where without interchain interaction $\rho(\omega) \Rightarrow \infty$ when $\omega \Rightarrow 0$, while with the interchain interaction $\rho(*) \Rightarrow 0$ when $\omega \Rightarrow 0$. The low frequency part of the calculated spectrum is in a very good accord with data of the inelastic scattering of the thermal neutrons [4, 20]. Hence, the acoustic vibrational spectrum of the interacting chains up to $100\,\text{cm}^{-1}$ is determined by the interchain interaction (a monotonous drop of $\rho(\omega)$), while in the region between 100 and $500\,\text{cm}^{-1}$ it is practically completely determined by the chemical structure of chains. Therefore, the low temperature heat capacity (up to 50–70 K) should depend both on the interchain and intrachain interactions. Above this temperature region one can always find such a temperature interval where the heat capacity is determined mainly by the intrachain interaction.

The calculation of the complete vibrational spectra considered above for PE can, in principle, be carried out for polymers of any chemical structure. This calculation, however, is normally so time-consuming that up to now such calculation has been carried out only for simple chains such as PE, POM, PEO [20].

As mentioned above, in order to study the low frequency acoustical part of the complete spectrum, which gives the main contribution to the low temperature heat capacity, there is no need to consider all the vibrational degrees of freedom of the repeating unit of chains, because the characteris-

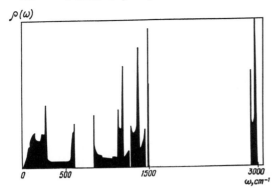

Fig. 1.3. Spectral density of crystalline PE [20]

tic vibrations of their skeletons and the side groups lie in different regions of their spectra. Generally speaking, this conclusion, which has been formulated for the first time by Kirkwood [21] and Pitzer [22] due to analysis of the vibrational spectra of hydrocarbons, is well grounded only for those macromolecules which contain "light" side groups. For the chains with "heavy" side groups this approach needs further justification because these "heavy" groups may have low characteristic frequencies. Cheban [23, 24] has analysed the problem of the skeleton approximation theoretically and showed that the simple skeleton approximation, in which the total mass of the repeating unit can be considered to be concentrated in a point, is rigorously valid only in the narrow region of spectra in the vicinity of zero frequency. If one considers the total low frequency area of the spectrum such approximation is possible only under the assumption that this point mass is dependent on frequency. This model, which has been termed the normalized skeleton approximation, allows us to fix the frames of the usual skeleton approximation. For example, an estimation showed that for PE the usual skeleton approximation is justified for the vibrational area 0–$50\,cm^{-1}$ for the torsional vibrations and 0–$200\,cm^{-1}$ for the deformational vibrations. The normalized skeleton approximation is especially useful in that cases when the side groups only interact with the atoms of their repeating unit. In this case a real macromolecule can be replaced with an equivalent linear chain consisting of the point masses $M(\omega)$ and the normal mode spectrum of such a chain can be obtained.

For the theoretical consideration of the low temperature heat capacity, some simplified models are very useful which facilitate the analytical solutions and describe rather satisfactorily the low frequency area of spectra. The first significant attempt to consider such a model was that of Stockmayer and Hecht [25]. Genensky and Newell [26], using the Born–van Karman method, analysed in considerable detail the vibrational spectrum and low temperature heat capacity of this model. In spite of some oversimplifications this model was very useful for considering the thermal properties of polymers in general (see Chaps. 2 and 3). The model is illustrated in Fig. 1.4. Hypothetical linear parallel chains form the simple rectangular lattice characterized with four empirical force constants. Two of the force constants, β_y and χ_y, refer to intrachain forces, being associated with C-C stretching and bending respectively. The other two constants, α_y, γ_y and δ_y, characterize the interchain forces α_y being associated with nearest neighbor interactions while γ_y and δ_y with next nearest neighbor interactions. The force constants were chosen in such a way as to display the two most important features of the polymers: a large anisotropy of the lattice and flexibility of chains. To obtain this polymerlike behavior one assumes that β_y and χ_y are much larger than the other force constants. The large anisotropy is introduced due to the condition $\beta_y \gg \alpha_y$, while the flexibility of the chains is characterized by the force constant χ_y.

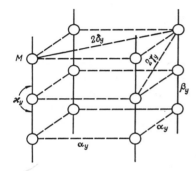

Fig. 1.4. Crystal structure and force constants of the Stockmayer-Hecht model [25]

It has been established that the low temperature heat capacity, determined by the acoustical vibrations of the model lattice, is characterized by three temperature regions (starting from the very low temperature): $C \sim T^3$, $C \sim T^{5/2}$ and $C \sim T^{1/2}$. These features are to be expected of typical highly anisotropiy crystals made of extended-chain macromolecules including macromolecules with plane zig-zag structure. One can see that these results are consistent with the Lifshitz model. Hence, although the Stockmayer-Hecht model is obviously oversimplified, nevertheless, it takes into account two most important features of polymers: the large anisotropy of the elastic forces in the crystalline lattice and the flexibility of macromolecules. Therefore, one can assume that it predicts qualitatively correctly the low temperature heat capacity behavior of polymer crystals.

It is a well known fact that the degree of anisotropy of crystalline macromolecules decreases significantly when the polymer chains are crystallized in the crystalline helix rather than in extended-chain plane zig-zag. The problem of the low temperature heat capacity of the helix-like macromolecules has been examined by Telezhenko [27–29]. Using a simple model of the integer and noninteger helices and the computational technique of lattice dynamics he has examined in particular detail the low frequency vibrational spectra of the helices and the corresponding temperature dependence of the heat capacity. The most important conclusions were consist for the following: i) the transformation of the plane structure to the helical structure is accompanied by a considerable compressing of the acoustic part of the spectrum (especially for noninteger helices) and the appearance of a large number of low frequency optical vibrations; due to such compression the low temperature heat capacity (at $T \approx 5\,\mathrm{K}$) happens to be insensitive to the particular structure of the spectrum of the acoustic vibrations; ii) the contribution of the low frequency optical vibrations to the heat capacity is essential at rather low temperature because of a large number of such optical branches and of their low frequencies close to zero; due to the low frequency optical modes the T^3-Debye law for heat capacity must be valid only at very low temperature $T \leq 2\,\mathrm{K}$; iii) the temperature interval, where the bending waves are dominated, for the helical structures is significantly narrower than for the plane structure. Thus, these calculations have demonstrated that the application of the extended chains Stockmayer-Hecht model to the chains of helical structure can lead to erroneous conclusions concerning the low temperature heat capacity behavior.

Finally, it is necessary to consider briefly the recent theoretical studies dealing with the anomalous heat capacity behavior of polymer glasses at very low temperatures. It has been experimentally established that, in contrast to the crystalline solids, the dielectric disordered solids including polymeric glasses at very low temperature behave anomalously: besides the normal Debye contribution to the heat capacity there is an additional contribution which depends linearly upon temperature and then a hump occurs (Fig. 1.5). Therefore, the low temperature heat capacity of the disordered materials may significantly exceed, at the same temperature, the heat capacity of the same materials in the crystalline state.

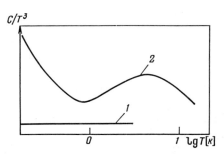

Fig. 1.5. Low temperature heat capacity of amorphous solids [30, 31]: *1* – The Debye heat capacity; *2* – experimental curve

Two groups of models have been suggested for explanation of this anomalous low temperature behavior of heat capacity [30–40]. The first group of models is based on the suggestion of the existence in the Debye matrix of some isolated low frequency vibrational modes, which, in turn, is closely related to the suggestion of the heterogeneous structure of glasses on the molecular level. As an example of such a model, one can mention the model assuming the existence of some internal submicrovoids in the disordered systems [33]. Similar models are also described in [34, 35]. Although this group of models can predict the appearance of the linear dependence of the heat capacity at very low temperature, these models usually involve parameters which can be identified with difficulty experimentally.

The other group of models invokes a distribution of localized vibrational modes. The model which has received the most attention is the phonon assisted tunneling model [37–40], in which it is assumed that due to the disordered structure some groups of atoms may have more than one equilibrium position and their tunneling between these states with the help of phonons may provide a substantial contribution to the heat capacity and explain both the linear dependence and the existence of the hump.

Finally, Montgomery [41] examined a simple model of the transformation of the Debye spectrum in disordered solids due to fluctuations of elastic constants around the values existing in the crystalline state. The low frequency Debye spectrum is widened due to such fluctuations and some very low frequency vibrational modes occur. Their appearance leads to the excessive values of low temperature heat capacity and change its temperature dependence. A very attractive side of this approach is that the model deals with the collective state of the disordered systems rather than the localized modes accessible only to a certain number of atoms or groups of atoms.

A very interesting model for the specific heat of amorphous polymers has been developed recently by Allen [42]. In the model it is assumed that the normal mode spectrum is given by $\rho(\omega) \sim \omega^{d-1}$, where d is the fractal, or spectral, dimension. By introducing this $\rho(\omega)$ into the Debye model [Eq. (1.6)] one can arrive at $C(T) \sim (T/\theta_d^d)$ for $T \ll \theta_d$, where θ_d is the characteristic temperature. This characteristic temperature is related to the formula weight FW by $\theta_d^d = A_d/FW$, where A_d is a constant that depends upon the dimension d and the structural characteristics, such as the velocity of sound, which are assumed to be equal for all polymers. In this model d is a characteristic of the polymer conformation. Every polymer is described by the same value of d resulting in the same power law dependence of temperature. The value of the heat capacity depends linearly on the second parameter FW. The general temperature dependence can be expressed by $C(T) = B_d FW\, T^d$, where B_d is inversely proportional to the constant A_d. The model predicts T^2 dependence at low temperatures and a crossover from T^2 to $T^{5/3}$ dependence.

1.2.2 Experimental Results and Comparison with Theoretical Concepts

According to the theoretical frameworks considered, the most interesting temperature dependencies of the heat capacity of polymers should occur at low temperatures. According to Wunderlich and coworkers [3, 42–48], who have measured and examined numerous data of their own and literature data on the temperature dependence of heat capacity of various polymers, one can indicate some temperature regions in the temperature dependence of the heat capacity. The first interval is below approximately 50 K. In this temperature region the heat capacity depends not only upon the chemical structure of macromolecules, but also upon the physical structure of polymers, i.e. the intermolecular interaction. In particular, in this temperature region, C is very sensitive to the degree of crystallinity. The behavior of crystalline and amorphous polymers in this temperature region differs significantly, because, in contrast to the crystalline polymers, the amorphous polymers do not follow the T^3-Debye law.

All measurements of heat capacity are normally done at constant pressure giving C_p. In all the theoretical considerations, heat capacities are, in contrast, in terms of constant volume C_v. For conversion of C_p to C_v the rigorous thermodynamical relation $C_p - C_v = TV\alpha^2/\chi$ or the Nernst-Lindemann equation $C_p - C_v = A_0 C_p^2 T/T_m$ can be used. However, for any solids at low temperatures this difference is small and as a first approximation can be neglected [4]. Therefore, in this section we do not use any subscripts for C.

Between approximately 50 K and the glass transition temperature is a wide temperature interval where the intermolecular interaction has practically no influence on C and it is determined by the skeletal vibrational spectrum and the motion of the side groups. In this temperature region one can wait for the appearance of the characteristic for chain-like structures laws in C, however, the superposition of the contributions arising from the low frequency part of the acoustic spectrum and the low frequency motion of the side chains do not allow the direct comparison of the theoretical predictions with the experimental data. In the vicinity of the glass transition of the amorphous polymers and amorphous part of the crystalline polymers with typical values of the degree of crystallinity, a jump in the heat capacity resulting from the segmental motion, occurs. In the vicinity of the melting point and other first order transitions, characteristic peaks in C occur. However, the absolute value of the heat capacity before and after the first order transitions are very similar. Finally, in the molten state the heat capacity is a very weak function of temperature. Following this general picture of the temperature dependence of the heat capacity, in this section we will consider typical experimental results and discuss to what extent the experimental results follow the theoretical predicitions.

Temperature Region Below 50 K

We start our consideration with PE, which is, in many respects, an ideal subject for studying the heat capacity of polymers due to the simple repeating unit not possessing any of the low-frequency optical modes which make it difficult to interpret the heat capacity of most polymers. Figure 1.6 shows the temperature dependence of C_p for completely crystalline and amorphous PE [4, 49]. The data for the completely amorphous sample has been obtained by extrapolation of data for samples of various crystallinity to zero crystallinity rather than direct calorimetric measurements of the completely amorphous PE samples. The heat capacity C_p of the crystalline PE in the temperature range $0 < T \leq 9\,\mathrm{K}$ follows the Debye law: $C = 0.1104\,10^{-3}T^3$ J/(mole K) with the characteristic Debye temperature $\theta_D = \theta_3 = 260\,\mathrm{K}$. The frequency ω_D, corresponding to θ_D, is equal to 3.4×10^{13} Hz ($181\,\mathrm{cm}^{-1}$), while the sound velocity c_l at $0\,K$ is equal to $2 \times 10^5\,cm/s$ what is in a good agreement with the data of acoustic measurements at helium temperatures [51]. According to the theories of Stockmayer-Hecht and Lifshitz the excitement of the bending waves in the chain-like structures should accompany the $T^{5/2}$ temperature dependence. In early studies [4, 52] such a dependence was not found. Later, however, Baur [50] considered the temperature dependence of heat capacity of PE above $10\,K$ and more thoroughly and came to the following conclusions (Fig. 1.7). In a narrow temperature interval 15–25 K C really follows $T^{5/2}$ and then one can indicate a temperature interval when $C \sim T^{3/2}$. Such behavior is also predicted by the Stockmayer-Hecht theory according to the totally geometrical consideration, while $T^{5/2}$ behavior is not a consequence of any geometrical reasons and must be characteristic of any fibrilar crystals [26]. Of course, these temperature intervals are not very long, therefore, the values of the characteristic temperatures found from the experimental data have a low accuracy. However, although qualitatively, they show the existence of the temperature regions where C follows the theoretical predictions. On the other hand, the width of the temperature intervals where these dependencies occur differs considerably from that predicted by the Stockmayer-Hecht theory at the reasonably chosen force constants. According to the theory $C \sim T^{5/2}$ should take place in the temperature interval $5 \leq T \leq 90\,K$, while the real temperature interval is very narrow. Hence, for a real crystalline lattice this theory can predict the temperature dependence of C only qualitatively.

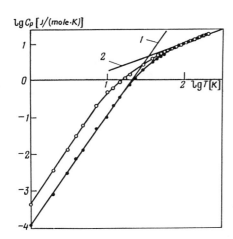

Fig. 1.6. Double logarithmic plot C vs T for crystalline ($\bullet$) and amorphous ($\circ$) PE according to the generalized data [4]. Curve *1* corresponds to $\lg C_v = 3.0 \lg T - 3.95$, curve *2* – $\lg C_v = 0.75 \lg T - 0.54$ [see Eq. (1.32)]

In spite of the fact that now the temperature dependence of the heat capacity of many crystalline polymers has been measured, the examination of these data according to the theoretical frameworks is difficult because normally crystalline polymers are semi-crystalline and possess also the amorphous disordered part. In the early interpretations of the low temperature heat capacity it is usually assumed that there is no principal difference between the amorphous and crystalline polymer at very low temperatures, therefore, one can apply fully the theories developed for crystalline solids also for disordered amorphous solids. However, during the last two decades it has been established that this conclusion is completely wrong. Hence, at very low temperatures the behavior of crystalline and amorphous materials needs separate consideration. The data shown in Fig. 1.6 for amorphous PE was obtained by extrapolation from a series of measurements of the heat capacity on samples of varying density (crystallinity), in which it was found that the heat capacity is a linear function of crystallinity. The data for amorphous PE below about 10 K should be reexamined after thorough measurements of the influence of the degree of crystallinity in PE. Above 10 K these data seem to be correct.

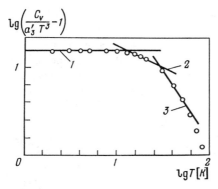

Fig. 1.7. Double logarithmical plot lg $(C_v / a_3' T^3 - 1)$ vs lg T [50]. *Points* – experimental values; *lines* correspond to $C_v = a_3' T^3 + a_n T^n$ with n = 3, 5/2 and 3/2

According to the schematic diagram shown in Fig. 1.5 the low temperature behavior of amorphous materials has three important features: first, considerably higher absolute values of heat capacity, that is exceed, in comparison with crystalline polymers; second, existence of a wide maximum; finally, below 1 K, an additional term linear in temperature is observed. This linear contribution is weakly dependent upon the nature of the disordered material and its purity. However, some admixture can influence the heat capacity without any influence on the thermal conductivity. Formally, this dependence can be described by the following expression

$$C = c_1 T + c_3 T^3 + c_E E(\theta_E / T) \qquad (1.30)$$

where c_1, c_3, c_E and θ_E are constants which are determined from experiment. Below 1 K the heat capacity is measured only for some polymers. In this temperature interval the temperature dependence of C can be represented by only the first two terms of Eq. (1.30). Therefore, the experimental data are usually represented as C/T vs T^2. Typical dependencies for three amorphous polymers are shown in Fig. 1.8. Main parameters of Eq. (1.30) for these polymers are listed in Table 1.1. Values of c_1 are rather similar for all the polymers, while values of c_3 exceed significantly the corresponding constant in the Debye equation calculated using the acoustical data.

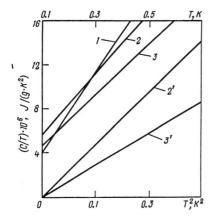

Fig. 1.8. Heat capacity of amorphous polymers below 1 K [30, 31]. *1* – PC, *2* – PS, *3* – PMMA; 1, 2, 3 – experimental dependencies; *2′*, *3′* – the Debye heat capacity calculated by means of acoustical data

In semi-crystalline polymers the additional linear contribution to the heat capacity should depend upon degree of crystallinity. Such data below 1 K have only been published for PE so far [53]. Two samples of superhigh molecular weight PE, namely isotropic and extruded with degree of crystallinity about 40%, were studied. Although below 0.2 K some difference in C for isotropic and extruded samples was observed, above 0.2 K their C was equal. The data below 1 K are described by the first two terms of Eq. (1.30). However, these results can be considered as qualitative and further accurate measurements below 1 K are necessary for a more definite conclusion.

Table 1.1. Parameters of Eq. (1.30) for various polymers [30]

Polymer	$c \, 10^{-5}$ cm/s	θ_D K	$c_D \, 10^7$ J/(gK^4)	$c_1 \, 10^7$ J/(gK^2)	$c_3 \, 10^7$ J/(gK^4)
PMMA	1.79	256	177	48	292
PS	1.67	223	262	53	457
PC	1.53	210	284	38	410

Very important information concerning the nature of the excess heat capacity was obtained in studies of the effect of degree of crystallinity on C above 1 K [53, 54, 55]. Especially valuable are the data for PET [55], because the variation of crystallinity was in the range 0–60%. Moreover, the values of the Debye heat capacity were calculated by means of independent acoustic data. Figure 1.9 shows that for all degrees of crystallinity experimental values of C are above those calculated according to the Debye formula. It means that the excess heat capacity exists in all samples. The peak on the plot C/T^3 versus T is diminished with degree of crystallinity and displaced to the low temperatures. All the experimental curves demonstrate also the tendency of linear increasing of C at the temperatures below approximately 1.5 K. For the samples of low degree of crystallinity the excess heat capacity approximates well by the Einstein heat capacity function with parameters listed in Table 1.2. The number of oscillators N and θ_E were chosen in such a way to describe the maxima completely. As is seen from Table 1.2 and Fig. 1.9 in order to reach this, only one Einstein term with $\theta_E = 15\,K$ is needed. Hence, the maxima on the curves for the samples with low degree of crystallinity are well described by the last term in Eq. (1.30).

Table 1.2. Parameters of low temperature heat capacity PET [55]

Degree of crystallinity w, %	$(C/T)^3)_D$ $J/(gK^4)$	θ_E K	$N\,10^{19}$ g^{-1}	$\frac{N}{1-w}10^{19}$ g^{-1}
2	4.21	15	10.4	10.6
17	4.05	15	9.6	11.6
41	3.82	15	5.2	8.8
59	3.64	12	1.01	2.5

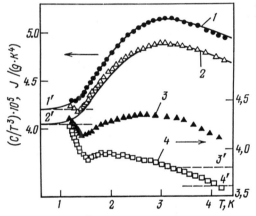

Fig. 1.9. C/T^3 vs T plots for samples of PET of various degree of crystallinity [55]. $1 - w = 2\%$; $2 - w = 17\%$, $3 - w = 41\%$, $4 - w = 59\%$. - - - corresponds to the Debye values calculated according to acoustical data. The two *solid curves* are calculated from $C/T^3 = (C/T^3)_{Debye} + [Nk_BE(15/T)]/T^3$

The behavior of samples with degree of crystallinity 41 and 59% can be described only assuming that the increase in C at $T < 1.5\,K$ reveals the appearance of the additional contribution linear in temperature (Fig. 1.10). However, this contribution is still small, therefore the accuracy of determination of c_1 is low. The behavior of the sample with the highest degree of crystallinity could be described only under assumption a smaller value $\theta_E = 12\,K$. The number of low frequency oscillators is proportional to the amorphous part, however, for the sample with the highest degree of crystallinity 59% it is considerably smaller. This fact, as well as the lower value of θ_E for this sample seems to demonstrate the influence of crystallinity on the heat capacity of the amorphous phase. It is remarkable that, with increasing degree of crystallinity, the influence of the linear contribution to C is revealed at higher temperatures. The changes in C due to crystallinity shown in Figs. 1.9 and 1.10 have been found also in PE [52] and oxacyclobutene [54], which allows us to consider it as very characteristic of semi-crystalline polymers.

The physical interpretation of the nonacoustic contributions with rather low frequencies and small number leading to the appearance of the "hump" on the C-T dependencies is often based on the idea [33] that there are cavities in the amorphous structure with one or more vibrating units (such as pendent groups or short branches) loosely bound inside the cavity [56, 57]. But in some cases this suggestion meets with some difficulties. The dimensions of the cavities may be smaller than the pendent groups or branches (for example, Ph group). Sometimes for the explanation of the "hump" it is assumed that the vibrating units are rather long parts of macromolecules between entanglements [58]. A more particular suggestion concerns a cooperative motion such bulky substitutions as Ph or F [59]. A more general idea is based on the results of neutron scattering in glasses [32]. It has been established that the "hump" in C/T^3 versus T dependencies is a consequence of the displacement of the peak responsible for the transverse acoustical vibrations in the vibrational spectrum of amorphous materials to the lower frequencies in comparison with the crystalline state. For the crystalline state this peak is usually in the vicinity of $70\,cm^{-1}$, while for the amorphous state – at $40\,cm^{-1}$. This displacement seems to be a result of a smaller density of the amorphous materials and must be a general characteristic of glassy state. Hence, according to this assumption the excess heat capacity of glasses in comparison with the crystalline state is a direct consequence of the transformation of the density distribution of the low frequency vibrations in the crystalline state due to the loss of range order and decrease of density.

Keeping this in mind let us return to Fig. 1.6 and compare the behavior of crystalline and amorphous PE at higher temperatures where C for both polymers is identical. The difference in C between the crystalline and amorphous polymers must be a result of only the intermolecular (various densities) contribution since the skeletal (intramolecular) contribution must be identical for both samples. Both the early estimations [56] and more recent calculations [60] show that all changes in the heat capacity can be described satisfactorily only by changing the characteristic temperature θ_3, which is characteristic of the intermolecular interaction.

With increasing temperature the difference between the crystalline and amorphous heat capacity of PE disappears and at $\sim 50\,K$ they are practically identical, since the dominating role belongs to the intramolecular vibrations. Hence, the idea of the transformation of the low frequency region of the crystalline vibrational spectrum is rather universal for the explanation of the heat capacity behavior at low temperatures.

Boyer et al. [61] studied the temperature and pressure dependence of the heat capacity in the temperature range 0.4–20 K and pressure up to 5.2 kbar for PTFE. The "hump" on C/T^3 versus T dependencies occurs at 2.8 K at normal pressure and at 4.4 K at 5.2 kbar. At very low temperatures C

$(C/T^3) \cdot 10^5$, $J/(g \cdot K^4)$

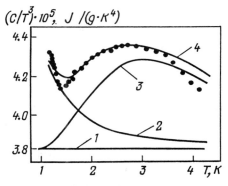

Fig. 1.10. C/T^3 against T for the sample of PET with degree of crystallinity w = 41% [55] *1* – the Debye contribution (see Table 1.2); *2* – the linear contribution [see Eq. (1.30)]; *3* – the Einstein contribution; *4* – $C/T^3 = 1 + 2 + 3 = 382 + 58/T^2 + 7.2 \times 10^3 E(15/T)/T^3$ $(\times 10^{-5} J/gK^4)$

is considerably more sensitive to pressure than at higher temperatures: the decrease of C with pressure in the vicinity of a few K is 2–3 times while at 20 K it is only some percent. The linear and cubic contributions to the heat capacity of PTFE were also considered.

Allen [42] has applied his new theory of the heat capacity of amorphous polymers to a number of polymers in the temperature range 5–40 K using the C-T data published by Wunderlich and coworkers [43–48]. More than 10 polymers including some biopolymers have been examined. Figure 1.11 shows typical results for three polymers. It is seen that the heat capacity follows the predicted temperature dependencies: T^2 from ~ 4 to ~ 12 K, and $T^{5/3}$ from ~ 12 to ~ 40 K. The data for simple polymers (low FW) fit the temperature dependence better than for more complex (higher FW) polymers. The relation $C(T_0)/FW = $ constant is found to hold well for all polymers in both the T^2 region ($T_0 = 10$ K) and $T^{5/3}$ region ($T_0 = 20$ K). Above 40 K some deviations especially for the simple polymers occur. In addition, the normalized dependence $C(T)/FW$ versus T for all amorphous polymers can be roughly represented by a single curve, showing that the heat capacity can be estimated using only the formula weight of any linear polymer.

Temperature Region Above 50 K

Above ~ 50 K the dominating role in the heat capacity starts to be played by intramolecular vibrations, rather than intermolecular ones. They include the skeletal motion and the vibrations of the pendent groups. Numerous experimental investigations of the heat capacity of linear polymers [4, 43–48, 56, 62] show that above ~ 50 K, C vs T dependencies for various polymers are rather uniform. Typical behavior is shown in Fig. 1.12. In this temperature range one can always find a part of the curve which is close to a linear one. However, this linear dependence is not a result of the validity of the Tarasov model, which predicts such behavior, as it is often assumed. There are many arguments against such identification, these arguments being not just arbitrary [the linear dependencies are not extrapolated to C = 0 at 0 K, as Eq. (1.13) predicts, but intersect C > 0]. More important physical arguments are that

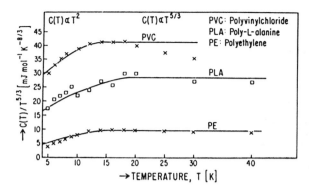

Fig. 1.11. Temperature dependence of heat capacity of three representative polymers with the data plotted as $C(T)/T^{5/3}$ to emphasize the $T^{5/3}$ dependence. The fits are $T^{1/3}$ (giving $C \sim T^2$) and $C/T^{5/3} =$ constant [42]

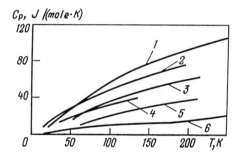

Fig. 1.12. Temperature dependencies of the heat capacity for lienar polymers [62]. *1* – PMMA; *2* – PS; *3* – PVC2; *4* – PTFE; *5* – PVAl; *6* – PE

the linear dependencies occur due to summation of various non-linear contributions from temperature resulting from low frequency skeletal and side group contributions [62, 63]. Therefore, it is quite natural that these linear sections cannot be useful in examination of the model dealing only with the chain skeleton. To perform such an examination, it is necessary to pick out from the total heat capacity only that part which belongs to the skeleton vibrations.

In the simplest case, that of PE, the vibrations of CH_2-groups are rather high-frequency (Table 1.3) and give very small contribution to C below 150 K. Therefore, one can consider the experimentally determined heat capacity as completely resulting from the skeleton motion. Most thoroughly this problem was discussed in [49, 50, 62, 63]. In the temperature region $100 \leq T \leq 200$ K the heat capacity of crystalline PE can be represented by (Fig. 1.6) [49]

$$\lg C_v = 0.75 \lg T - 0.54 \tag{1.31}$$

$$C_v = a_1 T + a_{1/2} T^{1/2} \tag{1.32}$$

where $a_1 = 0.045$ J/(mole K^2) and $a_{1/2} = 0.496$ J/(mole $K^{3/2}$). The slope of the line corresponding to Eq. (1.31) is 3/4, which is deviate from the prediction of both the Tarasov and Lifshitz models. In Eq. (1.32) the first term seems to be identified with the contribution from the torsional skeletal vibration and the second one – from the bending mode. Both terms contribute practically equally to C_v in this temperature interval. However, it seems better to consider both these equations as empirical expressions.

Table 1.3. Vibration frequencies of CF_2 and CH_2 groups [20, 64]

Vibration mode	CF_2		CH_2	
	ω, cm^{-1}	θ_E, K	ω, cm^{-1}	θ_E, K
Twisting	289	416	1173	1688
Bending	413	594	1433	2062
Rocking	430	619	1022	1471
Wagging	611	879	1394	2006
Symmetrical stretching	975	1403	2973	4278
Asymmetrical stretching	1350	1943	2998	4314
C-C stretching	1059	1524	1059	1524

In the above treatment, we tried to find a part of experimental C vs T curves which can be identified with the theoretical predictions. One more procedure which is widely used for describing the skeleton vibrations consists of using the 3-dimensional Debye function to characterize the intermolecular skeletal vibrations. The remaining intramolecular skeletal vibrations are then approximated by a box-distribution as proposed by Tarasov. The ratio of the number of vibrations corresponding to the 3-dimensional Debye function to the number of vibrations described by the box-distribution is fixed by the ratio of θ_3/θ_1 [see Eq. (1.17)]. The experimental procedure consists of finding the optimal values of the parameters θ_3 and θ_1 which describe C in the chosen temperature interval with the highest accuracy. Normally this accuracy is about 5%. These two parameters then are considered as the characteristic temperatures introduced in the Tarasov model and it is concluded that this model describes the experimental data. Sometimes θ_3 and θ_1 are differentiated by introducing the transverse and longitudinal components (θ_{3l} and θ_{3t}, θ_{1l} and θ_{1t}), which can lead to a better analytical description. However, it seems quite obvious that such a procedure is an empirical selection of either two or four parameters by means of which we can describe experimental data. Therefore, the formula which one arrives at as a result of this procedure can be considered only as the normal empirical relation. Any conclusions concerning the physical features of the Tarasov model, however, are not unequivocal.

The procedure described above has been applied to many polymers including PE. A recent variant of the scheme has been described by Wunderlich and coworkers [43–48]. They applied it to the heat capacity of fully crystalline PE obtained in 45 separate investigations on widely different samples with crystallinities between 0.42 and 0.97. The heat capacity below 150 K can be described well by means ot two theta temperatures $\theta_3 = 158\,K$ and $\theta_1 = 519\,K$ [45]. These values are only slightly different from the earlier values [4].

The same calculation scheme has been applied to many polymers, including aliphatic polyoxides, fluoropolymers, polybutadienes and others. Figure 1.13 shows a plot of θ_3 and θ_1 values for a series of polyoxides for which the low temperature heat capacity measurements are available [45]. The most drastic change with increasing the oxygen ratio O/CH_2 occurs in θ_1. It has been emphasized that the presently available normal mode calculations for crystalline PE and POM in the low frequency intermolecular vibration region are not in agreement with the heat capacities. In this region, heat capacity is best reproduced by the Tarasov two-parameters θ_3 and θ_1 expression.

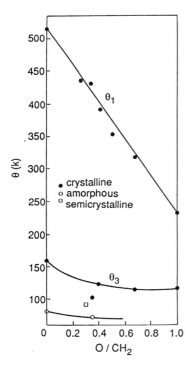

Fig. 1.13. θ_3 and θ_1 for homologous polyoxides as a function of oxygen ratio [45]. ($\bullet$) – crystalline; ($\circ$) – amorphous; ($\square$) – semicrystalline

For the chains which contain a large number of acoustic vibrations the procedure of the selection of the skeletal heat capacity of the total one is a more difficult. It can be examplified by PTFE behavior [46, 63–65]. Because F-atoms are rather heavy even at low temperatures a number of low frequency acoustic vibrations are excited. The heat capacity resulting from these low frequency optical vibrations must be extracted from the total heat capacity. The vibrational parameters of PTFE are listed in Table 1.3. During the calculations it is necessary to take into account the coupling of various vibrations. Figure 1.14 shows the various contributions to the total heat ca-

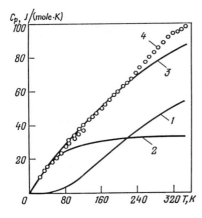

Fig. 1.14. Temperature dependence of heat capacity of PTFE [63, 65]. *1* – C_{optic}; *2* – C_{acoust}; *3* – $C = C_{optic} + C_{acoust}$; *4* – experimental values

pacity of PTFE. In the vicinity of 60 K the contribution of the low frequency optical vibrations to the total heat capacity is about 10%, while in the vicinity of 160 K it reaches about 50%. One can assume that a similar ratio is typical for chains with heavy pendent groups. For the side chains of more complex structure it is also necessary to take into account the possibility of some restricted internal rotation [62, 66].

Sochava [62] drew attention to the fact that the change in the absolute values of C_p in a large temperature interval from about 50 K up to T_g can be explained by the change in the mass of the repeating units because the corresponding vibrational frequencies are inversely proportional to the square root of the mass (see Fig. 1.12). Wunderlich and Jones [68] suggested an additive scheme based on this principle for estimating the heat capacity of linear macromolecules. Heat capacity contributions of various chemical groups have been tabulated [68, 69].

1.3 Temperature Transitions and Heat Capacity

1.3.1 Amorphous Polymers

Glass Transition (α-Relaxation)

The transition of any amorphous polymer from the glassy into the liquid state similar to low molecular substances is accompanied by sharp jump in the heat capacity. In an early treatment [67] based on the hole theory it has been assumed that this increase ΔC_p results mainly due to the hole formation. However, this treatment is certainly restricted because the similar sharp increase is also found under the constant volume condition. Therefore, for this increase of the heat capacity at the glass transition not only is the formation of the holes responsible but some other effects as well.

According to a later point of view, the following three contributions to the heat capacity jump at glass transition should be taken into account [70, 71]

$$\Delta C_p = \Delta C^c + \Delta C^h + \Delta C^v \tag{1.33}$$

where ΔC^c is the conformational contribution to the heat capacity change resulting from the change in conformational equilibrium in the glass transition region; ΔC^h is the contribution resulting from the increase of free volume at the glass transition and appearance of the new additional holes; finally, ΔC^v is the contribution arising from the change of the vibrational frequencies and the anharmonicity of the vibrational modes resulting, in turn, from the jump-like change of the compressibility and thermal expansivity. The rigorous determination of the relation between these contributions is a very complex experimental and theoretical problem. Below we will briefly consider the main approaches for determination of all the terms of Eq. (1.33).

According to the free volume concept, the change in heat capacity at the glass transition is a result of the rapid increase of the number of holes during unfreezing [67]. The corresponding heat capacity change is

$$\Delta C^h = E_h (\partial N_h / \partial T)_p \tag{1.34}$$

where E_h is the mole energy of the holes being the difference between the mole energy of a body which includes the holes and that of a body without holes. N_h is the number of moles of the holes. It is assumed that the equilibrium distribution is controlled by the Boltzmann equation

$$N_h = N_0 (V_0 / V_h) \exp(-E_n / RT) \tag{1.35}$$

where N_0 is the number of moles of the repeating unit, V_0 is their volume and V_h is the mole volume of the holes. The joint consideration of Eq. (1.34) and Eq. (1.35) gives at T_g

$$\Delta C_p^h = R(V_0/V_h)(E_h/RT_g)^2 \exp(-E_h/RT_g) \ . \tag{1.36}$$

Using the reasonable ration $V_0/V_h \not\subset \cong E_0/E_h = 5\text{–}6$ (E_0 is the cohesion energy) Wunderlich has arrived at

$$\Delta C_p' \cong \bar{M} \Delta C_p = 12.4 \ \frac{J}{\text{mol K}} \tag{1.37}$$

where $\bar{M}$ is the molecular weight of the "bead", the bead being the backbone unit which can be considered rigid. PE and PP have 1 and 2 beads, respectively. The analysis of experimental results for numerous glassy polymers (and other materials) [67, 69] shows that the heat capacity increase $\Delta C_p'$ is about $11\,\text{J K}^{-1}\,\text{mol}^{-1}$. In spite of some uncertainty as to the number of beads in complex macromolecules, the rule of constant heat capacity increase is a very important thermodynamic characteristic of the glass transition.

Normally macromolecules are characterized by the existence of some rotational isomers with the energy difference of $0.5\text{–}8\,\text{kJ/mole}$ [72, 73]. In the most widely distributed case of the carbochain macromolecules they possess one low-energy *trans*-isomer and two *gauche*-isomers with a higher energy.

If the conformational equilibrium changes with temperature, then the additional contribution to the heat capacity occurs, which is usually called "conformational heat capacity". This conformational heat capacity for the macromolecules with the simplest side groups can be calculated using the Gibbs-DiMarzio theory [74, 75]. According to this theory the fraction of the flexible bonds in the high energy state f is given by

$$f = \frac{2\exp(-E_c/k_B T)}{1 + 2\exp(-E_c/k_B T)} \qquad (1.38)$$

where E_c is the energy difference between the high-energy *gauche* state and the low-energy *trans* state. The statistical mechanical calculations show that for a low-energy *trans* state and two high-energy *gauche* states the conformational heat capacity is

$$C^c = R(E_c/k_B T)^2 \frac{2\exp(-E_c/k_B T)}{[1 + 2\exp(-E_c/k_B T)]^2} \cdot \qquad (1.39)$$

Combining Eqs. (1.38) and (1.39) we arrive at

$$C^c = R(E_c/k_B T)^2 f(1 - f) \,. \qquad (1.40)$$

The fraction of high-energy bonds and the conformational specific heat as a function of $E_c/k_B T$ are shown in Fig. 1.15. The apperance of the maximum as a function of temperature is the consequence of the monotonous increase of the fraction of high energy bonds as the temperature increases. The temperature, at which the maximum occurs, depends upon E_c and as E_c increases from 0.5 to 8 kJ/mole the maximum shifts approximately from 25 to 200 K.

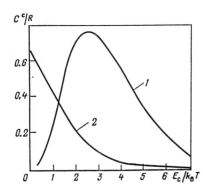

Fig. 1.15. Conformational specific heat for two states (*trans*, *gauche*) model (*1*) and fraction of bonds in high-energy state (*gauche*) (*2*) as a function of $E_c/k_B T$ (E_c is the energy difference between *gauche* and *trans* state) [76]

At what temperatures can one expect the manifestation of the conformational heat capacity under real conditions? Below T_g the fraction of *trans* and *gauche* bonds is constant due to their frozen state and the conformational contribution to the heat capacity is negligible. At T_g the segmental motion is unfrozen, therefore this transition is accompanied by the conformational isomerization. If one assumes that under constant volume the hole contribution is absent, then the basic premise is that the total change in heat capacity at constant volume, $\Delta C_v(T_g)$, is principally due to the change in conformational equilibrium with temperature [76–78]. Hence, this hypothesis is based on the assumption that the term C^v in Eq. (1.33) gives a considerably smaller contribution in comparison with C^c. The change in the heat capacity at constant volume ΔC_v can be readily calculated from the thermodynamic equation

$$C_p = C_v + TV\alpha^2/\chi .\tag{1.41}$$

At glass transition

$$\Delta C_v(T_g) = \Delta C_p(T_g) - T_g V_g \left(\frac{\alpha_l^2 - \alpha_g^2}{\chi_l - \chi_g} \right)\tag{1.42}$$

where the subscripts l and g refer to the liquid and glassy states above and below T_g. According to Eqs. (1.33) and (1.42) the second term in the right part of Eq. (1.42) corresponds to the contribution of the free volume (holes) to the heat capacity at T_g. According to the main hypothesis

$$R(E_c/k_BT)^2 f(1-f) = \Delta C_p(T_g) - T_g V_g \Delta \left(\frac{\alpha^2}{\chi} \right) .\tag{1.44}$$

O'Reilly [76, 77] has demonstrated that the energy difference E_c derived from the constant volume specific heat $\Delta C_v(T_g)$ for simply substituted vinyl chains [H, CH_3, Cl, $(CH_3)_2$ substitutes] are in good agreement with E_c calculated from the Gibbs-DiMarzio theory. For the chains with more complicated substitutes, conformational degrees of freedom in the side groups must be also taken into account.

Roe and Tonelli [70] in their treatment of the conformational heat capacity have also used the rotational isomeric states theory. Their calculations are of a more general character, because they are based on the generalized determination of the conformational energy

$$E_c = k_BT^2 \, d\ln Z_c/dT\tag{1.45}$$

where Z_c is the partition function. The conformational heat capacity can be immediately calculated according to

$$C^c = dE_c/dT .\tag{1.46}$$

Using this approach the conformational energy E_c of a large number of macromolecules above and below T_g has been calculated and their conformational heat capacity has been determined. The results are listed in Table 1.4. They will be discussed later.

Table 1.4. Ratio of various contributions to $\Delta C_p(T_g)$ [in J/(gK)] for amorphous polymers

Polymer	$\Delta C_p(T_g)$ experiment	ΔC_p^c Equations				ΔC_p^h Equations			ΔC_p Eq.
		(1.44) [76]	(1.39) [76]	(1.46) [70]	(1.49) [71]	(1.42) [70]	(1.47) [70]	(1.51) [71]	(1.52) [71]
PE	0.60	0.54	–	0.238	0.428	–	–	0.143	0.02
PP	0.48	0.20	–	0.176	0.279	–	–	0.156	0.075
PIB	0.40	0.22	0.34	0.125	0.214	–	–	0.163	0.05
PVC	0.30	0.15	0.235	0.054	0.192	0.213	0.155	0.096	0.074
PBA	0.41	0.22	0.381	0.062	0.209	0.363	0.382	0.107	0.110
PMMA	0.30	0.19	0.257	0.038	0.180	0.184	0.155	0.118	0.100
PS	0.34	0.24	0.260	0.071	0.115	0.176	0.155	0.118	0.078
α-PMS	0.32	0.20	0.251	0.046	0.102	–	–	0.125	0.082
PDMS	0.42	0.24	0.254	0.134	–	–	–	–	–
PC	0.24	0.06	0.198	0.084	0.049	–	–	0.097	0.135
PET	0.33	0.22	0.269	0.021	0.125	–	–	0.088	0.078
PPO	0.24	0.05	–	0.07	–	–	–	–	–

In contrast to O'Reilly's approach, Roe and Tonelli have assumed the difference $\Delta C_p - \Delta C^v \neq \Delta C^c$, i.e. it does not correspond only to the contribution of the free volume. In the glassy state as the temperature changes under the constant volume condition both the free and occupied volumes are constant. In the liquid state at the constant total volume between the free and occupied volumes, some redistribution resulting from the minimization of the free energy may occur. Therefore, the reliable values of ΔC^h from the experimental values of ΔC_p can be obtained only on the condition that both the total and free volume are constant. One needs to calculate $\Delta C_{v,v_h}$. It is impossible to reach this requirement experimentally. It is assumed that it can be calculated using the following relation

$$\Delta C_{v,v_h} = \Delta C_p - TV \frac{(\Delta \alpha)^2}{\Delta \chi} \tag{1.47}$$

which follows from the hole theory developed by Nose [79]. Similar relation was also obtained by Goldstein [80]. Tanaka [81] has considered the partition function which includes the conformational term and the term related to the free volume and has arrived at

$$\Delta C_p = \frac{d}{dT} \left[\left\{ \left(RT^2 \frac{d \ln Z_c(T)}{dT} \right) \Big/ p \right\} \right] - \left[\frac{d}{dT} - \left(RT^2 \frac{d \ln V_h}{dT} \right) \right.$$
$$\left. + \frac{dH_0}{dT} + \frac{3}{2} R \right] \tag{1.48}$$

where Z_c is the conformational partition function of the isolated chain at temperature T; V_h is the free volume corresponding to one structural unit; p is the degree of polymerization; H_0 is the intermolecular cohesion energy per a mole of the structural units. The first and second terms in the right part of Eq. (1.48) are characterized by the contributions of ΔC^c and ΔC^h, respectively, while two other terms reflect the change in the vibrational heat capacity at T_g.

The most complete theory of the heat capacity change at T_g for glassy polymers has been developed by DiMarzio and Dowell [71]. They have incorporated into the Gibbs-DiMarzio configurational entropy theory the lattice vibrations to reflect the changes in the vibrational heat capacity during the glass transition. The main result of their calculations is

$$\Delta C_p = R \left(\frac{E_c}{k_B T} \right)^2 f(1 - f) + 4RT\Delta\alpha(1 - 4.17T\Delta\alpha)$$
$$+ 0.5T\Delta\alpha C_p(T_{g-})$$ (1.49)

where $C_p(T_{g-})$ denotes the heat capacity in the glassy state below T_g. The first two terms are configurational and arise from the change of number of flexible bonds and of number of holes with temperature. The third term is a vibrational contribution. It is remarkable that the final expression has no adjustable parameters and only the chemical structure needs to be known to calculate the increase in the heat capacity at T_g. DiMarzio and Dowell have estimated the typical value of the three terms in Eq. (1.49) and arrived at the following results

$$R \left(\frac{E_c}{k_B T} \right)^2 f(1 - f) \cong 6 \text{ J/mole of flexible bonds K}$$ (1.50)

$$4RT\Delta\alpha(1 - 4.17\,T\Delta\alpha) \cong 2.0 \text{ J/mole of beads K}$$ (1.51)

$$0.5T\Delta\alpha C_p(T_{g-}) \cong 1.25 \text{ J/mole of beads K}$$ (1.52)

which give 9.25 J/mole K for all three terms. This value differs from that given by Wunderlich by about 20%. This accuracy can be considered as rather high keeping in mind that in the theory no adjustable parameters are involved.

Equation (1.49) allows us to make some qualitative conclusions concerning the heat capacity behavior of polymers in the liquid state. As the temperature increases the conformational contribution to C_p should decrease. The contribution resulting from the holes formation is a weak function of temperature, while the vibrational contribution increases although not so quickly as below T_g. Thus, one can expect that the temperature coefficient of C_p in the liquid state will be less than in the glassy state. This qualitative prediction agrees with experiment.

Table 1.4 gives a comparison of various contributions to the heat capacity change at T_g predicted by the equations considered above. The table summarizes the experimental values for 12 well studied polymers together with the computational data. First of all, a large difference in the conformational contribution should be indicated. DiMarzio and Dowell have concluded as a result of a comparison of their theoretical predictions with experiment, that the conformational contribution is about 50% of the total increase of C_p at

T_g, the contribution from the volume expansion is about 30% of the total and, finally, the vibrational contribution is about 20% of the total. According to estimations given by O'Reilly and Havlichek et al., the conformational contribution is higher. However, Havlichek et al. have not taken into account the vibrational contribution at all, therefore, if it is taken into account the conformational contribution should be decreased. Roe and Tonelli believe that ΔC^h should be between 1/3 and 1/2 of the total change.

Summarizing this discussion, one can emphasize that the most important conclusion of all these results is that all the contributions are comparable in their values and non of the contributions can be neglected in calculations of ΔC_p. This conclusion refutes the widely accepted suggestion of the free volume theory that the heat capacity increase at glass transition is mainly due to the rapid increase of numbers of holes with temperature. On the other hand, this conclusion does not give grounds for overestimating the role of the conformational contribution. Finally, the recognition of the important role of the vibrational contribution arising from the change of the characteristic frequencies with temperature demonstrates some important differences of the vibrational spectra of polymers in the glassy and liquid states.

Goldstein [82, 83] has examined in considerable detail the possible nature of the change in vibrational characteristic during the glass transition. If the change in the vibrational spectrum (characteristic frequencies) does not contribute to the total heat capacity jump at T_g then the variation of the entropy between two glasses with different thermal history should remain constant as temperature is lowered to 0 K. If, however, there are appreciable changes in vibrational frequencies, so that the structures of higher energy include more softer vibrational modes, the entropy difference between the two samples will decrease with temperature. According to Goldstein this decrease will resemble a Debye heat capacity curve. On the other hand, if some changes in the anharmonicities of the vibrational modes occur, there must be a rapid fall in the entropy with decreasing temperature and then a slower change with a further temperature decrease. Goldstein has emphasized that it is a very difficult problem to identify experimentally the nature of the changes, because of the secondary relaxations which can contribute to the heat capacity in a similar manner. Fortunately, the heat capacity is not very sensitive to the secondary relaxations (see below).

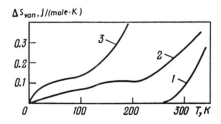

Fig. 1.16. The vanishing part of the entropy difference ΔS_{vp} as a function of temperature [82, 83]. 1 – PVC; 2 – PEMA; 3 – cis-1,4-PI

In order to answer the questions raised, Goldstein examined very carefully the very accurate heat capacity measurements of some polymers (PVC, PEMA, cis-1,4-PI) and some low molecular organic glasses. Figure 1.16 shows the vanishing part of the entropy difference as a function of temperature. An examination of these results led to the following conclusions. The conformational fraction of the heat capacity change ΔC^c is 0.79 for PI, 0.15 for PEMA and 0.58 for PVC. The later value is in a good agreement with the estimation of DiMarzio and Dowell, while for PEMA – of Roe and Tonelli. For all three polymers just after the freezing in this plot a rapid drop occurs (Fig. 1.16). A less pronounced entropy decrease also takes place below 50 K for PI and below 100 K for PEMA, however, the drop just below T_g is considerably higher. For PVC, only the high temperature drop exists. These results seem allow us to conclude that the larger part of ΔC_p is a result of the change in the anharmonicity.

Subglass Secondary Relaxations

It is generally accepted now that in the glassy state some secondary relaxational processes occur, which are referred to δ-relaxation, γ-relaxation and β-relaxation [84, 85]. These relaxations are usually related to the molecular rearrangement processes that do not freeze until much lower temperatures than the glass transition temperature. These motions, as shown by their activation energy, are hindered by potential barriers much lower than those that are associated with the main glass transition. Motions of this type are normally explained by the existence in macromolecules of some additional internal degrees of freedom other than those which are connected with α-relaxation. Such degrees of freedom can be associated, for example, with the motion of side groups attached to the main chain, which are capable of the hindered internal motions.

Very important results concerning the β-relaxation were obtained in the 1970s, when it was established that the β-relaxation is a characteristic property of amorphous packing rather than being associated with polymer structure [83–85] and it was suggested that β-processes arise as a liquid-like motion in the places where less dense packing exists. It was also shown that the β-relaxation and α-relaxation are closely related. For many polymers $T_\beta/T_\alpha \cong 0.75$ [84]. This relationship has been interpreted as an indiciation that the motion occurring in T_β is a precursor of that which will occur at T_g. A number of theoretical estimations [86–88] allow one to assume that the β-relaxation corresponds to the quasi-independent non-cooperative localized modes, while the glass transition (α-relaxation) corresponds to the cooperative motion of the same modes.

All these relaxational motions below T_g should give some contribution to the heat capacity. In general, the changes in physical properties resulting from these relaxations are much smaller than those accompanying the glass transition. Heat capacity measurements seem to be not very sensitive to these relaxational processes. No sudden changes similar to those occurring at T_g are detected at the temperatures where mechanical or dielectrical measurements indicate that these motions should freeze out. Therefore, in the early measurements of C_p, especially when carried out by means of adiabatic calorimetry, these relaxational transitions were not fixed at all. However, during the last decade Bershtein and Egorov have successfully used DSC for investigating the β-relaxation in many polymers. These experimental results have been summarized recently in their book [88]. In contrast to the discontinuity in C_p at the glass transition, the behavior of C_p in the vicinity of β-relaxation has, unfortunately, no uniform shape and is strongly dependent upon the thermal history of samples. Depending on the thermo-mechanical treatment of the samples it may be revealed either as a small hump or discontinuity, or even as a change in slope of C_p. Therefore, its detection and indication as β-transition is still "an art". Of course, there is no generalized quantitative calorimetric characteristic of the β-relaxation. Typical examples of C_p behavior in the vicinity of β-relaxation are examined in Chaps. 7 and 9 in connection of the energy state of deformed polymers.

T_{ll}-Transition

Above T_g in the liquid state in many polymers a temperature transition often occurs, which is referred to as T_{ll} – or "liquid 1–liquid 2" transition [90–93]. This term indicates that at the T_{ll} the transition of the liquid with a fixed structure into the really liquid state occurs. Many experimental methods including DSC are sensitive to the T_{ll}-transition. The shape of C_p vs T anomalous behavior in the vicinity T_{ll}-transition depends strongly upon a number of factors, such as the form of the samples, their thermal history, existence of the volatile additives etc. [94–96]. Normally, the T_{ll}-transition is accompanied by only the change of the slope ΔT vs T curve. A comprehensive discussion of the calorimetric characteristics of the T_{ll}-transitions in polymers is given by Bershtein and Egorov [89].

In conclusion, it is necessary to point out that we have considered here only the equilibrium aspects of the heat capacity behavior at the glass transition and β-relaxation and have not touched the non-equilibrium kinetic aspects of these relaxational transitions which are discussed in [4, 88, 89, 98].

1.3.2 Crystalline Polymers

Glass and Secondary Relaxations

According to the two-phase model of crystalline polymers, a simple linear relationship between the heat capacity change and the degree of crystallinity should exists

$$w = 1 - \frac{\Delta C_{p,c}}{\Delta C_p} \tag{1.53}$$

where $\Delta C_{p,c}$ is the heat capacity change at T_g of the semi-crystalline polymers. It is assumed that this relationship is valid for all the values w between

0 and 1. However, normally such linear dependence exists really only in the range of low and intermediate values of w, while at high values of w the heat capacity change is diminished considerably and even identification of T_g of the crystalline polymers using C_p-T dependence turns to be a problem.

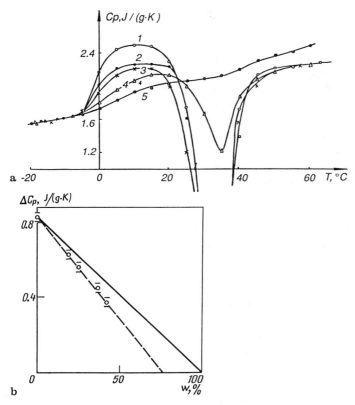

Fig. 1.17. Temperature dependence of heat capacity of PU (hexamethylene diizocianate and diethylene glycole) crystallized up to various degrees of crystallinity (**a**) and dependence of ΔC_p at T_g as a function of the degree of crystallinity (**b**) [97, 98]. Degree of crystallinity: *1* – 19%; *2* – 23%; *3* – 37%; *4* – 42%

Typical behavior of the heat capacity in the vicinity of the glass transition of a crystalline polymer with increasing degree of crystallinity is shown in Fig. 1.17 [97, 989. It is seen that when w increases ΔC_p decreases. However, the ΔC_p vs w extrapolates not to 1, but to about 0.75. It means that for the polymer with a high degree of crystallinity, ΔC_p will be absent. Such behavior is really characteristic of highly crystalline polymers [99, 100]. A non-proportional decrease of ΔC_p with w leads to some overestimated values of w evaluated using Eq. (1.53) in comparison with other methods including the method where the heat of fusion of crystalline polymers is used. It is necessary to emphasize that the disappearance of ΔC_p for highly crystalline polymers reflects a non adequate character of the two-phase model rather than a lack of measurement sensitivity.

The difficulties of the unequivocal identification of the glass transition of highly crystalline polymers using the C_p vs T relationship is seen especially clearly for PE – the most widely studied crystalline polymer. The discussion concerning the glass transition temperature of PE has been continuing for many years, nevertheless, any unanimous opinion is still absent. Usually, three temperature intervals are considered as the glass transition temperature of the amorphous part of PE: $150 \mp 10\,$K, $195 \mp 10\,$K and $240 \mp 10\,$K [100–103]. The most careful examination of the glass transition temperature of PE using the temperature dependence of heat capacity was carried out by Beatty and Karasz [100]. From their analysis they concluded that according to calorimetric measurements the only temperature interval $150\mp10\,$K may be identified as the glass transition of linear PE, because the discontinuity in C_p corresponding to the glass transition occurs only in the temperature region $150 \mp 10\,$K and is absent in the two other temperature intervals (Fig. 1.18). Figure 1.19 demonstrates the even more striking behavior of C_p below and above this T_g. Above it has been shown that below the glass transition the heat capacity of crystalline and amorphous polymers are practically identical. The horizontal line in Fig. 1.19 shows it very obviously. Above T_g, due to the additional sensitivity to the amorphous part contributions, C_p begins to depend upon w. This behavior is one more additional indication that T_g of linear PE is below $180\,$K. Finally, it should be mentioned that the cold crystallization in the highly amorphous PE occurs in the temperature region 160–$180\,$K [104], which is consistent with $T_g = 150 \mp 10\,$K and inconsistent with two other temperature intervals because below T_g crystallization cannot take place. Nevertheless, the temperature interval 150 ∓ 10 is often considered as the interval of the β-relaxation, while two next temperature intervals are identified as the multiple glass transition.

Figure 1.20 shows the discontinuous C_p behavior of other highly crystalline polymers for which the identification of T_g is also a problem. A more detailed consideration of the behavior of C_p at the glass transition in highly crystalline polymers can be found in [89].

Hence, ΔC_p at the glass transition of the amorphous regions of crystalline polymers falls with w quicker than the two-phase model predicts for crystalline polymers and at a high degree of crystallinity the discontinuity may disappear completely. The molecular reasons for such behavior are not clear yet, therefore it is difficult to estimate rigorously the changes in various contributions in C_p at T_g. One can suggest that the conformational contribution ΔC_p is the most sensitive to crystallinity. In contrast to the statistical coils in amorphous polymers, amorphous regions in crystalline polymers include loops of various sizes, chain ends, and especially highly stretched tie-molecules. It is quite obvious that in their conformational distributions low energy (*trans*, assumed) conformations dominate. Such chains seem to be similar to highly stretched parts of highly elastic rubber-like chains and for examination of their conformational state one can use the Volkenstein-Ptitsin

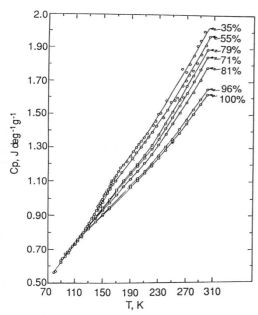

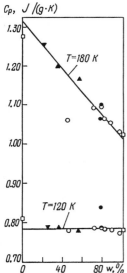

Fig. 1.18. Heat capacities of linear PE of various crystallinities shown near the curves [100]

Fig. 1.19. Heat capacities of linear PE at fixed temperatures as a function of degree of crystallinity [100]. Various marks correspond to various sources. The extrapolated w = 0 and 100% are also shown

[71] approach. They predicted theoretically that the conformational heat capacity of stretched macromolecules should be smaller than unstretched. Corresponding estimations show that the drop in the conformational heat capacity at the degree of stretching equal to 10 is quite comparable with the value of the conformational heat capacity, while at stretching up to 20–25 the conformational heat capacity is completely frozen. Hence, the absence of the proportionality between ΔC_p and w for highly crystalline polymers may

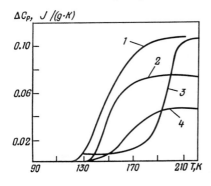

Fig. 1.20. Heat capacity difference plot for various crystalline polymers [100]. *1* – linear PE (w = 35%); *2* – linear PE (w = 55%); *3* – PTFE; *4* – POM

partly be a result of exclusion of the coil conformations in amorphous regions of crystalline polymers. To estimate the character of changes for two other contributions to ΔC_p with w is difficult.

A broad area of the calorimetric behavior of crystalline polymers on melting and other phase transitions has been discussed in a number of books [4, 89, 98, 105–107], therefore, it will not be discussed here.

References

1. Einstein A (1907) Ann Physik *22:* 180, 800
2. Maradudin AA, Montroll EW and Weiss G H (1963) Solid state physics, Suppl 3, Academic New York
3. Leibfried G and Ludwig W (1961) In: Solid state physics, vol 12, Seitz F, Turnbull D (ed), Academic New York
4. Wunderlich B and Baur H (1970) Fortschritte der Hochpolymeren Forschung *7:* 151
5. Debye P (1912) Ann. Physik *39:* 789
6. Tarasov VV (1945) DAN SSSR *46:* 22
7. Tarasov VV (1950) Zh Fiz Khim *24:* 111
8. Tarasov VV (1953) Zh Fiz Khim *27:* 1430
9. Tarasov VV (1979) Physical Problems of Glasses (in Russian), Stroiisdat Moscow
10. Lifshitz IM (1952) Zh Eksper Teor Fiz *22:* 471, 475
11. Lifshitz IM (1953) Zh Fiz Khim *27:* 294
12. Stepanov PE (1952) Zh Fiz Khim *26:* 1642
13. Landau LD and Lifshitz IM (1976) Statistical Physics, Part 1, Nauka Moscow
14. Kosevich AM (1981) Physical Mechanics of Real Crystals (in Russian), Naukova Dumka Kiev
15. Syrkin ES and Feodosyev SB (1982) Fiz Nizkich Temp (FNT) *8:* 761
16. Syrkin ES and Feodosyev SB (1982) Fiz Nizkich Temp (FNT) *8:* 1115
17. Kosevich YuA and Syrkin ES (1983) Fiz Nizkich Temp (FNT) *9:* 1196
18. Vilk YuP and Peresada VI (1985) Ukr Fiz Zh *30:* 1098
19. Kopinga K, Van der Leeden P and De Jonge WJW (1976) Phys Rev B *14:* 1519
20. Kitagawa T and Miyazawa T (1972) Adv Polymer Sci *9:* 336
21. Kirkwood JG (1939) J Chem Phys *7:* 506
22. Pitzer KS (1940) Chem Phys *8:* 711

23. Mikhaylov ID and Cheban YuV (1977) In: Chislenye Metody Resheniya Sadach Matemat Fiz i Teor System, p 73–76, Universitet Druzhby Narodov Moscow
24. Cheban, YuV (1979) Thesis, Universitet Druzhby Narodov Moscow
25. Stockmayer W and Hecht CE (1953) J Chem Phys *21:* 1954
26. Genensky SM and Newell CF (1957) J Chem Phys *26:* 486
27. Telezhenko YuV and Sucharevskii BYa (1982) Fiz Nizkich Temp *8:* 188
28. Telezhenko YuV (1982) Fiz Nizkich Temp *8:* 1228
29. Telezhenko YuV (1985) Thesis, Institute of Low Temperature Physics, Ukranian Academy of Sciences Kharkov
30. Stephens RB (1973) Phys Rev B *8:* 2896, (1976) *13:* 852
31. Pohl RO and Salinger GL (1976) Ann N Y Acad Sci *279:* 150
32. Bottger H (1974) Phys Stat Solidi, Ser B *62:* 9
33. Rosenstock HB (1972) J Non-Crystall Solids *7:* 123
34. Baltes P (1973) Solid State Comm *13:* 225
35. Tantilla WH (1977) Phys Rev Lett *39:* 554
36. Watkins GM and Fenichel H (1980) J Non-Crystall Solids *37:* 433
37. Anderson PW, Halperin BI and Varma CM (1972) Phil Mag *25:* 1
38. Phillips WA (1972) J Low Temp Phys *7:* 351
39. Smoljakov BP and Khaimovich EP (1982) Uspekhy fizicheskich nauk (UFN) *136:* 317
40. Goldanskii VI, Trachtenberg LI and Fleurov VN (1986) Tunneling Events in Chemical Physics (in Russian), chap 6, Nauka Moscow
41. Montgomery CG (1980) J Low Temp *39:* 13
42. Allen, JP (1986) J Chem Phys *84:* 4680
43. Gaur U and Wunderlich B (1981) J Phys Chem Ref Data *10:* 119, 1001, 1051; *11:* 313; Gaur U, Lau SF, Wunderlich BB and Wunderlich B (1982) J Phys Chem Ref Data *11:* 1065; (1983) *12:* 29, 65, 91
44. Cheban YuV, Lau SF and Wunderlich B (1982) Colloid Polym Sci *260:* 9
45. Grebowicz J, Suzuki H and Wunderlich B (1985) Polymer *27:* 561
46. Loufakis K and Wunderlich B (1985) Polymer *26:* 1875
47. Loufakis K and Wunderlich B (1986) Polymer *27:* 563
48. Grebowicz J, Aycock W and Wunderlich B (1986) Polymer *27:* 575
49. Baur H (1970) Kolloid Z Z Polym *241:* 1057
50. Baur H (1971) Kolloid Z Z Polym *244:* 293
51. Perepechko II (1980) Low Temperature Properties of Polymers, Pergamon Oxford
52. Tucker JE and Reese W (1967) J Chem Phys *46:* 1388
53. Finlayson DM, Mason P and Rogers JN (1980) J Phys C *13:* 185
54. Yosida S, Suga H and Seki S (1973) Polymer J *5:* 11
55. Choy CL, Hug M and Moody DE (1975) Phys Lett, Ser A *54:* 375
56. Reese W (1969) J Macromol Sci *A3:* 1257
57. Choy CL, Hunt RG and Salinger GL (1970) J Chem Phys *52:* 3629
58. Zoller P, Fehl DL and Dillinger JR (1973) J Polym Sci, Polym Phys Ed *11:* 1441
59. Madding RP, Fehl DL and Dillinger JR (1973) J Polym Sci, Polym Phys Ed *11:* 1435
60. Swaminathan K and Tewari SP (1980) J Polym Sci, Polym Phys Ed *18:* 1707
61. Boyer JD, Lasjauntias JC, Fisher RA and Phillips NE (1983) J Non-Crystall Solids *55:* 413
62. Sochava IV (1964) Vestn Leningrad Univ *19:* 56
63. Gotlib YuYa and Sochava IV (1962) DAN SSSR *147:* 580
64. Lee WK and Choy CL (1975) J Polym Sci, Polym Phys Ed *17:* 619
65. Choy CL, Leung WY and Chen FC (1979) J Polym Sci, Polym Phys Ed *17:* 87

66. O'Reilly JM and Karacz FE (1966) J Polym Sci, Polym Phys Ed *14:* 49
67. Wunderlich B (1960) J Phys Chem *64:* 1052
68. Wunderlich B and Jones LD (1969) J Macromol Sci *B3:* 67
69. Wunderlich B and Gaur (1980) Pure Appl Chem *52:* 445
70. Roe R-J and Tonelli AE (1978) Macromolecules *11:* 114
71. DiMarzio EA and Dowell F (1979) J Appl Phys *50:* 6061
72. Birshtein TM and Ptitsin OB (1964) Conformations of Macromolecules (in Russian) Nauka Moscow
73. Flory PJ (1969) Statistical mechanics of chain molecules, John Wiley New York London Sydney Toronto
74. Gibbs JH and DiMarzio EA (1958) J Chem Phys *28:* 373, 807
75. DiMarzio EA, Gibbs JH, Fleming PD III and Sanchez IC (1976) *9:* 763
76. O'Reilly JM (1977) J Appl Phys *48:* 4043
77. O'Reilly JM (1978) J Polym Sci, Polym Symp *N63:* 165
78. Havlicek I, Vojta V, Ilavsky M and Hrouz J (1980) Macromolecules *13:* 357 (1980)
79. Nose T (1971) Polym J *2:* 437
80. Goldstein M (1963) J Chem Phys *39:* 3369
81. Tanaka N (1978) Polymer *19:* 770
82. Goldstein M (1976) J Chem Phys *64:* 4767; (1977) *67:* 2246
83. Goldstein M (1976) Ann N Y Acad Sci *279:* 68
84. Boyer R (1975) J Polym Sci, Polym Symp *50:* 189
85. Boyer R (1976) Polymer *17:* 996
86. Gotlib YuYa (1968) J Polym Sci, Polym Symp *16:* 3365
87. Johari G and Goldstein M (1970) J Chem Phys *53:* 2372
88. Rostiashvili VG, Irzhak VI and Rosenberg BA (1987) Glass Transition of Polymers (in Russian), Khimiya Leningrad
89. Bershtein BA and Egorov VM (1990) Differential scanning calorimetry in physical chemistry of polymers (in Russian), Khimiya Leningrad
90. Boyer R (1980) J Macromol Sci-Phys *18:* 461, 563
91. Gillham J and Boyer R (1977) J Macromol Sci-Phys *13:* 497
92. Lobanov AM and Frenkel SYa (1980) Vysokomol Soed *A22:* 1045
93. Enns J and Boyer R (1987) In: Order in the amorphous state of polymers, p 221, Plenum New York
94. Kumler, Machajevski A, Fitzgerald J et al (1987) Macromolecules *20:* 1060
95. Chen J, Kow C, Fetters L and Platzek D (1985) J Polym Sci, Polym Phys Ed *23:* 13
96. Bershtein VA, Egorova LM, Egorov VM and Sinani AB (1989) Vysokomol Soed *B31:* 457
97. Godovsky YuK and Lipatov YuS (1968) Vysokomol Soed *A10:* 32
98. Godovsky YuK (1976) Thermophysical Methods of Polymers Characterization (in Russian), Khimiya Moscow
99. O'Reilly JM and Karacz FE (1964) ACS Polymer Prepr *5:* N2, 351
100. Beatty CL and Karacz FE (1979) J Macromol Sci, Part C *17:* 37
101. Stehling FC and Mandelkern L (1970) Macromolecules *3:* 242
102. Boyer R (1973) Macromolecules *6:* 288
103. Boyer R (1973) J Macromol Sci-Phys *7:* 487
104. Hendra PJ, Jobic HP and Holland-Moritz K (1975) J Polym Sci, Polym Lett Ed *13:* 365
105. Wunderlich B (1980) Macromolecular Physics, vol 3, Crystal Melting, Academic New York
106. Turi EA (ed) (1981) Thermal characterization of polymeric materials, Academic New York
107. Wunderlich B (to be published) Thermal analysis, Academic New York

2 Thermal Conductivity

2.1 Basic Concepts of Thermal Conductivity of Solids

Most polymers are insulating systems, therefore, any electronic effects are absent in them and heat conduction occurs as a result of lattice vibrations. Theoretical consideration of the thermal conductivity of the crystalline dielectrics where the lattice vibrations can be resolved into normal modes which can then be treated as phonons leads to the Debye equation [1]

$$\chi = \tfrac{1}{3}C(T)\bar{c}\bar{l} \tag{2.1}$$

where C is the heat capacity per unit volume, $\bar{c}$ is the average phonon velocity and $\bar{l}$ is the phonon mean free path. The temperature dependence of the thermal conductivity of crystalline solids (Fig. 2.1) can be understood according to this equation as follows. At high temperatures where most of the phonons are excited $\chi \sim T^{-1}$ since C constant and $\bar{l}$ is proportional to T^{-1}. With decreasing temperature the number of interacting phonons decreases exponentially and correspondingly both χ and $\bar{l}$ should increase exponentially. The further decrease of temperature must lead to the situation when $\bar{l}$ will be compared with the dimension of the sample and therefore its temperature dependence should disappear. The phonons then will be scattered only by the boundaries of the sample which characterized the mean free path. Hence, the thermal conductivity reaches a maximum after which its temperature dependence follows that of the specific heat, i.e. χ being proportional to T^3 at low temperatures.

Thus, according to the two suggested mechanisms of phonon scattering – on phonons (Umklapp process) and on the boundaries of the crystals the temperature dependence of χ of the crystalline dielectrics should have a maximum. A comparison of the theoretical values of χ in the maximum, estimated with these two suggestions, with experimental ones for most perfect crystals showed [1] that the experimental values are considerably lower than the theoretical ones. The reason for such behavior is the existence in the real crystals of various defects such as dislocations, point defects, impurities, isotopes and so on. Therefore the measurements of low temperature conductivity of solids turns out to be a very sensitive method of characterizing various defects of crystalline solids.

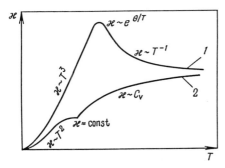

Fig. 2.1. Schematic drawing of the temperature dependence of the conductivity of crystalline (*1*) and amorphous (*2*) dielectric solids

The modern theory of low temperature thermal conductivity of crystalline solids predicts quantitatively both the temperature dependence and values of χ. However, at intermediate and elevated temperatures the theory can only predict its temperature dependence ($\chi \sim T^{-1}$), while for evaluating of the quantitative values of χ semiempirical relations are usually used [2].

The behavior of the thermal conductivity of amorphous solids is quite different in both the magnitude and temperature dependence (Fig. 2.1). The absence of the translation symmetry of the amorphous solids excludes the phonon scattering due to Umklapp process[1]. At normal and high temperatures the length of the free path can be considered as a constant, independent of temperature. In fact, simple estimations of the $\bar{l}$ according to Eq. (2.1) show that above 100–200 K its value is a few ångstroms, i.e. corresponds to the short range order in glasses. As the temperature is reduced the length of the mean free path changes at first very slowly. Therefore the temperature dependence of the heat conductivity in this temperature range follows the temperature dependence of C_V. Such behavior was found for many glasses up to about 20–50 K (see Fig. 2.1).

In the temperature range of 5–20 K, for many glasses including polymer glasses, it has been experimentally established [3–6] that χ does not follow the temperature dependence of heat capacity. In this temperature range, a plateau occurs where χ is independent of temperature and then χ becomes proportional to T^2. This very surprising fact that not only the temperature dependence of the heat conductivity for glasses of different nature is the same but also the absolute values of χ are also rather close has recently attracted considerable attention.

For an interpretation of the thermal conductivity of amorphous solids over a wide temperature range Klemens [7] introduced the idea about the scattering of phonons on the elastic disordered regions in the amorphous structures. Because of the absence of the long range order in glasses the propagation of elastic waves in such structures is accompanied with their scattering due to the change of their velocity. The main parameter in the mechanism of the "structural scattering" is the relation between the wavelength and the

---

[1] Inspite of the fact that the meaning of "phonon" for the amorphous solids can be introduced rather rigorously only for the longwave acoustic vibrations it is often used for the whole vibration spectrum

correlation length of the microscopic disorder [1]. At comparative values of these characteristics, intensive scattering occurs. The further development of the mechanism of the structure scattering has led to the hypothesis about a more effective scattering of the transverse waves than longitudinal. According to this idea the low temperature conductivity is governed first of all by the longitudinal modes because the transverse modes are characterized by small values of the mean free path. As the temperature is arised, the probability of interaction of the longitudinal and transverse waves increases considerably which leads to a very intensive scattering. Although this theoretical model could explain the temperature behavior of the thermal conductivity of amorphous solids in some temperature range it fails to explain the plateau and the following T^2 decrease of χ at very low temperatures.

Two approaches have been suggested for explaining this low temperature behavior of χ. According to the first approach [8,9], in which the general idea of Klemens concerning the scattering of phonons due to spatial fluctuations in the elastic properties of the amorphous solids was more quantitatively elaborated, T^2 dependence appears due to the assumption of a long correlation length for these fluctuations, while the plateau is explained by the increasing contribution with rising temperature of the scattering on the short-range correlation. The second approach is based on the idea that some atoms or groups of atoms may occur in energetically close states separated by an energy barrier [10, 11]. Resonant scattering by the tunneling between these states makes the principle contribution and this can lead to T^2 dependence and to the plateau region. Both these approaches have stimulated intensive theoretical and experimental investigation of the low temperature behavior of amorphous solids (see, for example, [12–14]).

Using these general concepts of the thermal conductivity of solid dielectric, we will consider the temperature dependence of the thermal conductivity of amorphous and crystalline polymers, an influence of various molecular parameters of polymers and some external parameters on the heat conductivity of polymers and polymer materials and, finally, the behavior of the thermal diffusivity which is closely related to the thermal conductivity.

2.2 Temperature Dependence of Thermal Conductivity of Polymers

In the study of the thermal conductivity of polymers one can distinguish two periods. Most of the studies up to about the middle of the 1970s were confined mostly to the temperature range around room temperature and only a few measurements were performed at low temperatures. These data were discussed in several reviews [15–21]. Starting from the middle of the 1970s, some new measurements performed at low temperature for various polymers were published. These measurements were initiated and stimulated, first of all, by a great interest in the anomalous thermal properties of amorphous dielectrics at low temperatures. For a number of both amorphous and crystalline polymers data are now available in the wide temperature range of 1 to 400 K. In our consideration of the temperature dependence of the thermal conductivity of polymers we follow the most comprehensive interpretation of the problem given by Choy [21].

2.2.1 Amorphous Polymers

Temperature dependence of χ is rather similar for various amorphous polymers. It has been most widely studied for PMMA [6, 16, 18, 21, 22, 23]. Figure 2.2 shows the generalized curve for PMMA obtained on the basis of these publications. It is suitable to discuss separately the behavior of χ above and below 100 K.

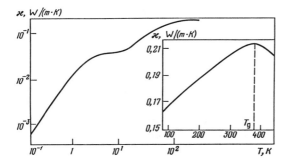

Fig. 2.2. Temperature dependence of the thermal conductivity of PMMA (the generalized curve drawn according to the literature experimental data)

Temperature Region Above 100 K

In this temperature range below T_g χ follows the temperature dependence of C because $\bar{l}$ is practically independent of temperature being a few ångstroms. Hence, below T_g the thermal conductivity will increase slightly with temperature as is seen from Fig. 2.3. However, in the temperature interval of glass transition a change of the slope of the dependence occurs (Fig. 2.2); below the glass transition the slope is positive while above T_g it is negative. The appearence of the break of χ at T_g is not obvious according to Eq. (2.1). According to this equation and the behavior of the heat capacity at T_g it is more likely to arrive at a jump of χ at T_g. Moreover besides the jump at the static T_g one could also expect a jump-like decrease of χ in the range of a jump-like decrease of $\bar{c}$.

Because of the absence of such jump-like changes in χ Eiermann [18, 23] has assumed that the heat conduction in the amorphous polymers near the glass transition both below and above T_g occurs due to a non-correlated heat transfer by inter- and intramolecular interactions, i.e. according to the

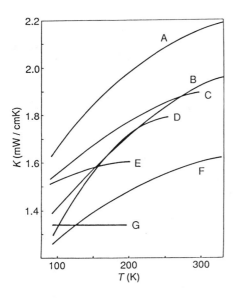

Fig. 2.3. Thermal conductivity of amorphous polymers above 100 K [21]. *A* – PET; *B* – PMMA; *C* – PBMA; *D* – PP (atactic); *E* – NR; *F* – PVC; *G* – PIB. Only data below the glass transitions are shown. Above the glass transitions the slopes of the curves change from positive to negative (see Fig. 2.2)

mechanism responsible for the heat conduction in liquids. In order to explain the experimental dependence of χ upon temperature below and above T_g, Eiermann performed the analytical analysis of the structural model taking into account a large local anisotropy of heat transfer along the chains and between them and arrived at the following relation

$$\Delta\left(\frac{1\mathrm{d}\chi}{\chi\mathrm{dT}}\right) = -5.8\Delta\alpha \tag{2.2}$$

where $\Delta\alpha$ corresponds to the jump of the thermal expansivity at T_g. Hence the physical reason of the break in the temperature dependence of χ is the difference in the coefficients of the thermal expansions below and above T_g. Equation (2.2) describes quite satisfactorily the behavior of a number of amorphous polymers of various chemical structure including PMMA [18, 19, 23].

Temperature Region Below 100 K

Two problems have attracted considerable attention in the low temperature thermal conductivity of amorphous solids: the physical nature of the low temperature plateau at 5–15 K and the temperature behavior of χ below 5 K (see Fig. 2.1 and 2.2). Recent experimental studies have established that the behavior of χ at very low temperatures (< 1 K) can be described by the following relationship: $\chi = AT^m$ with m = 1.9 ∓ 0.1 and A = 3×10^{-2} W/mK (variation of A for all amorphous solids studied does not exceed 3).

For explanation of this behavior, both the phonon scattering by disordered structure of glasses and the resonant scattering of phonons by tunneling-state approaches were used [21]. Analyzing the applicability of the first approach, Choy came to the conclusion that with the appropriate choice of 4 parameters of the model, reasonable agreement with the experimental results can be obtained. These parameters include the long and the short correlation lengths of the order of $3000\,\text{Å}$ and $10\,\text{Å}$ and the corresponding velocity fluctuations. While short correlation length can really exist in amorphous polymers the question of the long correlation length is still open.

According to the second approach the whole frequency spectrum can be divided into three regions, in which the phonon mean free path has different temperature dependence. In the first most low temperature region $\omega < \omega_R$ (where ω_R is the lower limit of the band of localized vibrations) and $\bar{l} = D\omega^{-1}$ (D is a parameter characterizing the material). The contribution of the phonons in the frequency region is

$$\chi_1 = \frac{k_B D}{2\pi^2 \hbar^2 \bar{c}^2} T^2 \int_0^{\theta/R} \frac{x^3 e^x}{(e^x - 1)^2} dx \qquad (2.3)$$

where $\theta_R = \hbar\omega_R/k_B$ and $x = \hbar\omega/kT_B$. In this region the Debye spectrum is used. The next two regions of the spectrum are characterized by the following relations $\omega_R \leq \omega \leq \omega_c$ and $\omega_c \leq \omega \leq \omega_1$, where ω_c is the upper limit of the band of localized vibrations and ω_1 is the maximum frequency of the acoustic phonon spectrum. In both these regions the mean free path is expected to be of the order of inter-atomic distances. The total contribution from these two regions is

$$\chi_2 = \frac{1}{3}\bar{c}\bar{l} \int_{\theta_R/T}^{\theta_1/T} \rho(\omega)\frac{x^2 e^x}{(e^x - 1)^2} dx \qquad (2.4)$$

where $\rho(\omega)$ is proportional to the density of the acoustic modes $\theta_1 = \hbar\omega_1/k_B$. Hence, in these two regions χ is proportional to C_v.

Choy [21] applied Eq. (2.3) and (2.4) to analyze the low temperature thermal conductivity of PMMA and some other amorphous polymers (Fig. 2.4 and Table 2.1). As is seen from Fig. 2.4 there is an reasonable agreement throughout the whole temperature range. The values of θ_R which have been used at the calculations of χ are also in a good coincidence with the results obtained during the analysis of the exceed heat capacity (Table 2.1, see also Table 1.2). Table 2.1 also shows that the values of D and θ_R are rather close for various polymers. This seems to be the reason that the thermal conductivity of all amorphous polymers are rather close at any temperature and above $15\,\text{K}$ it is within 15% of the average value.

Table 2.1. Parameters D and θ_R obtained from the analysis of the low-temperature thermal conductivity of amorphous polymers [21]

Polymer	$D\times10^{-5}$, m/s	θ_R, K	θ_a^*, K	θ_b^*, K
PMMA	13.5	4.7	4.9	17.5
PS	6.02	5.2	5.5	16
PC	5.61	4.5	5.0	14
PET	5.43	4.3	–	15

*Obtained from the analysis of the low-temperature heat cpacity (see Chapter 1)

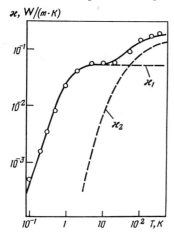

$\varkappa$, $W/(m\cdot K)$

Fig. 2.4. Comparison of the experimental data for the thermal conductivity of PMMA with the theoretical prediction according to Eqs. (2.3) and (2.4) [21]

In spite of the very close relation of the anomalous behavior of the heat capacity and thermal conductivity of glasses at low temperatures there are some experimental facts which cannot be explained from a general point of view. In particular, the thermal conductivity of glasses is completely insensitive to the chemical purity and thermal history of the samples while the heat capacity is very sensitive to these parameters [4]. Although between the excess heat capacity and the anomalous behavior of the thermal conductivity some correlation takes place (see Table 2.1), however, the suggestion that they relate to the same centers leads to the conclusion about the anomalously large (in comparison with the known defects in crystals) scattering ability of these centers. Further investigations with the combination of several experimental techniques need for elucidating the structure of these centers.

2.2.2 Crystalline Polymers

The temperature dependence of the thermal conductivity of crystalline polymers differs considerably from that of amorphous polymers. At low temperatures, no plateau region is observed and the temperature dependence of the thermal conductivity is highly sensitive to degree of crystallinity. Even with a similar degree of crystallinity the thermal conductivity of various polymers can vary considerably. In one of the group of crystalline polymers (HDPE, POM, POE) the thermal conductivity at first increase with temperature then reaches a maximum after which it begins to decrease (Fig. 2.5, curve 1). For another group of polymers (LDPE, PET, PP) the thermal conductivity increases monotonously up to the glass transition temperature after which the slope of the dependence changes from positive to negative (Fig. 2.5, curve 2), i.e. this behavior is similar to amorphous polymers in the vicinity of glass transition. The absolute values of χ for the first group of polymers is higher than for the second one. For both groups above $30\,K$, χ increases with degree of crystallinity. Below $20\,K$, reverse behavior takes place: with increasing degree of crystallinity χ drops sharply. For example, for PET at $2\,K$, χ for the amorphous sample is more than an order of magnitude higher than for the sample with 75% crystallinity. This unusual relation between the thermal conductivity of crystalline and amorphous polymers seems to indicate the existence in crystalline polymers of some extra mechanism of phonon scattering at low temperatures. Because of the different influence of crystallinity of the thermal conductivity below and above approximately 30 K the behavior of χ in these regions will be considered separately.

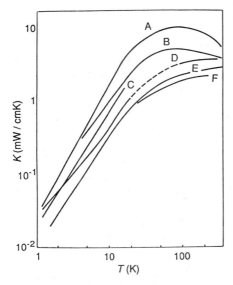

Fig. 2.5. Thermal conductivity of crystalline polymers [21]. A – PE ((w = 0.81); B – POM (w = 0.70); C – PE (w = 0.71); D – PE (W) = 0.43); E – PET (w = 0.51); F – PP (w = 0.62)

Temperature Region Above 30 K

In this temperature region χ is dependent upon the degree of crystallinity and shape of the crystallites. If the last parameter is not taken into account, then the thermal conductivity of the two phase system can be represented by the simple additive relation

$$\chi = w\chi_c + (1 - w)\chi_a \tag{2.5}$$

However, one can expect that the form of the crystallites and a very large anisotropy of the thermal conductivity will play an important role in the heat transfer process in crystalline polymers. If one assumes that such a polymer consists of the amorphous matrix with the crystallites embedded statistically then the Maxwell relation can be used for determination of the thermal conductivity of the system [18, 24]

$$\frac{\chi - \chi_a}{\chi + 2\chi_a} = w\frac{\chi_c - \chi_a}{\chi_c + 2\chi_a} \tag{2.5a}$$

If χ, χ_a and w are known one can estimate χ_c. Such calculations have been performed [18, 24] for some crystalline polymers (PE, POM) for that temperature interval where w can be considered to be constant. It turned out that χ is proportional to T^{-1} in quite good accord with the behavior of crystalline dielectrics in the temperature region where three-phonon umklapp scattering processes are responsible for the thermal conductivity.

A weak point of the model even for small values of the degree of crystallinity, when such a polymer can be considered as the amorphous matrix with the crystallites as some inclusions (at high values of w this suggestion can hardly take place), consists of ignoring the microanisotropy of the crystallites. It has already been mentioned that the conductivity along chains is at least one order of magnitude higher than through the van der Waals bonds perpendicular to the chains. Some early theoretical estimations give the value equal to 50 [25]. Recent model calculation for PE crystalline lattice led to an anisotropy factor as high as 1000 [26]. Therefore, the next step in the analysis of the thermal conductivity of crystalline polymers on the basis of the Maxwell model was the consideration of the thermal anisotropy of crystallites [27]. Two parameters have been introduced

$$k_{\parallel} = \chi_{c_{\parallel}}/\chi_a \gg 1 \quad \text{and} \quad k_{\perp} = \chi_{c_{\parallel}}/\chi_a \cong 1 \qquad (2.6)$$

By incorporating these parameters into the Maxwell equation and taking into account $k_{\parallel} \gg 1$ for polymers one can arrive at

$$\frac{\chi - \chi_a}{\chi + 2\chi_c} = w \left(\frac{2k_{\perp}^{-1}}{3k_{\perp} + 2} + \frac{1}{3} \right) \qquad (2.7)$$

Because $k_{\parallel} \gg 1$ this relation is practically insensitive to $k_{\parallel}$. Moreover, because the thermal conductivity of all amorphous polymers above $30\,\text{K}$ are rather close, the only parameter in Eq. (2.7) is $\chi_{c_{\parallel}}$. It can be estimated using the dependence of χ upon the degree of crystallinity.

Such an analysis has been performed for the polymers with a large enough variation of degree of crystallinity [21, 27]. Figure 2.6 shows the resulting $\chi_{c_{\perp}}$ values for four polymers as a function of temperature. It demonstrates obviously the above-mentioned separation of the crystalline polymers into two distinct groups with the values of χ strongly decreasing with rising temperature and with the values of χ a weak function of temperature being smaller in the whole temperature range than the values of the first group. The much lower χ values for PET and PP may be partly explained by weaker interchain interactions in these polymers. However, a surprisingly weak temperature dependence of χ according to Choy [21] seems to be the result of higher phonon scattering by defects in the crystallites of the polymers of this group due to more complicated and bulky repeating units. The phonon mean free path in these polymers is comparable with the intermolecular distances, therefore, the model analysis suggested for amorphous polymers [24] can be applied in this case. A larger conductivity of the crystalline phase is a result of the higher force constants of the interchain interaction, i.e. their lower thermal resistance. Therefore, the thermal conductivity may be connected with density. It follows from the model analysis

$$\frac{\chi_c - \chi_a}{\chi_a} = 5.8 \frac{\rho_c - \rho_a}{\rho_a} \tag{2.8a}$$

$$\frac{\chi - \chi_a}{\chi_a} = 5.8 \frac{\rho - \rho_a}{\rho_a} \tag{2.8b}$$

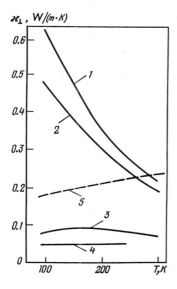

Fig. 2.6. Temperature dependence of the thermal conductivity of the crystallites normal to the chain axis [21]. 1 – PE; 2 – POM; 3 – PET; 4 – E – PP; 5 – the temperature dependence of the thermal conductivity of amorphous PE obtained by extrapolation of the data for the melt; the dependence is typical for amorphous polymers (for comparison)

Keeping in mind all these ideas, the temperature dependence of χ for crystalline polymers above 30 K can be explained as follows [21]. For polymers with simple chemical structure (PE and POM) $\chi_{c_\perp} \gg \chi_a$ near 30 K therefore Eq. (2.7) further reduces to

$$\chi = \chi_a \left(\frac{1 + 2w}{1 - w} \right) \tag{2.9}$$

Hence, χ is proportional to χ_a and thus increases with rising temperature. Above 30 K χ will depend on both $\chi_{c_\perp}$ and χ_a because these values are comparable in this region. For samples with high crystallinity (w > 70%) temperature dependence of their thermal conductivity will decrease with increasing temperature due to the dominating role of $\chi_{c_\perp}$. Hence, for the polymers with high crystallinity χ at first increases reaching a maximum in the vicinity of 80 K and then starts to decrease with temperature. With an increasing degree of crystallinity the maximum is shifted to low temperatures and more sharply. For polymers of the first group with low crystallinity (w < 0.4) χ is determined mainly by χ_a. Therefore, for such polymers a slight increase of χ with temperature occurs up to about 200 K when a shallow maximum occurs. For such polymers $\chi_{c_\perp}$ is practically independent of temperature and, therefore, χ exhibits a temperature dependence similar to χ_a.

Temperature Region Below 10 K

The most striking feature of the the low temperature conductivity of crystalline polymers consists of a very unusual influence of the crystallinity on the temperature dependence and absolute values of χ. Although Eq. (2.7) predicts an increase of χ in whole temperature range, the experimentally found dependencies at low temperature are quite different. Figure 2.7 shows that below 10 K the increase of crystallinity is accompanied by a drastic (by 10-fold and even more) decrease. A very large difference is also found for the temperature dependence χ of the samples with various crystallinity.

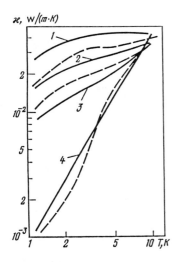

Fig. 2.7. Low temperature thermal conductivity of PET of various crystallinity (*dashed lines* – experimental data; *solid lines* – the modified Klemens theory) [28]. Degree of crystallinity: *1*– 0; *2* – 0.05; *3* – 0.14; *4* – 0.75

Even in early studies [16, 29] attention was paid to the important role of crystallites in additional phonon scattering and the accompanying decrease in crystallinity. It has been postulated that in the system of crystallites–amorphous phase besides the structural scattering in the amorphous phase an additional mechanism of phonon scattering takes place and some correlation of the scattering with the thickness of crystallites has been found. Today the main idea of this approach can be formulated as follows [9, 21, 25, 28]. The presence of crystallites introduces a correlation length of the order of the size of the crystallites (10–15 nm) leading additional phonon scattering and an increase in conductivity. Such an approach should strictly be valid when the mean free path is larger than the correlation length, i.e. at temperatures of a few K.

In [25, 28, 30] this idea has been checked quantitativley. Assfalg [28] obtained the structure factor for velocity fluctuation from SAXS measurements and using this and some other parameters calculated the thermal conductivity of the samples with various crystallinity. As can be seen from Fig. 2.7 where these results are shown the main features of the experimental curves, i.e. the large decrease of conductivity with increasing w can be reproduced by this model. Burgers and Greig [25] have analysed their experimental data above 2 K for isotropic and extruded PE using the following equation

$$l^{-1}(\chi) = l_b^{-1} + l_L^{-1}(\chi) + (1-w)l_s^{-1}(\chi) \qquad (2.10)$$

where $l_L(\chi)$ and $l_s(\chi)$ denote the mean free paths for long-range and short-range correlations, respectively, and l_b is the constant corresponding to the boundary dimension. The l_L^{-1} term describes the scattering of phonons due to velocity fluctuation in alternate amorphous and crystalline regions and the term $(1-w)l_s^{-1}$ corresponds to phonon scattering in amorphous regions. Figure 2.8 shows a comparison of experimental results with the theoretical calculations. The values of 100 and 8 Å are for long-range and short-range correlations, respectively. Because a large number of suggestions used and some arbitrary chosen parameters a very good agreement between the experimental data and theoretical predictions seems to be fortuitous. Nevertheless, the model calculations display rather obviously the main features of the low temperature behavior of the thermal conductivity of semi-crystalline PE and one can hope that the approach dealing with phonon scattering by disordered structures happen to be very useful for describing the thermal conductivity both amorphous and semi-crystalline polymers.

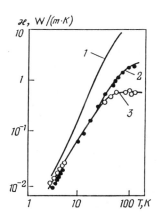

Fig. 2.8. Thermal conductivity of PE [25]. *1* – theoretical calculation according to Eq. (2.10) for the single correlation length equal to 10 nm (w = 0); *2* – theoretical calculation according to Eq. (2.10) for two correlation lengths equal to 10 nm and 0.8 nm (w = 0.6); *3* – theoretical calculation according to Eq. (2.10) taking into account umklapp processes; ○ – data for the isotropic sample; ● – data for the extruded sample

This idea was supported recently by Finlayson and Mason [30] who measured the thermal conductivity of the isotropic and extruded PE samples over the range 0.08–1.8. It was established that χ of the isotropic sample differed significantly both in magnitude and temperature dependence from χ of the extruded samples. The difference has been explained using different values of the correlation length related to the dimension of crystalline lamellae and their orientation. Finlayson and Mason also performed a joint consideration of their data and the data obtained by Burgess and Greig [25] and concluded that the structure scattering approach is essential to any explanation of the thermal conductivity, since only one type of scattering mechanism is required to explain the conductivity of a wide range of amorphous and semi-crystalline materials over a wide range of temperature (0.1 to 100 K).

In semi-crystalline polymers, the acoustic mismatch between the crystallites and the amorphous phase exists which introduced an additional thermal resistance due to the difference in the densities and sound velocities in these phases. This problem has been considered for the model of the structure of crystalline polymers in which amorphous and crystalline layers are arranged alternately and this structure is additionally arranged in parallel with amorphous region [21, 31]. The main relation for conductivity of such a model is

$$\frac{\chi}{\chi_a} = \frac{\kappa^2}{2w\chi_a(r_b/d) + (\kappa - w)} + (1 - \kappa) \tag{2.11}$$

where κ is the fraction of the crystallites in the sectional area perpendicular to the direction of heat flow, d is the thickness of crystalline lamellae and r_b is the thermal resistance of the interphase amorphous region-crystallite. Equation (2.11) includes three parameterss: κ, $r_b(T)$ and $\chi_a(T)$ (the thickness of the crystalline lamellae can be obtained from SAXS-measurements). For the polymers which can be quenched $\chi_a(T)$ can be obtained experimentally. Because $r_b(T)$ cannot be measured directly it can be estimated from Little's theory [32] which at low temperature gives $r_b = AT^3$ where A includes the sound velocities and densities of the amorphous and crystalline phases.

According to Eq. (2.11) the unusual behavior of the thermal conductivity of semi-crystalline polymers at low temperatures (Fig. 2.7) can be explained as follows [21]. At high temperatures the interphase thermal resistance is negligibly small and the conductivity of crystalline polymers is determined by crystallites possessing a higher χ. As the temperature drop r_b increases and at a certain temperature its contribution not only overcomes any increase of the conductivity resulting from its increase due to the presence of crystallites, but also leads to a greater decrease of the thermal conductivity of the samples with higher crystallinity. Therefore, below some temperature a cross-over in the thermal conductivity curve occurs. The increase in slopes of the curves with crystallinity below the cross-over can also be explained due to a combination of the resistance of the amorphous phase and amorphous-crystalline interphase.

Thus, Eq. (2.1) can be used immediately for those crystalline polymers for which the degree of crystallinity can be changed over a wide range. Besides PET there is one more series of low temperature thermal conductivity data for PE (1.2–20 K) with the degree of crystallinity ranged from 0.43 to 0.81 [29]. These data can also be understood and described quantitatively above 3 K to an accuracy of about 10% on the ground of the above approach with some simplified suggestions which, however, cannot change the main conclusion concerning the applicability of this approach for semi-crystalline polymers. Finally, Choy [21] estimated the interphase resistance for some other polymers of various chemical structure, crystallinity and lamellae thickness. The data are listed in Table 2.2. It is seen that the thermal resistance does not depend critically upon the degree of crystallinity and the chemical structure of polymers.

Hence, both mechanisms, namely, phonon scattering due to correlation in sound velocity fluctuation and thermal interphase resistance can, in principle, explain why the conductivity of crystalline polymers at very low temperatures is always below the conductivity of the same polymers in the amorphous state. However, the parameters involved, (r_b, the correlation lengths related to structures in amorphous regions and to crystalline structure) cannot be directly measured. It makes the performance of the quantitative analysis more difficult and requires the introduction of some crude assumptions.

Table 2.2. Parameters obtained from the analysis of the low temperature conductivity of crystalline polymers according to Eq. (2.11) [21]

Polymer	w	$\chi 10^2$ W/m K	$\chi_a 10^2$ W/m K	d^* nm	$(r_b T^3/d)10^3$ m · K/W	$(r_b T^3)10^5$ m^2 · K/W
PE	0.81	2.85	3.2	28.2	1.95	5.5
PE	0.43	1.2	3.2	11.3	4.9	5.5
POM	0.71	2.8	3.8	28.2	1.27	3.6
PP	0.47	1.2	3.8	17.0	5.0	8.5
PA 66	0.42	1.0	3.8	14.5	7.5	10.2
PET	0.39	0.95	3.75	14.1	8.7	12.2
PCTFE	0.34	1.0	3.8	28.2	3.2	9.1

*d = $\sqrt{2}\times$ average lamellae thickness

2.3 Effect of Molecular Parameters

The theoretical models considered are to elucidate the physical mechanisms responsible for the low temperature dependence of the thermal conductivity and cannot be used immediately for estimating the absolute values of χ because the main phenomenological characteristics of these models (short-range and long-range corelation length, the thermal resistance of the interphases) are the variable parameters of the models and their direct relation to the structure of acrtual glassy polymers is not really obvious. There have been some attempts in the literature to estimate the thermal conductivity of amorphous and crystalline polymers using simple molecular models. Eiermann [18, 23, 24] used his model briefly considered above and arrived at the following main relation for the thermal conductivity at normal temperatures

$$\chi = K_p \frac{C_v}{l_{ml}} (f_i/m)^{1/2} \tag{2.12}$$

where K_p is a temperature independent constant, f_i is the force intermolecular constant, l_{ml} is the length of the valence bond and m is the mass of the repeating unit. Unfortunately, although the weak molecular interaction is assumed to be the only restriction for the phonon free path, however, in this equation the total heat capacity is used rather than its intermolecular contribution (see Chapter 3). Therefore, this model can be used only for qualitative estimations. Eiermann [23, 24] made such an estimation of the temperature dependence of the thermal conductivity, the influence of the side chains on the thermal conductivity and some other factors.

Bondi [33] considered a similar model to estimate the thermal conductivity of polymers in the molten state. In the model, the total conductivity is considered as a combination of the conductivity along macromolecules and between chains. For the thermal conductivity along macromolecules χ_m he arrived at the following equation

$$\chi_m = \frac{\Sigma C_D \rho_0}{M} \left(\frac{E_m}{\rho_0} \right)^{1/2} 0.5 \varphi_b l_m \tag{2.13}$$

where ΣC_D is the mole heat capacity of the acoustic vibrations in the Debye approximation, E_m is the spectroscopic modulus of the chain, φ_b is the part of the cross-section of macromolecules accessible for phonons and l_m is the length of the mean free path which is assumed to be equal to the persistent length. The thermal conductivity through the van der Waals bonds can be estimated according the equation

$$\chi_w = \frac{R}{Na^2} \left(\frac{E^0}{M} \right)^{1/2} \left(3.22 - \frac{8.8 \, c_{in} RT}{E^0} \right) \tag{2.14}$$

where E^0 is the standard vaporization energy of the repeating unit, a is the interchain distance and c_{in} is the number of internal freedom of the repeating unit. The total thermal resistance is

$$\frac{1}{\chi} = \frac{l_m}{l_m + d} \frac{d}{\chi_m} + \frac{d}{l_m + d} \frac{1}{\chi_w} \tag{2.15}$$

where

$$d = a + \frac{V}{A} \left(\frac{M}{\rho V} - 1 \right) \tag{2.16}$$

where A is the area of one mole of the repeating units.

The calculations performed with these formulas [33, Table 10.2] showed that the thermal conductivity can be estimated with the accuracy of about 50% relative to the values measured. The model can also be modified for the calculation of the thermal conductivity of polymers with macromolecules having relatively long side chains. The corresponding estimations showed that the thermal conductivity of such macromolecules is smaller than without side chains and that the negative thermal coefficient $d\chi/dT$ is larger. Both these predictions agree with experimental findings.

One of the consequences of the large difference in the inter- and intramolecular thermal resistance in polymers must be the dependence of the thermal conductivity upon molecular weight. Some studies have demonstrated such a dependence [18], however, various models suggest a different character of the dependence. Eiermann [23] on the basis of the model suggested, arrived at the following dependence of an amorphous polymer upon molecular weight

$$\frac{1}{\chi} = \frac{1}{\chi_*} + \frac{A}{M} \qquad (2.17)$$

where χ_* is the thermal conductivity of the infinite molecular weight and A is a constant. Checking of this relation for PDMS and PEO of various molecular weights has shown that the experimental dependence does not follow this prediction because a real dependence takes place only for rather low molecular weights and then a leveling occurs. Lohe [34] predicted the $M^{-1/3}$ dependence, while Hansen and Ho [35] – $M^{-1/2}$. Figure 2.9 shows that experimental data for two polymers follow rather accurately the $M^{-1/2}$ dependence. The physical reason of the leveling of the thermal conductivity with the molecular weight seems to be a result of the fact that in long enough macromolecules the interchain heat transfer occurs more often and, therefore, the total conductivity seems to become insensitive to the chain length. Two more model calculations [36, 37] also arrived at $M^{-1/2}$ dependence. Hence, one can consider this dependence as the theoretically most acceptable.

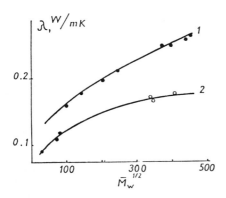

Fig. 2.9. Molecular weight dependence of thermal conductivity [35]. *1* – PS; *2* – PE

Lohe [34] studied the influence of the chain branching on the thermal conductivity for a number of polyolefins in the molten state (Fig. 2.10). It is seen that with increasing of the number of methylene groups in chains the thermal conductivity decreases significantly. According to the model calculations this decrease is a consequence of the increase of the molecular weight of the repeating unit (see Eq. 2.12). Similar behavior was also found for acrylic polymers.

Statistical incorporation of various single atoms and groups of atoms into macromolecules is accompanied by a large variety of the thermal resistances. For two-component copolymers both the linear and non-linear relationships for the thermal conductivity versus copolymer content can be observed [18].

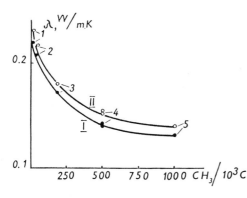

Fig. 2.10. Thermal conductivity of branched macromolecules [34]. *1* – HDPE; *2* – LDPE; *3* – EP copolymer; *4* – P; *5* – PIB. *I* – pressure 0.1 MPa; *II* – pressure 30 MPa

2.4 Anisotropy of Thermal Conductivity – Effect of Orientation

Drawn crystalline and amorphous polymers as well as stretched elastomers exhibit anisotropy of the thermal conductivity resulting from the tendency of the chain molecules and crystallites to align along the orientation direction. The thermal conductivity along the draw or stretching direction is always higher than in the isotropic state or perpendicular to this direction. The anisotropy of conductivity has been investigated for several polymers PMMA and PE being the most thoroughly studied amorphous and crystalline polymers. Below the anisotropy of the thermal conductivity will be considered first of all for these polymers.

Amorphous Polymers.   Drawing of the amorphous glassy polymers and stretching of non-crystallizable elastomers is accompanied with an increase of the part of the segments along the draw direction. Therefore, the only physical reason of the appearence of the anisotropy in conductivity is the increase of the part of the covalent bonds aligned along the stretching direction and their decrease to the perpendicular direction. Using this simple assumption Eiermann [18, 23] with the help of his model arrived at the following simple relation which can be valid for any degree of orientation

$$\frac{1}{\chi_{\|}} + \frac{2}{\chi_{\perp}} = \frac{3}{\chi} \tag{2.18}$$

Similar equation is also followed from the aggregate model in which an unoriented polymer is regarded as a random aggregate of axially symmetric units whose thermal conductivities were those of the fully oriented material [38]. The thermal conductivity of the partially oriented polymer may be calculated using either series or parallel model alignment of the unchanged units. The physical reason of the existence of such aggregates in the amorphous glassy polymers is unclear.

   Hansen and Ho [35] considered the anisotropy of the thermal conductivity of amorphous polymers using their model which treats the heat transfer in polymers similar to the mechanism assumed for liquids and arrived at

$$\frac{\chi}{\chi_\perp} = \left(\frac{\chi_\parallel}{\chi}\right)^{1/2} \tag{2.19}$$

Although Eqs. (2.18) and (2.19) are very different in form, the experimental checking of their predictions for some polymers as is seen from Table 2.3 has shown that the accuracy of the prediction in both cases is practically identical. This seems to be the result of a small orientation which can be reached for amorphous polymers in practice.

Table 2.3. Thermal conductivity of drawn polymers [18]

Polymer	λ	Thermal Conductivity, W/m K				$\chi/\chi_\perp$	$(\chi_\parallel/\chi)^{1/2}$
		$\chi_\parallel$	$\chi_\perp$	χ_{calc}	χ_{exper}		
PS	5	0.173	0.154	0.160	0.162	1.06	1.03
PMMA	2.57	0.238	0.181	0.200	0.197	1.08	1.10
PMMA	3.75	0.280	0.168	0.195	0.197	1.17	1.19
PVC	1.85	0.228	0.149	0.168	0.168	1.13	1.17
PVC	2.65	0.279	0.140	0.167	0.168	1.20	1.29

Wallace, Moreland and Picot [39] have studied the shear dependence of thermal conductivity in PE melt at 160° C up to $400 \, s^{-1}$. The results expressed in terms of a normalized thermal conductivity, namely, a ratio of the thermal conductivity measured normal to the flow direction to that under conditions of zero shear. A sharp decrease at low rates of strain (in the vicinity of $50 \, s^{-1}$), indicating that alignment of the polymer chains along the flow direction, was observed. After shearing at the maximum shear rate, $400 \, s^{-1}$ approximately 90 min was required to recover the values corresponding to the zero shear condition. Such behavior has been explained by orientation with stretching of the molecular chains in the direction of flow at low shear rates, and formation of rotating units of entangled clusters at higher shear rates.

Crystalline Polymers. The relationships for anisotropy of conductivity of crystalline polymers are significantly more complex than for amorphous polymers. They were intensively studied [40–49]. For amorphous polymers the anisotropy coefficient $\chi_\parallel/\chi_\perp$ is normally not higher than 2, while for crystalline polymers it can reach such high values as 50 and even higher as is seen from Fig. 2.11. These studies have revealed two important features. At normal and high temperatures $\chi_\parallel$ increases continuously with increasing draw ratio λ, while $\chi_\perp$ at first decreases in comparison with the thermal conductivity of the undrawn state and then reaches a constant value independent of λ. Besides the data for HDPE shown in Fig. 2.11, similar behavior was observed also for PP, POM, PET and PVF_2 [41–43]. The second feature is that at low temperature ($< 30 \, K$) the anisotropy of conductivity even for samples with the highest draw ratio is extremely small. For example, at low temperatures ($< 10 \, K$) HDPE samples of all draw ratios show $\chi_\parallel \simeq \chi \simeq 1.5\chi_\perp$.

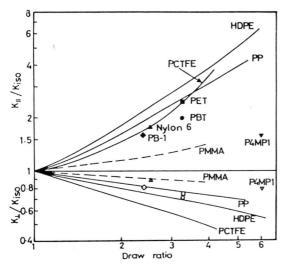

Fig. 2.11. Draw ratio dependence of thermal conductivity of various oriented polymers [47]

In the anisotropic behavior of drawn crystalline polymers a very important role play crystallites, therefore, the correct estimate of the thermal conductivity parallel and perpendicular to the chain direction in crystallites is of great importance. Early estimations [25] arrived at $\chi_\parallel/\chi_\perp > 50$ at room temperature. In 1985, Choy and coworkers [26] performed model calculations of the magnitude and temperature dependence of the thermal conductivity of PE crystals parally and perpendicular to the chain direction. They used the linear Stockmayer–Hecht lattice model (see Chapter 1). The model shows that heat transfer occurs principally by phonons polarized transverse to the chain direction, while phonons polarized longitudinal to the chain direction contribute only about 20% to the heat transport along the chain direction, and negligibly to the heat transport perpendicular to the chain direction. As a result of the model calculations they arrived at the following data: $\chi_\parallel = 465\,\mathrm{W\,m^{-1}K^{-1}}$, $\chi_\perp = 0.16\,\mathrm{W\,m^{-1}K^{-1}}$ at $300\,\mathrm{K}$ in agreement with experimental results for ultradrawn PE samples. Figure 2.12 shows the temperature dependence of the thermal conductivity of PE crystals along the chain axis and perpendicular to the chain axis for various mean free paths for boundary scattering. Both sets of dependences show a $1/T$ behavior down to about $100\,\mathrm{K}$, below which they drop with decreasing temperature. Hence the axial thermal conductivity of PE crystals is extremely high and quite comparable to that of copper. These data lead to an extremely large anisotropy ratio $\chi_\parallel/\chi_\perp = 3 \times 10^3$.

Three model approaches have been used for explaining the large anisotropy of thermal conductivity of drawn semi-crystalline polymers at normal temperatures [21]. If the only reason of the anisotropy is the orientation of the crystallites towards the draw direction resulting from the drawing, then the orientation may be characterized by the function f_c given by:

$$f_c = \tfrac{1}{2}\left[3\langle \cos^2 \varphi \rangle - 1\right] \quad 0 < f_c < 1 \qquad (2.20)$$

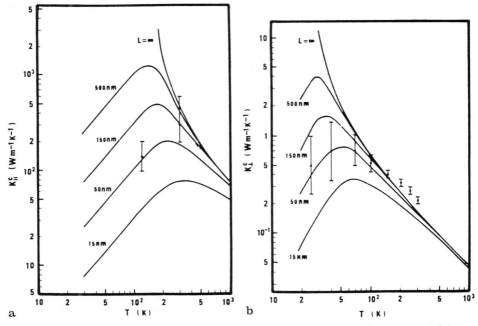

Fig. 2.12. Temperature dependence of the thermal conductivity of PE crystal (**a**) along the chain axis and (**b**) perpendicular to the chain axis [26]. L corresponds to the mean free path for boundary scattering. The data points for $\chi_{c_\parallel}$ and $\chi_{c_\perp}$ are deduced from measurements on ultradrawn and isotropic samples, respectively

WAXS diffraction measurements show (see Chapter 5) that for various crystalline polymers f_c increases very fast at low draw ratios and arrives at a constant value 0.9 at $\lambda \simeq 4$–5. The orientation of the chains in the amorphous phase is rather low at these draw ratios. Taking into consideration these results and the small orientation effect in amorphous polymers the modified Maxwell model can be applied for describing the anisotropy effect in drawn crystalline polymers. Assuming $k_\parallel = \chi_c/\chi_a \gg 1$ [see Eq. (2.7)] one can arirve at

$$\frac{\chi_\perp - \chi_a}{\chi_\perp + 2\chi_a} = w \left\{ \frac{k_\perp - 1}{k_\perp + 2} \frac{1 + \langle \cos^2 \varphi \rangle}{2} + \frac{1}{2} \langle \sin^2 \varphi \rangle \right\} \qquad (2.21)$$

$$\frac{\chi_\parallel - \chi_a}{\chi_\parallel + 2\chi_a} = w \left\{ \frac{k_\parallel - 1}{k_\parallel + 2} \langle \sin^2 \varphi \rangle + \langle \cos^2 \varphi \rangle \right\} \qquad (2.22)$$

$\langle \sin^2 \varphi \rangle$ and $\langle \cos^2 \varphi \rangle$ can be estimated from f_c and $\chi_\perp$ – from data on isotropic materials (see Fig. 2.3), therefore, there are no adjustable parameters in Eqs. (2.21) and (2.22). Figure 2.13, in which the predicted and experimental behavior for PE is compared, shows that the model gives the correct trends of the anisotropy of drawn PE. The quantitative agreement takes place for $\chi_\perp$ for all values of λ and f_c. For $\chi_\parallel$ such an agreement occurs only at $\lambda < 4(f_c \simeq 0.9)$ while at higher draw ratios experimental values of

$\chi_\parallel$ increase rapidly. This seems to indicate that some other factors rather than orientation of crystallites contribute significantly to the thermal conductivity along the draw direction. Nevertheless, at $\lambda < 4$ this model displays correctly the dependence of the anisotropy of thermal conductivity upon degree of crystallinity and a gradual change of the slope of the curve of $\chi_\parallel$ versus temperature from negative to positive as f_c increases from 0 to 1. Finally it shows that temperature dependence of $\chi_\perp$ should follow that of χ again in agreement with experimental findings.

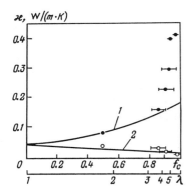

Fig. 2.13. Thermal conductivity of drawn PE at 323 K against the crystalline orientation function f_c [21]. The theoretical curves $\chi_\parallel$ (*1*) and $\chi_\perp$ (*2*) are calculated according to Eqs. (2.21) and (2.22)

For describing the rapid increase of $\chi_\parallel$ for highly drawn crystalline polymers, first of all, super-drawn super-high-modulus PE the model of "statistical intercrystalline bridges" was used [21] (see Chapter 5). According to this model the rapid increase of the axial Young's modulus as well as the thermal conductivity and some other properties is a result of a series arrangement of amorphous and crystalline regions with taut tie molecules which has been termed intercrystalline bridges connecting the crystallites. Assuming that the intercrystalline bridges possess the same thermal conductivity as the crystallites one can arrive at the following relation

$$\frac{1}{\chi_\parallel} = \frac{1-a}{\chi_{c_\parallel}} + \frac{a}{b\chi_{c_\parallel} + (1-b)\chi_a} \tag{2.23}$$

where a and b are the geometrical characteristics of the model b being the fraction of the intercrystalline bridges. Above 30 K if b is not too small ($> 1\%$) $b\chi_{c_\parallel} \gg (1-b)\chi_a$, so Eq. (2.23) transforms to

$$\frac{\chi_\parallel}{\chi_{c_\parallel}} = \frac{b}{b(1-a)+a} \tag{2.24}$$

Similar analysis can be performed for the modulus of elasticity $E_{c_\parallel}$ which leads to

$$\frac{\chi_\parallel(T)}{\chi_{c_\parallel}(T)} = \frac{E_\parallel(T)}{E_{c_\parallel}} \tag{2.25}$$

for the plateau region of modulus of PE. Since values of $\chi_\parallel(T)$ and $E_\parallel$ (200 K) are known from experiment and as $E_{c_\parallel} = 240\,\mathrm{GPa}$ one can estimate $\chi_{c_\parallel}$. Such estimations lead to $\chi_{c_{\parallel}} = 31\,\mathrm{W\,m^{-1}K^{-1}}$ at $100\,\mathrm{K}$ which is about 200 times higher than χ_a.

The thermal conductivity perpendicular to the draw direction in this model is

$$\chi_{\perp} = (1 - a)\chi_{c_{\perp}} + a\chi_a \tag{2.26}$$

i.e. $\chi_{\perp}$ is practically independent of draw ratio which agrees with experimental findings (see Fig. 2.11). Assuming $\chi_a = 0.18\,\mathrm{W\,m^{-1}K^{-1}}$ we arrive at $\chi_{c_{\parallel}} \simeq 0.6\,\mathrm{W\,m^{-1}K^{-1}}$, which coincides with the value obtained from the modified Maxwell model (see Fig. 2.6).

The third model of anisotropic behavior of drawn crystalline polymers, the aggregate model, was suggested by Kilian and Pietrella [45, 46]. The model assumes the strong coupling of the crystalline and amorphous regions with a definite orientation correlation of neighboring crystalline lamellae which form clusters. These clusters can be considered as the main units of the aggregate. Introducing the anisotropy factor of the main unit $A_u = \chi_{u_{\parallel}}/\chi_{u_{\perp}}$ one can arrive at the following relation for anisotropy A

$$A = \frac{1}{2}\left[\frac{2A_u + 1}{A_u - (A_u - 1)\langle\cos^2\varphi\rangle} - 1\right] \tag{2.27}$$

Assuming that A_u depends on crystallinity and the orientation of the main units is adequately described by f_c, one can calculate the anisotropy A in terms of the adjustable parameter A_u. Figure 2.14 demonstrates the behavior of three samples of PE of different crystallinity. Although there is a good coincidence between the model and experiment, nevertheless the weak point of the model is the suggestion that the clusters remain unchangeable during drawing.

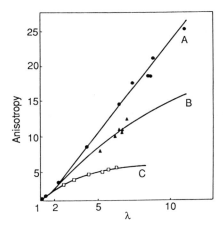

Fig. 2.14. Variation of the anisotropy of PE with draw ratio [45, 46]. $\bullet$ – $w = 0.74$; $\triangle$ – $w = 0.67$; $\square$ – $w = 0.45$. The theoretical curves are calculated according to Eq. (2.27) using $A_u = 26$, 16, 7, respectively, for the samples with $w = 0.74$, 0.67 and 0.45

Finally, concerning the second feature of the thermal conductivity anisotropy – a sharp decrease of the anisotropy below $30\,\mathrm{K}$ and its practical disappearence below $10\,\mathrm{K}$ for PE – Choy [21] supposed, that this is a result of insensitivity of the dominant resistive mechanisms, arising either due to the correlation in sound velocity fluctuation or boundary resistance, to the orientation of the chains. A very small anisotropy of the thermal conductivity at low temperatures has been observed recently for extruded samples of liquid crystal copolyesters [50].

2.5 Effect of Radiation

Radiation effects the thermal conductivity and other thermophysical properties of polymers and polymer material. The results concerning the influence of various types of radiation on polymers have been reviewed recently [51, 52]. Effects arising in polymers due to radiation can be reversible or irreversible. The latter seem to be a result of irreversible changes in the chemical and physical structure. Data for LDPE shown in Fig. 2.15 demonstrate typical behavior of crystalline polymers. It is seen that below the melting point i.e. in the solid state the increase of the radiation dose is accompanied with a significant decrease of χ, first of all, due to gradual decreasing up to complete disappearing of crystallinity (at doses above $10\,\mathrm{MGy}$). Such a behavior is in agreement with the influence of crystallinity on the thermal conductivity. In contrast to the solid state, in the molten state χ increases with the radiation dose. With large doses chemical destruction occurs. Similar behavior is a characteristic of i-PP. For PTFE, radiation has a small effect on χ.

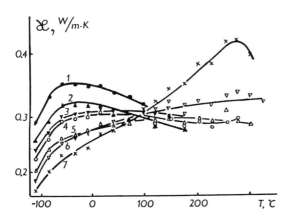

Fig. 2.15. Temperature dependence of the thermal conductivity of LDPE irradiated with electrons [52]. Dose (MGy): $1 - 0$; $2 - 1$; $3 - 2$; $4 - 5$; $5 - 10$; $6 - 20$; $7 - 40$

    The effect of radiation on the thermal conductivity of amorphous phases is closely related with the chemical changes occurred: radiational cross linking or destruction. Although in early investigations there were found no radiation effects in PS, recent studies have established such effects in the temperature interval from -100 to $+20°$ C after radiation by both electron and neutron radiation. In the region of 1–$2\,\mathrm{MGy}$ a small minimum on the dependencies χ versus dose absorberd occurs after which a very weak increase of χ takes place. At high values of absorbed doses cracking of the irradiated samples occurred.

2.6 Effect of Pressure

The effect of pressure on the thermal conductivity of polymers has not been extensively investigated. Early data was summarized by Knappe [18]. Lohe [34] studied the pressure dependence of the thermal conductivity for a number of polymers (Fig. 2.12) in the molten state up to 30 MPa. Similarly to liquids, χ increases under pressure the average pressure coefficient $1/\chi(d\chi/dP)$ being $\simeq 1.6 \times 10^{-3} \mathrm{MPa}^{-1}$. The network model of Eiermann [23] gives the possibility of estimating the pressure dependence of χ for a linear amorphous polymer in the limit of low pressure according the following relation $1/\chi(d\chi/dP) = 5.25/K$ (K is the bulk modulus). A very similar relation was also obtained in [53, 54]. Dietz [55] also studied the influence of pressure on the thermal conductivity of a number of polymers in the molten state. Figure 2.16 shows the dependencies of relative values of χ upon pressure. The increase of χ with pressure depends on the chemical structure of polymers being 10–20% in the vicinity of 10^8 Pa. Similar results have also been obtained recently on a number of the molten polymers in the range of pressure 0–1×10^8 Pa [56].

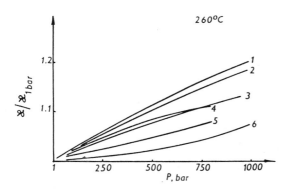

Fig. 2.16. Relative variation of the thermal conductivity of various polymers as a function of pressure [55]. *1* – PP; *2* – HDPE; *3* – LDPE; *4* – PS; *5* – PC; *6* – PA-6

Frost, Chen and Barker [57] investigated the effect of pressure on χ theoretically, assuming the interchain vibrations give the main contribution. They formulated the theory of thermal conductivity in terms of a quasi lattice model and showed that the anharmonic contributions to the quasi-lattice vibrations lead to a rather simple pressure dependence for the conductivity

$$[\chi(P) - \chi(0)]/\chi(P) = \gamma(T)f_0(0, T)[1 - \exp(-PV_0/RT)] \qquad (2.28)$$

where $f_0(0, T)$ is the fractional free volume and V_0 is the volume of a lattice cell. They applied this theory to the thermal conductivity data of PE in the pressure range up to 1.25×10^8 MPa and temperature range $-50\ +50°$ C.

Although their theory was developed for amorphous polymers Eq. (2.28) describes these data surprisingly well. However, this treatment gives an extremely large increase of χ under pressure. It is hardly believable that under pressure of 1.25×10^8 Pa the thermal conductivity of PE at room temperature increases about 3 times.

Privalko [58] has shown that the relative pressure dependence of the thermal conductivity of polymer glasses can be expressed as follows

$$[\chi(P) - \chi(0)]/\chi(P) = \gamma(T) f_0 A \ln(1 + P/B_\chi) \qquad (2.29)$$

where A and B are empirical constants. Using this Eq. (2.29) he treated successfully the pressure dependence of the thermal conductivity for a number of glassy polymers (Fig. 2.17).

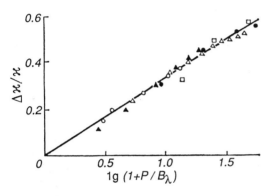

Fig. 2.17. Relative variation of the thermal conductivity of various polymers as a function of the pressure parameter Eq. (2.29) [58]. $\circ$ – PMMA; $\bullet$ – PC; $\blacktriangle$ – PEMA; $\triangle$ – PS; $\square$ – PBMA

Anderson and Sundquist [53, 54] have measured the thermal conductivity and diffusivity of HDPE, LDPE, PTFE, PS, atactic and isotactic PP, PMMA at 300 K in the pressure range 0–30×10^8 Pa. The thermal conductivity strongly increases with pressure, the pressure dependence being most pronounced at low pressure. The values of χ at 25×10^8 Pa are higher than those at atmospheric pressure by a factor of 2.75 for HDPE, 2.19 for LDPE, 1.95 for PMMA, 1.93 for PS, 1.71 and 1.56 for atactic and isotactic PP, respectively.

Summarizing the consideration of the effect of pressure on the thermal conductivity of polymers it is worth of mentioning that agreement between the experimental results obtained at various laboratories is, as a rule, poor and the variation of the values may reach 2–3 times.

2.7 Thermal Conductivity of Filled Polymer Materials

The main approach to an effective improvement of the low thermal conductivity of polymers consists of filling them with powdered or fiber fillers of high thermal conductivity. A number of studies concerning the improvement of χ of a polymer by adding various types of particles have been published

[59–69]. Most of these investigations dealt with filler concentration of less than 30%, however, in some studies the concentration of fillers has been varied in a wide range from a low to a very high volume content. The literature abounds in various attempts to relate experimental results of the thermal conductivity and the conductivity of the components. Many semi-empirical expressions have been proposed besides the early fundamental treatment by Maxwell and Rayleigh (see, for example, review articles [70, 71]). The thermal conductivity of the composite materials in which the conductivity of both phases are very close can be described satisfactorily by various formulas. The shape of the filler particles is not a very important parameter in this case. However, if the conductivity of the filler, χ_f, is much higher (say, two orders of magnitude) than that of the matrix, χ_m, then the shape of the particles can become important. In order to characterize the shape of the particles the sphericity parameter can be introduced, which is the ratio of the surface area of a sphere with the same volume as that of the particle to the surface area of the particle. The particle size of the filler is also important, primarily at low temperatures.

Rossenberg and coworkers [63, 64] studied the thermal conductivity of polymer composites made from an epoxy resin with powder filler of glass spheres, quartz, corundum, diamond, metal-powder fillers Cu, Ag, Au, Al, Sn, PB, stainless steel and bronze. In many cases the particles were spherical and rounded, while in some cases were random, irregular or platelets. Figure 2.18 shows the dependence of χ for epoxy–metal composites upon the volume concentration of metallic spherical particles at 300 K. It is seen that χ is independent of the particle size, the thermal conductivity of the particles and is in a good agreement with the values predicted by one of the mixtrue equations (Meredith–Tobias equation). Similar behavior was also established for other fillers mentioned above for temperature region 20–300 K. Below about 20 K the thermal conductivity of the composites is lower than predicted by the theory, especially for specimens containing smaller particles, and it can even be less than the conductivity of the unfilled resin. This is shown to be due to the acoustic mismatch of phonons at the resin–particle interface (see Sect. 2.2). Similar behavior in low temperature conductivity was observed also for epoxy resins filled with copper powders by Schmidt [65], who studied χ as a function of grain size (5–30 microns) and filler concentration (up to $\simeq 50\%$ by volume) in the temperature range 1.5–20 K. He also measured the thermal boundary layer resistance (Kapitza resistance) between the epoxy matrix and copper particles and established that the thermal conductivity is strongly dependent on grain size in the low temperature this dependence being the consequence of the thermal boundary resistance. Below a characteristic temperature dependent on particle size the thermal conductivity of the composite is below the resin. This possibility may be attractive for some applications at very low temperatures.

Kusy and Corneliussen [60] studied the thermal conductivity of PVC filled with Cu and Ni particles ranging from 0 to 15% (by volume). The thermal conductivity was found to increase continuously over an order of

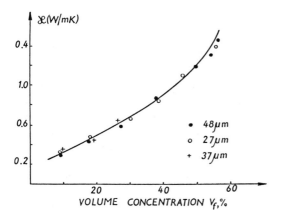

Fig. 2.18. The dependence of the thermal conductivity upon the volume concentration of spherical metallic particles [64]. Diameter of particles: • − 48 μm Ag; ○ − 27μm Ag; + − 37μm Au. The *solid line* is calculated according to the Meredith–Tobias equation

magnitude for PVC–Cu (15%) composition and about 4 times for PVC–Ni (15%) composition. In contrast the electrical resistivity decreases discontinuously at a concentration of the particles of some percent. Bigg [61, 62] has established that i-PP filled with short aluminum fibers of some mm length and the values of aspect ratios 15–30 become electrically conductive at a fiber concentration below 10% while the thermal conductivity increases monotonically with concentrations up to 30% of filler in accordance with Nilson's formula. On the other hand, Godovsky et al. [69] studied the concentration dependence of the thermal conductivity of HDPE filled with powdered two-component Cu/Ni and C/Ni particles. The mean size of the particles was 1.37 and 0.46 μm, respectively. It was unexpectedly established that in the compositions with filler concentration of about 12–13%, a rapid increase in the thermal conductivity occurred and in a rather narrow concentration interval, the values of χ are doubled. It has been suggested that this behavior is closely related to the percolation state and is a consequence of the formation of the infinite cluster. The values of the critical concentration are rather close to the theoretical value for the three-dimensional system. It is quite possible that during pressure molding the particles were arranged not randomly but formed needle-like structures with a predominant orientation. Nevertheless, this behavior is in contrast to a well established fact that the thermal conductivity of polymers filled with metallic particles is practically independent of the thermal conductivity of the particles due to formation of the interphase layers. Therefore, this result needs further experimental confirmation. The percolation approach to the problem of the thermal conductivity of heterogeneous composite materials and filled polymers was considered theoretically [72–75].

In some studies of the thermal conductivity of polymers filled with particles in the wide range from low to super-high volume content was studied

[63, 64, 68]. The conductivity of the systems containing a high or super-high volume of particles is greatly dependent on the powdery properties of particles, i.e. their ability to aggregate, the limit of packing, etc. The fractional void volume essentially occupied by filler particles when high or super-high content of filler is used can be left unfilled. In order to reach the maximum content of the filler, a mixing of particles of various size can be used. In polymers filled with such an optimum mixture of fillers the expected monotonous increase in the thermal conductivity in the wide range from low to super-high filler content can be obtained. Agari et al. [68] have recently demonstrated it for PE and PS filled with quartz and Al_2O_3 up to 80% by volume.

The thermal conductivity of a large number of glass-fiber and glass-cloth reinforced composites with epoxy, phenolic, polyester, silicone, phenylsilane, teflon and polybenzimidazole matrix in the temperature range 20–300 K was reviewed by Kasen [76] and some polymer engineering composites by Ziebland [77].

2.8 Thermal (Temperature) Diffusivity

The thermal diffusivity is the thermophysical parameter which characterizes the rate of the temperature diffusion in the material due to a heat flux in the unsteady-state heat transfer processes. It is closely related to the thermal conductivity according to the formula

$$\chi = \rho\, C\, a \qquad (2.30)$$

where a is the thermal diffusivity. Hence, the thermal diffusivity is an important transport property both from the theoretical and practical points of view. In contrast to metallic materials polymers have low values of thermal diffusivity. Similar to thermal conductivity this characteristic is connected with the chemical and physical structure of polymers being also dependent on temperature, pressure etc. The thermal diffusivity is the less frequently studied thermophysical characteristics of polymers. The early results were summarized in [17,19], more recent results were published in [47, 55, 78–80]. Below we will briefly consider the main features of the thermal diffusivity for typical polymers.

The temperature dependence of the thermal diffusivity for some amorphous polymers is shown in Fig. 2.19. In the glassy and molten states the value of a decreases monotonously with temperature. In the glass transition region a rapid drop occurs, which is the consequence of the rapid increase of heat capacity in this region because for the thermal conductivity only the change in slope for χ versus temperature occurs. This rapid drop in the thermal diffusivity can be used for identification of the glass transition of polymers which was demonstrated especially obviously on a series of PMMA [81]. Hence, one can conclude that the thermal diffusivity is almost the same for amorphous polymers.

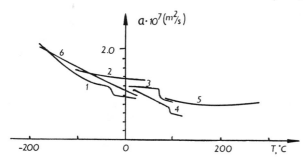

Fig. 2.19. Temperature dependence of the thermal diffusivity of amorphous polymers extracted from the literature [19, 55, 78]. *1* – PET; *2* – PC; *3* – PVC; *4* – PS; *5* – PS; *6* – PMMA

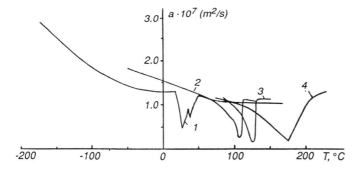

Fig. 2.20. Temperature dependence of the thermal diffusivity of crystalline polymers extracted from the literature [19, 55, 78]. *1* – PTFE; *2* – LDPE; *3* – HDPE; *4* – PP

The temperature dependence of the thermal diffusivity of crystalline polymers is shown in Fig. 2.20. The main feature of this behavior is the minimum in the temperature region of melting and crystal–crystal phase transition (PTFE) which can be predicted from the temperature change of other parameters in Eq. (2.30). It is clear from Fig. 2.20 that a decrease with rising temperature as T^{-m}, with $1 < m < 0.5$. For a number of crystalline polymers it has been shown that with increasing crystallinity a increases. However, some polymers (PE, PTFCE) exhibit a rather strong crystallinity dependence while others (PP, PB-1, P4MP1) exhibit weak crystallinity dependence. The change in crystallinity is displayed by the behavior of the thermal diffusivity considerably less than by the thermal conductivity. As a group, the fluoropolymers have very low thermal diffusivity close to that of amorphous polymers. The absolute values of the thermal diffusivity increase with increasing molecular weight. The effect of pressure on the thermal diffusivity was studied for PE, PP and PA-6 in a wide temperature range including the molten state. This effect is rather weak in the whole temperature range being about 10% per 10^8 Pa in the molten state. Similar to thermal conductivity, the thermal diffusivity is a very sensitive parameter to the molecular orientation both in the solid and molten states.

References

1. Berman R (1982) Thermal conduction in solids, Clarendon Oxford
2. Misnar A (1968) Thermal Conductivity of Solids, Liquids, Gases and Their Compositions (in Russian), Mir Moscow
3. Zeller RZ and Phol RO (1971) Phys Rev *B4:* 2029
4. Stephens RB (1973) Phys Rev *B8:* 2896
5. Stephens RB (1976) Phys Rev *B13:* 852
6. Phol RO and Salinger GL (1976) Ann N Y Acad Sci *279:* 150
7. Klemens PG (1960) In: Non crystalline solids, Frechette VC (ed), J Wiley New York
8. Oskotsky VS and Smirnov IA (1979) Defects in Crystals and Thermal Conductivity (in Russian), Nauka Moscow
9. Morgan GI and Smith D (1974) J Phys (C) *7:* 649
10. Philips WA (1972) J Low Temp *7:* 351
11. Anderson PW, Halperin BJ and Varma CM (1972) Phil Mag *25:* 1
12. Philips WA (ed) (1981) Amorphous solids: Low temperature properties, Springer Berlin Heidelberg New York
13. Smoljakov BP and Khaimovich EP (1982) Uspekhy fisicheskich nauk (UFN) *136:* 317
14. Phonon Scatter Condenced Matter 5 Proc 5 Intern Conf (1986), I11, Urbana Berlin
15. Andersen DR (1966) Chem Rev *66:* 677
16. Reese W (1969) J Macromol Sci *3A:* 1257
17. Godovsky YuK (1970) In: Uspekhy phys chem polym, pp 173–205, Khimija Moscow
18. Knappe W (1971) Adv Polym Sci *7:* 477
19. Godovsky YuK (1976) Thermophysical Methods of Polymers Characterization (in Russian), Khimija Moscow
20. Perepechko II (1980) Low Temperature Properties of Polymers, Pergamon Oxford
21. Choy CL (1977) Polymer *18:* 984
22. Reese W (1966) J Appl Phys *37:* 864, 3227
23. Eiermann K (1964) Kolloid Z Z Polym *198:* 5
24. Eiermann K (1965) Kolloid Z Z Polym *201:* 3
25. Burgess S and Greig D (1975) J Phys (C) *8:* 1637
26. Choy CL, Wang SP and Young K (1985) J Polym Sci Polym Phys Ed *23:* 1495
27. Choy CL and Young K (1975) Polymer *18:* 769
28. Assfalg A (1975) J Phys Chem Solids *36:* 1389
29. Kolouch RJ and Brown RG (1968) J Appl Phys *39:* 3999
30. Finlayson DM and Mason PJ (1985) J Phys (C) *18:* 1777, 1791
31. Choy CL and Greig D (1975) J Phys (C) *8:* 3121
32. Little WA (1959) Canad J Phys *37:* 334
33. Bondi A (1968) Physical Properties of Molecular Crystals, Liquids and Glasses, J Wiley New York
34. Lohe P (1965) Kolloid Z *203:* 115
35. Hansen D and Ho CC (1965) J Polym Sci *A3:* 659

36. Morgan GJ and Scovell PD (1977) J Polym Sci Polym Lett *15:* 193
37. Fesciyan S and Frisch HL (1977) J Chem Phys *67:* 5691
38. Hannig J (1967) Polym Sympos *C16:* 2751
39. Wallace DJ, Moreland C and Picot JJC (1985) Polym Eng Sci *25:* 70
40. Choy CL and Greig D (1977) J Phys (C) *10:* 169
41. Gibson AG, Greig D, Sabota M, Ward IM and Choy CL (1977) J Polym Sci Polym Lett *15:* 183
42. Choy CL, Luk WH and Chen FC (1978) Polymer *19:* 155
43. Choy CL, Chen FC and Luk WH (1980) J Polym Sci Polym Phys Ed *18:* 1187
44. Gibson AG, Greig D and Ward IM (1980) J Polym Sci Polym Phys Ed *18:* 1481
45. Pietralla M (1981) Colloid Polym Sci *259:* 111
46. Kilian GH and Pietralla M (1978) Polymer *19:* 664
47. Choy CL, Ong EL and Chen F (1981) J Appl Polym Sci *25:* 2325
48. Choy CL and Leung WP (1983) Polym Sci Polym Phys Ed *21:* 1243
49. Crispin AJ, Taraiya AK, Greig D and Ward M (1987) Polym Commun *28:* 95
50. Crispin AJ and Greig D (1986) Polym Commun *27:* 265
51. Briskman BA (1983) Uspechy Khimiji *52:* 830
52. Briskman BA (1988) In: Itogy nauky i techniki, serija "Khimija i tekhnologija polymerov" *25:* pp 136–194
53. Anderson P and Sundquist B (1975) J Polym Sci Polym Phys Ed *13:* 243
54. Anderson P (1976) Makromol Chem *177:* 271
55. Dietz W (1975) Colloid Polym Sci *265:* 755
56. Privalko VP and Rekhteta (1989) In: Proc XI-th AIRAPT Intern Conf, vol 4, pp 241–246, Naukova Dumka Kiev
57. Frost RS, Chen RYS and Barker REJr (1975) J Appl Phys *46:* 4506 (1975)
58. Privalko VP (1986) Molecular Structure and Properties of Polymers (in Russian), Khimija Leningrad
59. Sundstrom DW, Lee YD (1975) J Appl Polym Sci *16:* 3159
60. Kusy RP and Corneliussen RD (1975) Polym Eng Sci *15:* 107
61. Bigg, DM (1973) Polym Eng Sci *19:* 1188
62. Bigg, DM (1986) Polym Composit *7:* 125
63. Garret KW and Rossenberg HM (1974) J Phys (D) *7:* 1247
64. de Arauja FFT and Rossenberg HM (1976) J Phys (D) *9:* 665
65. Schmidt C (1975) Cryogenics *15:* 17
66. Agari Y and Uno T (1985) J App Polym Sci *30:* 2225
67. Agari Y, Tanaka M, Nagai S and Uno T (1986) J Appl Polym Sci *32:* 5705
68. Agari Y, Ueda A, Tanaka M and Nagai S (1990) J Appl Polym Sci *40:* 929
69. Godovsky YuK, Kiguradze OR and Lolua DG (1986) Vysokomol Soyed *28B:* 564
70. Godbee HW and Ziegler TZ (1966) J Appl Phys *37:* 56
71. Sundstrom DW and Chen SY (1970) J Compos Mater *4:* 113
72. Dul'nev GN and Zarichnijak YuP (1974) Thermal Conductivity of Mixtures and Composite Materials (in Russian), Energija Leningrad
73. Volkov DP and Zarichnijak YuP (1979) Mechanics of Compos Mater *N5:* 939
74. Dul'nev GN and Novikov VV (1983) Ingenerno fisicheskii zhurnal (IFJ) *45:* 443
75. Novikov VV (1988) Compos Polym Mater (in Russian), *N38:* 17, Naukova Dumka Kiev
76. Kasen MB (1975) Cryogenics *15:* 327
77. Zieband H (1977) In: Polymer Engineering Composites, chap 7, Richardson MOW (ed), Applied Science Publishers London
78. Choy CL, Leung WP and Ng YK (1987) J Polym Sci Polym Phys Ed *25:* 1779
79. Sterzynski T and Linster JJ (1987) Polym Eng Sci *27:* 906
80. Ueberreiter K and Naghizadeh J (1972) Kolloid Z Z Polym *25:* 927
81. Okuda M and Nagashima A (1989) High Temp – High Pressure *21:* 205

3 Thermal Expansion

3.1 The Basic Concept of Thermal Expansion of Solids

Thermal expansion, or more widely thermal deformation, is characterized by the changes of the dimensions of a body resulting from the temperature changes. Similar to the thermal conductivity, the thermal expansivity occurs due to anharmonicity of various modes of lattice vibrations [1, 2]. Any formulations concerning the thermal expansivity of solids are closely related to the main ideas of an equation of state for solids. According to the original Gruneisen assumption [1] the internal energy of a solid can be divided into a static and athermal portion. This assumption leads to an equation of state for the pressure which contains two terms: one of the terms corresponds to the static interaction (internal pressure) and the other represents the thermal pressure due to the expansivity of the lattice vibrations. Thus, the most widely used form of the equation of state for solids is

$$P = P_i + \gamma P_T \tag{3.1}$$

which corresponds to the usual Mie–Gruneisen approximation. In this equation $P_i = -dU_L/dV$ is the internal pressure, γ is the Gruneisen parameter, $P_T = U_T/V$ is the thermal pressure, and U_T is the thermal energy. Differentiation of Eq. (3.1) with respect to temperature at constant volume yields

$$\left(\frac{\partial P}{\partial T} \right)_V = \alpha K_T = \gamma \frac{C_v}{V} \tag{3.2}$$

where α is the thermal expansion coefficient, and K_T is the isothermal bulk modulus. It is well known that thermal expansivity results from anharmonic lattice vibrations. Therefore, the Gruneisen parameter in Eq. (3.2) indicates the degree of anharmonicity because all the other values in this equation can be obtained in the harmonic approximation. Hence, the magnitude of γ is an index of just how anharmonic system is. In arriving at Eq. (3.2), the Gruneisen constant is assumed to be independent of temperature. Moreover, it is assumed that the frequencies of all vibrational modes of a solid change similarly upon the change of volume during heating. The dependence of frequencies upon the volume can be expressed as follows [1]

$$\frac{-\partial \ln \omega_i}{\partial \ln V} = \gamma_i \qquad (3.3)$$

where γ_i is the mode Gruneisen parameter. If all γ_i are equal, Eq. (3.3) transforms to Eq. (3.2). For a Debye solid, the Gruneisen parameter is

$$\frac{-\partial \ln \theta_D}{\partial \ln V} = \gamma_D \qquad (3.4)$$

It is well known now that Gruneisen parameters for many solids are functions of temperature [4]. Therefore, Eq. (3.2) can be written in the following form

$$\alpha K_T = \gamma(T) \frac{C_v}{V} \qquad (3.5)$$

where

$$\gamma(T) = \frac{\Sigma_i \gamma_i \left(\frac{\hbar \omega_i}{k_B T}\right)^2 \exp\left(\frac{\hbar \omega_i}{k_B T}\right) \Big/ \left[\exp\left(\frac{\hbar \omega_i}{K_B T}\right) - 1\right]}{\Sigma_i \left(\frac{\hbar \omega_i}{k_B T}\right)^2 \exp\left(\frac{\hbar \omega_i}{k_B T}\right) \Big/ \left[\exp\left(\frac{\hbar \omega_i}{K_B T}\right) - 1\right]} \qquad (3.6)$$

In this case the macroscopic Gruneisen parameter $\gamma(T)$ is the average over all the microscopic Gruneisen parameters for corresponding modes of the vibrational spectrum weighted by the contribution of each mode to heat capacity. Thus, although there is only one $\gamma(T)$ for each solid at any specific temperature, nevertheless, there are many γ_i corresponding to each particular mode.

Taking into account these general principles in this section we will consider the thermal expansivity of solid polymers and elucidate the role of the chain structure of macromolecules and the large local anisotropy of the interchain and intrachain interactions in thermal expansion of polymers and polymer materials.

3.2 Equation of State for Polymers

In Chap. 1 it was demonstrated that the crystalline lattices of polymers are highly anisotropic due to a large difference of intra- and interchain interactions. Moreover, the crystalline lattices consist as a rule of a large number of atoms capable of independent motion. Therefore, thermal motion of polymers consists of interchain vibrations which are essentially governed by relatively weak van der Waals forces between the chains and intrachain vibrations governed by covalent forces along the chains. The interchain vibrations are characterized by low frequencies and long wave length and intrachain vibrations by high frequencies and short wave length modes. At low temperature, the interchain potential (low-frequency long wave vibrations) dominates while at normal and high temperatures, to a first approximation, intrachain potential dominates. Due to the large difference of inter-and intrachain interactions in polymers the anharmonicity of the interchain vibrations must be considerably larger than that corresponding to the intrachain vibrations. It is quite obvious that due to this large difference, the contribution of the interchain vibrations to many of the properties of polymers which are related to the interchain vibrations, such as thermal conductivity and thermal expansion must dominate. It, in turn, indicates that the thermal expansion of polymer crystalline lattices must be extremely anisotropic and the expansion normal to the chain axis must be considerably larger then along the chain. Hence, it is clear that the direct use of the ideas concerning the thermal expansion and Gruneisen parameter of normal solids with the central forces of interaction (for example, metals and inorganic crystals) on polymers can lead to incorrect conclusions. In fact, if, for example, for estimating of the Gruneisen parameter, one uses Eq. (3.2), which is a normal procedure for low molecular weight solids, then for a polymer one can arrive at quite different values of γ depending on what value of heat capacity C_V is used for the estimation – the whole value of heat capacity or only that part which corresponds to the interchain contribution. The concept of the main contribution of the interchain interactions to the properties of polymers determined by anharmonicity which is now widely used in polymer physics [5–9] has been formulated in the most clear form by Wada et al [9].

For the theoretical calculations of the equation of state and Gruneisen parameters of crystalline polymers, a number of models have been used. The most complete equation of state for crystalline PE has been formulated by Pastine [10], who analysed the internal pressure and the thermal pressure in the Mie–Gruneisen approximation [Eq. (3.1)]. Pastine assumed that 2Nf of the 9N possible modes of a polyethylene chain were Debye-like (N is the number of CH_2 groups per chain and f is some fraction). Having assumed that only these Debye-like modes had appreciable volume dependence, Pastine found the expression for the thermal pressure

$$P_T = (2fNk_BT\gamma_D/V)D(O_D/T) \tag{3.7}$$

Gruneisen parameter γ_D defined by Eq. (3.4) was obtained by using a semiempirical equation suggested by Slater for a two-dimensional lattice

$$\gamma_D = -1/2[1 + V(d^2P_0/dV^2)/dP_0/dV)] \tag{3.8}$$

Inserting Eq. (3.8) into Eq. (3.7) one can immediately arrive at the value of thermal pressure as a function of V. The total pressure can be estimated as the sum of the thermal pressure and the lattice pressure [Eq. (3.1)]. Wu et al. [11] have compared the theoretical curve of Pastine for a PE crystalline lattice and concluded that the agreement can be considered to be very good as is seen from Fig. 3.1. In principle Pastine's method could be applied to other crystalline polymers.

For calculations of equation of state and Gruneisen parameters of crystalline polymers Barker [5] and Broadhurst and Mopsik [12] proposed a model

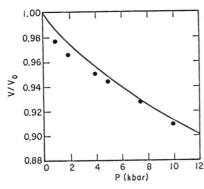

Fig. 3.1. Comparison of compression data for 100% crystalline PE (*circles*) with Pastine's equation of state (*solid line*) [7, 11]

of "a bundle of tubes". This model is based on Wada's hypothesis concerning the dominant role of the interchain potential in the equation of state. In this two-dimensional model the polymer chains are approximated by rigid rods and only the between-chain vibrational modes make a significant contribution to the Gruneisen parameter. In applying hydrostatic pressure to such a crystalline lattice it compresses only perpendicularly to the chains and the compressibility along the chains can be neglected. The Gruneisen parameter for this model can be estimated according to the relation

$$\gamma = -1/2(\Sigma \partial \ln K_T / \partial \ln V)_T - \text{const} \tag{3.9}$$

where the constant is 0.2–0.4 depending on various suggestions. Good agreement was found with experimental thermal expansion coefficients for the PE unit cell and paraffin hydrocarbons as a function of temperature [7, 9]. The values of the Gruneisen parameters obtained by using this approach will be considered in the next section.

The problem of the equation of state and the Gruneisen parameter for amorphous polymers, both liquid and glassy, is considerably more complex than for crystalline polymers due to the following reasons. Firstly, for amorphous polymers and for amorphous parts of crystalline polymers, the physical meaning of γ is a little bit more complex because in this case the compressibility is determined not only by anharmonicity of vibrational modes but also by the free volume. Besides, in amorphous polymers it is more difficult to estimate quantitatively the thermal pressure and the internal pressure due to some difficulties of correct calculation of the interchain distances and interchain interactions. Nevertheless, Pastine [10] made an attempt to obtain theoretically the equation of state of purely amorphous PE and partly crystalline PE, using some suggestions and empirical values. Therefore, such an approach could be considered as semi-empirical.

A separate direction in this area are those theoretical investigations of the equations of state of amorphous liquid and glassy polymers using cell and hole models [13]. This direction was initiated by Prigogine and coworkers [14] and was widely used in works by Simha and coworkers [15]. This approach is based on the principle of corresponding state. A comprehensive consideration of all these models were given by Curro [13]. As a result of a comparative analysis of various models, he concluded that the cell theories are able to adequately predict the PVT behaviour of polymer liquids over a limited range of temperature and pressure. The hole theory developed by Simha and coworkers [15] gave better agreement then the cell model for PVT data, but fails to explain the internal pressure behavior. It should be mentioned that often totally empirical equations of state for polymer liquids can be used more easily for practical purposes (see, for example [15, 16]).

Simha et al. [17] and Curro [18] developed a low-temperature equation of state for polymer glasses using a cell model with the harmonic oscillator approximation. The temperature dependent part of the free energy has been analysed by using both the Einstein and Debye models. For the Debye model the equation of state is

$$P\tilde{V} = \left(\frac{2}{\tilde{V}^2}\right)\left(\frac{A}{\tilde{V}^2} - B\right) + 3\gamma\tilde{\theta}_D\left[\frac{3}{8} + 3\left(\frac{\tilde{T}}{\tilde{\theta}_D}\right)^4 D\left(\frac{\tilde{\theta}_D}{\tilde{T}}\right)\right] \qquad (3.10)$$

where A and B are numerical coefficients. The reduced thermal expansion coefficient α has the form

$$\tilde{\alpha}_{0D} \sim (\tilde{T}/\tilde{\theta}_D)^3 \qquad (3.11)$$

where $\tilde{\alpha} = \alpha T^*$; $\tilde{\theta} = \theta/T^*$ and T^*, P^* is reduced temperature and pressure, correspondingly. In order to satisfy a principle of corresponding states, it is necessary to fulfill the condition $\tilde{\theta} = $ const. Comparison of experimental and theoretical reduced thermal expansivities as a function of reduced temperature for typical polymer glasses such as PS, PC, PPO and various polymethacrylates demonstrated a very good agreement below about 60 K. Above these temperatures experimental values are considerably above the theoretical curve.

3.3 Gruneisen Parameters of Polymers

In the thermal physics of polymers it is now generally accepted that for solid polymers it is necessary to distinguish two types of Gruneisen parameter, namely the thermodynamic γ_T and the lattice γ_L Gruneisen parameters. Warfield [6] has pointed out that in polymers due to two distinct modes of lattice vibrations there must be two different types of γ. Those measurements which relate to the interchain thermal motion determine the microscopic or lattice Gruneisen parameter γ_L. On the other hand, when calculations correspond to all vibrational modes, both interchain and intrachain, on average give the thermodynamic Gruneisen parameter γ_T. Much confusion has existed in the polymer literature by the failure to recognize the difference between the two types of Gruneisen parameter for polymers. For metals and ionic crystals $\gamma_L \simeq \gamma_T$. However, for polymers a large difference between these two Gruneisen parameters can exist. Normally, intrachain covalent bonds of high frequency vibrations have very low values of the Gruneisen parameter, while interchain low frequency anharmonic vibrations involving van der Waals bonds have very high values of the Gruneisen parameter. This difference must be taken into account first of all during the experimental determination of γ.

At present various experimental techniques can be used for determining the Gruneisen parameter. Generally, all experimental methods for determining γ can be divided into two categories [6]. One of the categories corresponds to those methods which are used for measurements of one of the anharmonic properties of a polymer, such as thermal expansion, thermal conductivity, thermoelasticity or the pressure dependence of the bulk modulus. The second group of methods is based on measurements of various dynamic properties in the bulk. Ultrasonic wave velocity measurements under pressure, X-ray and laser measurements have been employed to determine γ_L. Hence, if one determines γ according to the pressure dependence of the bulk modulus using the formula

$$\gamma_L = 1/2(\partial K_T/\partial P)_T \qquad (3.12)$$

then this value relates only to the interchain interaction and is the lattice Gruneisen parameter. Wada and coworkers [9] have demonstrated that

$$\alpha K_T = \gamma_L \frac{C_{inter}}{V} \qquad (3.13)$$

where C_{inter} is that part of the whole heat capacity related to the interchain vibrations. Although this value of C_{inter} is very difficult to estimate by direct calorimetric measurements, nevertheless, it can be evaluated in those cases when both γ_L and γ_T are known. In fact, the relation between the total heat capacity C_V and C_{inter} must be identical to the relation between γ_L and γ_T

$$\gamma_T = \gamma_L(C_{inter}/C_V) \qquad (3.14)$$

The theoretical ground of this formula was given by Curro [18] who calculated the Gruneisen parameter for glassy amorphous polymers in a very wide temperature interval using a cell model. It has been predicted that the limited value of $\gamma_L = 3.16$. One more very interesting conclusion concerns the relation between the Gruneisen parameter and the number of atoms in the repeating unit of polymers. At high temperatures the Gruneisen parameter should be inversely proportional to the number. Wide experimental checking of the relation (3.14) has not been made yet, although experimental data for PS and PMMA are in a good agreement with it [18].

According to the modern theoretical considerations [5, 7, 18] the temperature dependence of the Gruneisen parameter in glassy or semi-crystalline polymers is as follows. At low temperatures the long wavelength low frequencies acoustic modes dominate. As the temperature is raised the higher frequency optic modes begin to contribute. Due to the averaging over the whole vibrational spectrum according to their contribution as is described by Eq. (3.6) the Gruneisen parameter drops in value until at very high temperatures it approaches a direct average over the dominated optics modes owing to the large number of internal degrees of freedom in polymers. At normal temperatures γ is a weak function of temperature. Hence, as the temperature drops, γ_T increases and approaches γ_L Theoretically, γ_T becomes equal to γ_L at $0\,\text{K}$ [5, 7, 18].

Crystalline polymers are two-phase systems consisting of crystallites and amorphous regions. Gibbons [19] considered the relationship between the bulk Gruneisen parameter of a polycrystalline polymer and the Gruneisen parameter of the individual anisotropic crystallites. In spite of presence of the anisotropic crystallites, typical bulk crystalline polymers are macroisotropic due to the statistical orientation of the crystallites and existence of the amorphous regions. For calculations of γ in the framework of the "bundle-of-tubes" model the Reuss and the Voigt approximations were used. For polymer crystallites, which are characterized by a two-dimensional compressibility, the isotropic stress in the bulk state is reached due to their anisotropic volume change. It corresponds to the Reuss approximation and leads to the relation

$$\gamma = (\gamma_1 K_{T_2} + \gamma_2 K_{T_1})/(K_{T_1} + K_{T_2}) \tag{3.15}$$

where index 1 and 2 correspond to two directions perpendicular to the chain axis in crystallites.

The most widely used method of evaluating of the Gruneisen parameter is the calculation according to Eqs. (3.5) and (3.14). As a rule, using this approach one can estimate the value of γ at some particular temperature. The estimation of γ_T in a wide temperature range by means of calculations is more difficult because often all the necessary values of $K_T(T)$, $C_V(T)$ and $\alpha(T)$ for the same samples are absent. To overcome this difficulty Shen et al. [20] suggested using the following equation for the determination of (T)

$$\gamma(T) = \frac{C_P - C_V}{C_V \alpha T} \tag{3.16}$$

This equation can be obtained from the well known thermodynamic relation $C_P - C_V = \alpha^2 K_T VT$ by introducing Eq. (3.5) into it.

The most complete investigation of the Gruneisen parameters was carried out for PE. The data obtained with various methods have been summarized and analysed by Shen et al. [7, 8] (Table 3.1). Although these authors do not distinguish between the lattice and thermodynamic Gruneisen parameters we do it following the principles considered above. The data shown in Table 3.1 demonstrate a strong dependence of γ_L upon the experimental method used. It must be also kept in mind that the Gruneisen parameters obtained by means of the acoustic methods correspond to the adiabatic ones, while with the majority of other methods they are isothermic parameters. Normally adiabatic values are about 10% higher than isothermal ones. Such a relation has been actually observed for a number of polymers [6].

Table 3.1. Gruneisen parameters of PE [7]

Equation	Gruneisen parameter	Method
1. Lattice and thermodynamic Gruneisen parameter		
$\gamma_T = \alpha K_T C_V/V$	(1.1)	Calculation
$\gamma_L = -1/2(\partial \ln K_T/\partial \ln V)_T - 5/12$	7.5	Compression
	(3.5)	X-ray under pressure
$\gamma_L = K_T(\partial \ln \bar{c}/\partial P)_T + 1/3$	4.0	Ultrasonic
	7.7	Heat capacity
$\gamma_L = K_T(\partial \ln \kappa/\partial P)_T - 1/3$	11.0	Thermal conductivity
$\gamma_T = (C_P - C_V)/C_V \alpha T$	(1.1)	Heat capacity
	4.5*	Thermoelasticity
2. Mode Gruneisen parameter		
$\gamma_i = -d \ln \omega_i/d \ln V$		
$72\,cm^{-1}$ translational optical mode	1.28	IR
$730\,cm^{-1}$ rocking CH mode	0.07	IR
$1473\,cm^{-1}$ CH bending mode	0.0024	IR
$11\,cm^{-1}$ accordion mode	0.47	Raman
$1066\,cm^{-1}$ C-C stretching mode	0.092	Raman
$1133\,cm^{-1}$ C-C stretching mode	0.087	Raman
$1296\,cm^{-1}$ CH twisting mode	0.027	Raman
$2883\,cm^{-1}$ asymmetric CH stretching mode	0.070	Raman

Data in parentheses correspond to PE crystalline lattice.
*Data from [21]

The results obtained with the help of the second formula in Table 3.1 also demonstrate a dependence of γ_L upon the degree of crystallinity. In X-ray studies the bulk modulus of only the crystalline lattice is determined while in the usual measurements of the bulk modulus both contributions of the amorphous and crystalline parts are taken into account. These data, however, have been obtained on different samples and therefore it is quite possible that the difference is the result not only of the influence of the degree of crystallinity but some other factors as well, for example various types of defects. For a quantitative comparison it is important because it has been reported that the values of Gruneisen parameters for the quenched and annealed samples are different [22].

In order to check the possible influence of various defects in the crystalline lattice, Kijima et al. [23] studied the temperature dependence of the Gruneisen parameters over the temperature range 15–90° C and its pressure dependence (up to 550 MPa) on PE single crystals. In the temperature interval the value of γ_L increases from about 6.5 up to 9 at normal pressure being practically constant $\gamma_L = 6.5$ over the whole temperature range at pressures up to 200 MPa. Such behavior was explained by simultaneous influence of the lattice defects and the dependence of anharmonicity on temperature and pressure.

The Gruneisen parameters for various vibrational modes differ considerably (more than an order of magnitude) demonstrating quite understandable behavior: vibrational modes with higher frequencies are characterized by smaller values of γ because they are less dependent upon the increase during compression of the intermolecular interaction.

As it has been underlined above that a drop in the temperature must be accompanied with freezing of high frequency vibrational modes and as a result with an increasing contribution of low frequency modes. It must lead to an increase of γ_T when the temperature is decreased. The estimations made for the crystalline lattice of PE support this conclusion qualitatively (Fig. 3.2). In contrast to the thermodynamic Gruneisen parameter, γ_L is dependent on temperature only through a dependence of the volume upon temperature, i.e. it must be a weak function of temperature. Investigations in a narrow temperature interval from -50 up to $100°$ C support this conclusion [6, 9].

Recently, Engeln et al. [24] have studied the bulk Gruneisen parameter of PE with various degrees of crystallinity in the range between 0.44 and 0.98. For nearly completely crystalline PE below 10 K, a low temperature limit of the bulk Gruneisen parameter 2.0 ∓ 0.5 was reached. At about 25 K a maximum in γ occurs, which has been related to in-plane-bending vibrations of the PE chain. The maximum is followed by a continuous decrease to 0.6 at 300 due to an increasing influence of high frequency phonons with small Gruneisen parameters. However, the low temperature limit $\gamma = 2$ at T $\Rightarrow$ 0 K is smaller than expected from theory ($\gamma = 5.25$). The bulk-Gruneisen parameter of noncrystalline PE is constant between 40 and 100 K ($\gamma = 1.6$), whereas at lower temperature a rapid increase was observed, reaching $\gamma = 12.5$ at 5 K. This unusually high value was assumed to be due to nonidentified excess modes with extremely high Gruneisen parameters. Similar behavior has also been found for some glassy polymers [25].

In order to check the temperature dependence of the lattice and thermodynamic Gruneisen parameters Kato [26] studied Brillouin scattering and thermal expansion of amorphous PMMA between about 4 and 300 K. Experimental values of γ_L are shown in Fig. 3.2. At room temperature, the value of γ_L is about 4, which coincides with other published results [6, 9]. With decreasing temperature this value only decreases slightly. Such behavior seems to agree nearly with the behaviour theoretically predicted by Curro [10] and does not agree with the theoretical prediction of Barker [5]. These results also support the conclusion that at low temperatures γ_T must approach γ_L.

Although for other polymers the investigation of the Gruneisen parameters is less complete than for PE, nevertheless, it is possible to estimate values of γ_L and γ_T for some polymers. The corresponding data are listed in Table 3.2. One may conclude that there seems to be no visible correlation of the Gruneisen parameters with the chemical structure of polymers.

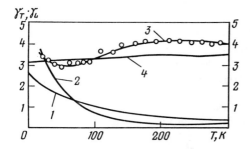

Fig. 3.2. Variations of the lattice and thermodynamic Gruneisen parameters γ_L and γ_T with temperature. *1* – the theoretical curve for γ_T calculated for 100% crystalline PE [7]; *2* – the theoretical curve for γ_T for glassy PMMA, calculated from Eq. (3.2); *3* – experimental values of γ_L for PMMA [26]; *4* – the theoretical line, predicted by Curro [18]

Table 3.2. Lattice and thermodynamic Gruneisen parameters for some polymers at 298 K [6]

Polymer	γ_L	γ_T	$\gamma_T/\gamma_L = C_{int}/C_V$
HDPE	4.1	0.52	0.126
LDPE	6.4	0.38	0.059
i-PP	9.0	0.96	0.094
PEO	5.0	1.02	0.204
PTFE	8.0	0.40	0.05
PMMA	4.0	0.82	0.205
PS	4.4	0.79	0.180

3.4 Thermal Expansivity of Polymeric Crystals

For the isotropic solids and crystals with cubic lattice the thermal expansivity is isotropic and may be described only by the single volume coefficient of thermal expansion $\alpha = \partial \ln V/\partial T)_P$ [2]. The thermal expansivity of all other crystals is anisotropic and during heating both the volume and shape of such crystals occur. However, the thermal expansivity must follow the symmetry of the crystal and, therefore, without the phase transitions the temperature change cannot change the symmetry of the crystal. The thermal deformation $[\varepsilon_{ij}]$, which occurs during heating of a crystal to ΔT is the tensor, all the components of which may be represented by

$$\varepsilon_{ij} = \beta_{ij}\Delta T \tag{3.17}$$

where β_{ij} is the coefficient of linear thermal expansion. Because $[\varepsilon_{ij}]$ is the tensor of rank two and because ΔT is a scalar, therefore, $[\beta_{ij}]$ is also the symmetrical tensor of rank two. The tensor of thermal expansivity may be transformed to the main axes

$$\varepsilon_1 = \beta_1 \Delta T_1 , \quad \varepsilon_2 = \beta_2 \Delta T_2 , \quad \varepsilon_3 = \beta_3 \Delta T_3 \tag{3.18}$$

where β_1, β_2 and β_3 are the main coefficients of thermal expansion.

For the majority of crystals the main coefficients of the thermal expansion are positive. However, some crystals (for example, tellurium, zinc, selenium) [2] possess a negative thermal expansivity in some directions over some temperature intervals. Thus, the tensor of thermal expansion of the ansotropic structures reflects the anisotropic forces acting in these structures.

As demonstrated above, the crystalline polymers are characterized by two essentially different types of interactions, namely the intrachain and interchain ones. The consequence of the difference is a striking anisotropy of physical properties including thermal expansivity. A very large difference in the potentials of inter- and intrachain interactions must lead to a large difference in the thermal expansion along the chains and perpendicular to them. Moreover, if one follows the suggestion concerning the succession of excitement of the vibrational modes in the chain structures according to their contribution to the heat capacity one may assume, that the thermal expansivity along the polymer chains must be completely absent at low and moderate temperatures because at these temperatures only acoustic modes associated with C-C bending and internal rotations are excited. However, the theoretical consideration of the thermal expansivity of chain-like and layer structures given by Lifshitz [28] showed that excitement of the bending waves in such structures with the dispersion relationship $\omega \sim k^2$ must lead to the appearance of the negative thermal expansion coefficients along the chains and in the plane of the layers due to a specific "membrane effect". The physical nature of the effect is that, in contrast to the usual case when during extension the distances between the particles are increased and correspondingly the frequencies are decreased, under extension the frequencies of the long bending waves are increased, i.e., the vibrations become more rigid. Due to this "membrane effect" in the chain-like and layer structures, the excitement of the bending waves is accompanied only with a shortening of the dimensions along the chains and in the plane of the layers.

The anisotropic tensor of the thermal expansion coefficients for chain-like structures

$$\beta_{ij} = \partial \varepsilon_{ij} / \partial T \tag{3.19}$$

(where ε_{ij} is the tensor of thermal deformation) is characterized by three components: β_{zz} along the chains and β_{xx} and β_{yy} normal to them. Because the free energy of solids is connected with the tensor of the thermal deformation with a simple relation

$$\varepsilon_{ij} = -\partial F / \partial \sigma_{ij} \tag{3.20}$$

(where σ_{ij} is the stress tensor), then

$$\beta_{zz} = -\frac{\partial^2 F}{\partial \sigma_{zz} \partial T} \tag{3.21}$$

$$\beta_{xx} + \beta_{yy} = -\frac{\partial^2 F}{\partial \sigma_{xx(yy)} \partial T} \qquad (3.22)$$

At low temperatures one can arrive at

$$\varepsilon_{ij} = -\frac{\partial F}{\partial \sigma_{ij}} = -\sum \frac{\left(\dfrac{\partial \omega_i}{\partial \sigma_{ij}}\right) \exp\left(\dfrac{\hbar \omega_i}{k_B T}\right)}{1 - \exp\left(\dfrac{\hbar \omega_i}{k_B T}\right)} \qquad (3.23)$$

Lifshitz demonstrated that for non-interacting chain-like and layer structures $\partial \omega / \partial \sigma > 0$ along the chains and layers, while perpendicular to the chains and layers $\partial \omega / \partial \sigma < 0$. According to Eqs. (3.21)–(3.23) it means that along the chains and layers the thermal expansivity must be negative, i.e. at heating of such structures in these directions they must become shorter, while in two other directions – longer. Hence, the main reason for the appearance of the negative thermal expansivity in chain-like and layer structures is a strong anisotropy of the elastic forces acting in the crystalline lattices of such structures. This differs principally from other solids possessing the negative thermal expansivities [2], where they appear due to, first of all, the long-acting forces resulting from their polarization.

The negative coefficients of thermal expansion along the axes of macro-molecules in crystalline lattices were discovered for many polymers (Table 3.3). Hence, the contraction of the crystalline macromolecules along their axes can be considered as rather general behavior, independent of their chemical structure. Normally, the contraction along the macromolecules is sometimes less than the expansion perpendicular to the chains. It means that the volume thermal expansion coefficient has a positive value. It is very interesting that the negative thermal expansivity along the chains not only occurs at low temperatures but at room temperatures as well. One more very important feature which can be drawn from Table 3.3, consists of the fact that the negative thermal expansivity is a rather weak function of the chemical structure of polymers. Therefore, one can suggest the existence of a rather universal molecular mechanism responsible for the thermal contraction. According to Lifshitz the thermal contraction of the chain-like structures is not connected with any molecular features of these structures but is a result only of a strong anisotropy of the forces along the chains and between them, which lead to the appearance of the bending waves with the unusual dispersion relation-ship in the long-wave part of the vibrational spectrum. Indeed, the negative thermal expansion coefficients has been found not only for the organic chain-like macromolecules but also for the helical inorganic macromolecules, such as tellurium [2].

Table 3.3. Thermal expansion coefficients of polymeric lattices

Polymer	$\beta_{c_\parallel}$ $\times 10^5\,\mathrm{K}^{-1}$	$\beta_{c\perp}$ $\times 10^5\,\mathrm{K}^{-1}$	Temperature interval, $^\circ C$	Reference
PE	-1.2	$\beta_a = 14.7$	$20-60$	30
	-1.8	$\beta_b = 6.0$	$20-120$	30
	$-(1.1-1.3)$		$-130+60$	22
PET	-2.2	$\beta_a = 17.1$	$-196+20$	29
		$\beta_b = 12.7$		
PA-6	-4.5	$\beta_a = 6.0$	$-196+20$	29
		$\beta_b = 23.4$		
Cellulose	-4.5	$\beta_a = 6.0$	$-196+20$	29
		$\beta_c = 2.8$		
PCP	-4.1	$\beta_a = 13.1$	$-196+20$	29
NR	-8.4	$\beta_a = 12.5$	$-196+20$	
		$\beta_b = 20.5$		

Although the Lifshitz's prediction of the negative thermal expansion of the chain-like structures was made in the early 1950s, nevertheless in the early 1970s, interest in the negative thermal expansivity behaviour of polymers increased considerably and some suggestions that this behavior was molecular occurred [30, 31].

They were based on the intuitive idea that the increase of the amplitude of the lateral motion of the atoms of the chains with increasing temperature must be accompanied by a shortening of macromolecules along their axes since the strong interchain forces keep the total length essentially unchanged. In particular, at first it was assumed for PE that shortening of the chains with rising temperature occurred due to increasing out-of-plane torsional modes of a planar zig-zag, which can be characterized by the rotation angle between consequitive planes defined by pairs of C-C bonds. The early value of the rotational angle equal to 3°, which was postulated by Kabayashi and Keller [30] and after a more elaborated consideration [31] increased to 13° to explain the experimentally found shortening of PE macromolecules in the temperature range 20–100°C.

Based on the temperature dependence of the heat capacity of chain-like structures, one can assume that shortening of the chains with rising temperature is not only the result of the torsional modes but bending modes as well. The attempts to consider theoretically the negative thermal expansivity of the planar zig-zag chains was undertaken by Kan [32], Choy et al. [33,34] and Barron et al. [35]. Here we will consider only the results of the more elaborate calculations, taking into account both the torsional and bending waves [33]. These calculations were conducted for crystalline lattices of two types: the Stockmayer–Hecht model lattice (see Fig. 1.4) and the lattice of planar zig-zag chains lying in parallel planes. The first model allows to consider only the torsional modes, while the second model allows to take into account both the torsional and bending modes.

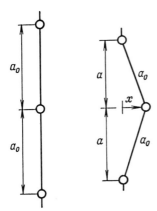

Fig. 3.3. Schematic diagram of the shortening of the longitudinal dimensions of an inextensible chain due to lateral motion [33]

The main idea is very simple and is demonstrated in Fig. 3.3. A linear chain made up of atoms joined together by inextensible bonds each of length a under heating is shortened due to a lateral displacement x of atoms in the middle. The length of bond projected along the chain direction changes from a to

$$a = (a_0^2 - x^2)^{1/2} \cong a_0 - x^2/2a_0 \tag{3.24}$$

where it is assumed that $|x| \ll a_0$. If an effective elastic constant restraining such lateral motion is K_{ef} then the relation for two lateral modes is

$$\frac{1}{2} K_{ef} \langle x^2 \rangle = k_B T \tag{3.25}$$

where $\langle x^2 \rangle$ is the average displacement. The mean projection is

$$\langle a \rangle = a_0 - \frac{k_b T}{K_{ef} a_0} \tag{3.26}$$

and the expansivity along the chain direction is

$$\beta_{c_\parallel} = \frac{1}{a_0} \frac{d\langle a \rangle}{dT} = -\frac{k_B}{K_{ef} a_0^2} \tag{3.27}$$

Hence, the only parameter which needs identification is K_{ef}. It is very interesting that in this model no reference to the anharmonicity is made which is the molecular origin of the thermal expansivity. However, in calculating the constant Kef the anharmonicity of lateral motion is taken into account, therefore, the reason for shortening of chains is the anharmonicity although not in the direction of shortening of chains. In the lattice consisting of parallel linear chains there are two force constants which determine K_{ef}, namely: the interchain force constant K_v and bending constant K_b. The expression for K_{ef} is

$$\frac{1}{K_{ef}} = \left(\frac{1}{K_g^3 K_w} \right)^{1/4} \frac{1}{2\sqrt{2}} (1.18 - 0.54t - 0.59t^2 + \ldots) \tag{3.28}$$

where $t^2 = K_w/16K_g$. Using $K_g = 35\,N/m$ and $K_w = 3.2\,N/m$, which is characteristic of PE, gives $\beta_{c_\parallel} = -1.3 \times 10^{-5}\,K^{-1}$. This value is quite consistent with the value listed in Table 3.3. Corresponding estimations for the planar zig-zag chain model gives $\beta_c = -1.31 \times 10^{-5}\,K^{-1}$.

The theoretical approach described above allows us to make some general conclusions. First of all, the thermal shortening is independent of the mass of atoms in the repeating unit of macromolecules. Further, the negative thermal expansivity is only slightly dependent on K_w. It means that for polymers with similar backbones the values of $\beta_{c_\parallel}$ will not differ significantly. Data in Table 3.3 do not contradict this conclusion. A comparison of the results obtained with these two models demonstrates that the torsional modes are responsible for about a half of the negative expansion, while the rest being attributed to the bending waves. This is the reason why the Stockmayer–Hecht model, which is not taken into account the torsional modes but with twice as many bending modes, also gives reasonable values of the expansion.

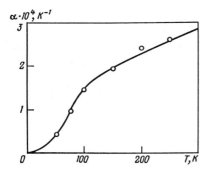

Fig. 3.4. Thermal expansion coefficient of PE unit cell as a function of temperature [7]. The *solid curve* – theoretical calculation; *circles* – experimental points from the temperature behavior of IR- and X-ray data

The model allows us to estimate the average negative thermal expansion in the classical temperature region. Such estimations showed [32] that for PE, thermal shortening must be weakly dependent on temperature in the temperature interval above 100 K. According to Table 3.3 it seems to take place in all the polymers studied. It is especially interesting to consider the temperature dependence of the thermal expansion of polymer lattices at low temperatures when unusual temperature behavior of thermal expansion is possible. Shen et al. [20] used the Gruneisen equation for the thermal expansion of solids [1] for estimating the volume thermal expansion of the PE lattice

$$\alpha = \frac{C_v}{Q(1 - \xi U_T^2/Q)} \tag{3.29}$$

where $Q = V_0K_0/\gamma_0$; $\xi = (m + n + 3)/6$, and m and n are the exponents in the Mie potential. U_T and C_v can be calculated from the Debye model. Figure 3.4 shows the thermal expansion coefficient of the PE unit cell as a function of temperature calculated from the Debye model. Eq. (3.29) was fitted to the data of Shen et al. [20] using the following values of parameters: $Q - 115\,kJ/mole$, $\xi = 3.75$, $m = 6$, $n = 13.5$ and the Debye temperature

$\theta_D = 315$ K. Although in this particular case the agreement between the calculated and experimental data is rather good one can hardly hope that Eq. (3.29) will be acceptable for a wide number of polymers, because the agreement with the experimental data was reached as a result of a rather arbitrary selection of the constants. Nevertheless, Fig. 3.4 shows that the main ideas of thermal physics of non-polymeric solids may be used quite successfully for the theoretical estimations of the thermal expansion of polymers if their structural features are taken into account. In the calculations considered above, the main feature of the polymeric lattices was taken into account due to the two-dimensional character of thermal expansivity of PE unit cell because it has been suggested that the thermal expansivity along the chain was equal to zero. Broadhurst and Mopsik [36] analysed the thermal expansion of hydrocarbons consisting of semirigid chains of beads in a quasiharmonic potential and demonstrated its sensitivity to the length of beads. The thermal expansion coefficient calculated for sufficiently long chains was found to resemble closely the behavior shown in Fig. 3.4.

Dadobaev and Slutzker [37] studied the thermal expansion behavior of three polymeric crystals over a wide temperature range of 5–400 K. The polymers studied were PE, PA-6 and PVAl. The temperature dependence of the crystal lattice parameters was determined using X-ray techniques. The main results are shown in Fig. 3.5. It can be seen from these data that the temperature dependence of thermal expansion both perpendicular to and along the chain direction is qualitatively similar, although they differ in sign and absolute values: the expansion parallel to the chain axis is negative, while perpendicular to the chain axis it is positive. Both decrease rather slowly to about 100 K and more rapidly below 50 K. The similarity of the behavior $\beta_{c\perp}$ vs T and $\beta_{c\parallel}$ seems to indicate that the negative expansion along the chains is a direct consequence of the lateral vibrations. For PE the dependence β vs T resembles the behavior of C_v vs T (see Chap. 1). With regard to this temperature dependence, Dadobaev and Slutzker suggested that below 100 K the main contribution to $\beta_{c\perp}$ for all three polymers studied is the torsional modes and that increasing of $\beta_{c\perp}$ above 100 K is a consequence of the increase of the anharmonicity of the torsional modes and the excitement of the bending modes. Simple estimations showed that in the region of about 200 K, $\beta_{c\perp}$ reaches the "classical" values. Indeed, according to the relation given by Frenkel [38]

$$\beta \cong k_B/El^2 \qquad (3.30)$$

where l is the linear characteristic dimension. For Pe $E_\perp = 4$ GPa and for PVAl $E_\perp = 9$ GPa. Assuming that $l = 0.3$–0.4 nm we arrive at $\beta_{c\perp} = 10 \times 10^{-5}$ K^{-1} for PE and 5×10^{-5} K^{-1} for PVAl. These estimated values are very close to the experimental results.

In spite of the similarity of the temperature behavior of the thermal expansivities $\beta_{c\perp}$ and $\beta_{c\parallel}$ for these polymers the difference in their absolute values is rather large, being 4–5 times for Pa-6 and PVAl which differ significantly in their intermolecular interactions. In this connection a very important problem arises: what factor is responsible for this difference — either the

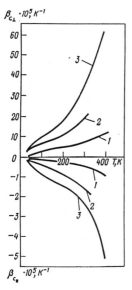

Fig. 3.5. Linear thermal expansion coefficients $\beta_{c\perp}$ and $\beta_{c\parallel}$ as a function of temperature for PVA (*1*), PE (*2*) and PA-6 (*3*) [37]

difference in the amplitudes of vibrations, or the difference in the coefficients of the anharmonicity. Dadobaev and Slutzker have concluded that because the similarity of the lateral expansion and linear contraction, the main factor responsible for this is the difference in the amplitudes. The variation of the degree of the anharmonicity plays a considerably smaller role.

The conclusion about the important role of the vibrational amplitude in thermal expansivity is quite consistent with peculiarities of the arrangement and interaction of macromolecules in their lattices. This is shown in Fig.3.6. The largest thermal deformation occurs in PA-6, where planar zigzag macromolecules are arranged parallel to each other with hydrogen bonds occurring at every seventh bond. Although the hydrogen bonding is absent in PE the planes of the molecule back bones are arranged crosswise. Therefore, the largest influence on the increase of the intermolecular distances is that of the macromolecules whose zig-zag planes are close to the plane (110). To such macromolecules belong half the total number of macromolecules. This ratio seems to be closely related with the fact that the thermal expansion of PA-6 is twice that of PE (see Figs. 3.5 and 3.6). Finally, although in PVAl lattice all the planes in which zig-zag molecules are arranged, are parallel, but a large linear density of the hydrogen bonds allows only a rather small vibrational amplitude. Such behavior corresponds the inverse proportionality between the lateral thermal expansion of the polymer lattices and the lateral moduli of elasticity.

Hence, the negative thermal expansion seems to be a universal phenomenon for polymers with planar zig-zag chains, because of the very strong elastic anisotropy of the lattices formed by this type of macromolecules. The situation is less clear for polymeric crystals with helical chain structure, since the elastic anisotropy in this case is not so high as for planar zig-zag chains. Therefore, recently Choy and Nakafuku [39] raised the question concerning

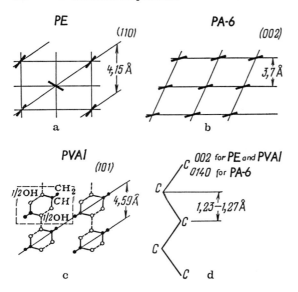

Fig. 3.6. Schematic diagrams of the unit cells of various crystalline polymers [37]. **a, b, c** – projections of the planar *trans*-zig-zag on the plane, perpendicular to the axes of the macromolecules (the interplane distances are indicated for room temperature); **d** – planar *trans*-zig-zag of the backbone of macromolecules

the negative thermal expansivity of helical chains structures in crystalline lattices. White et al. [40] studied the thermal expansivity of the single crystals of POM, which crystallize with a helical chain structure. The measurements were conducted starting from a very low temperature (Fig. 3.7). In spite of a large scattering of experimental points below 10 K it has been found that

$$\beta_\perp = (0.15 \pm 0.005) \times 10^{-8} T^3 (K^{-1}) \quad T < 15\,K$$
$$\beta_\parallel = (-0.028 \pm 0.001) \times 10^{-8} T^3 \quad (T < 10\,K) \qquad (3.31)$$
$$\alpha = 2\beta_\perp + \beta_\parallel = 0.27 \times 10^{-8} T^3 \quad (K^{-1})$$

First of all, these data show that in the vicinity of very low temperatures both the linear and volume coefficients of thermal expansion follow the cubic temperature dependence, which is in very good agreement with the behavior of the heat capacity measured on the same samples [$C \cong (13.5 \pm 0.5) T^3\,\mu J/gK$ in the temperature range 3–12 K].

As is seen from Fig. 3.7 the behavior of $\beta_{c_\parallel}$ is much more complex in comparison with the planar zig-zag chains. The negative thermal expansion of POM single crystals occurs only below approximately 100 K and above this temperature $\beta_{c_\parallel}$ changes its sign from the negative to positive, while for the planar zig-zag chains it remains negative in the whole temperature range. The reason for such behavior is unclear yet. The density of the single crystals studied was less than the crystallographic density of POM, therefore, one may assume that the degree of crystallinity was less than 100%. It is very likely

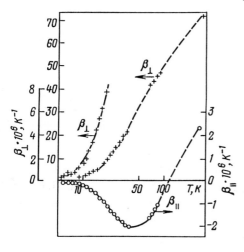

Fig. 3.7. Linear thermal expansion coefficients $\beta_\perp$ and $\beta_\parallel$ of POM single crystals as a function of temperature [40]

that the single crystals studied included some defects or amorphous regions which may change the sing of $\beta_{c_\parallel}$.

Choy and Nakafuku [39] studied the thermal expansion of the crystal lattice of POM using X-ray diffraction method. They have found that $\beta_{c_\parallel}$ is negative, with a value increasing from 0.22 to 0.85 $\times$ 10^{-5} K^{-1} as the temperature rises from 160 to 400 K. As a result of these findings one may conclude that amplitudes of the torsional and bending modes are sufficient for producing a thermal contraction of the helical chains. However, the thermal expansion measurements of another polymer crystal with helical chains – P4MP1 published recently showed that, although the $\beta_{c\perp}$ value is close to that of POM, $\beta_{c_\parallel}$ is not only positive in the temperature range -160 $+160°$ C but its value is larger by an order of magnitude. Choy and Nakafuku have analysed the thermal expansivity for these and other helical macromolecules. The result of their analysis is summarized in Table 3.4. It is seen that the behavior of $\beta_{c\perp}$ is similar for the polymer crystals with helical and zig-zag chains. On the other hand, $\beta_{c_\parallel}$ depends strongly on the magnitude of the elastic anisotropy. Indeed, POM is characterized by a large anisotropy factor $(E_{c_\parallel}/E_{c\perp})$ equal to 7 and possesses a negative thermal expansivity along the helical chains. P4MP1 has a considerably smaller anisotropy factor 2.3 which leads to a large positive value of $\beta_{c_\parallel}$. A similar relation between the elastic anisotropy and axial expansivity exists also in other crystalline solids with helical chain structure (see Table 3.4).

Now a natural question arises: what consequences for crystalline polymers may the negative thermal expansivity along the chain axes in crystalline lattices have? For the bulk undrawn crystalline polymers there will be no consequences because of the following reasons. First, the volume thermal expansivity in crystalline polymers will be generally positive, because the two other coefficients of thermal expansion are positive and considerably larger than $\beta_{c_\parallel}$. Taking into account the statistical arrangement of crystallites and correspondingly of chains axes in bulk crystalline polymers it is quite obvious, that the thermal expansivity along any direction happens to be positive. The second reason consists in the existence of the amorphous regions in crystalline polymers. According to the additive principle of structure of crystalline polymers

$$\alpha = \alpha_a(1 - w) + \alpha_c w \qquad (3.32)$$

At $T < T_g$ $\alpha_a \cong \alpha_c$. However, in contrast to crystallites there is no reason for the appearance of a significant local anisotropy of thermal expansion in microvolumes with the linear dimensions 10–20 nm typical of crystalline polymers with a normal degree of crystallinity. At $T > T_g$ $\alpha_a \cong (3-4)\alpha_c > 0$, therefore, the amorphous regions make a significant contribution to the total thermal expansion. These suggestions allow us to conclude that the negative thermal expansion along the chain axes in crystalline lattices will have no consequences in macroscopic thermal properties. This conclusion is in agreement with experimental data.

Table 3.4. Young's moduli (in GPa) and thermal expansion (in $10^{-5}\,\mathrm{K}^{-1}$), of several crystals with chain-like structure at 300 K [39]

| | Planar zig-zag chains | | | Helical chains | | | |
	PE	PTFE	PA − 6	POM	P4MP1	Se	Te
$E_{c\parallel}$	235	228	165	55	6.6	51	42
$E_{c\perp}$	3.5	10.7	11	7.8	2.9	13	24
$E_{c\parallel}/E_{c\perp}$	67	21	15	7.0	2.3	3.9	1.8
$\beta_{c\parallel}$	−1.3	−1.1	−2.3	−0.59	8.5	−1.34	−0.23
$\beta_{c\perp}$	14	17.5	14.8	8.4	10.4	7.0	3.0

The situation may change drastically for drawn crystalline polymers. It is well known that the drawn crystalline polymers may show a considerable contraction upon heating. However, because of the presence of the drawn amorphous regions the contraction is not the result of only the negative contribution of the crystallites alone, but also of the amorphous regions. Therefore, the thermal behavior of drawn crystalline polymers is much more complex.

3.5 Thermal Expansion of Drawn Polymers

3.5.1 Negative Thermal Expansion of Drawn Crystalline Polymers

Above T_g most drawn crystalline polymers often exhibit a shrinkage on heating, i.e. a negative thermal expansion along the orientation axis [29, 33, 34, 41]. Because $\beta_{c\parallel}$ is negative, one can suggest that this shrinkage is the consequence of that fact. However, a comparison of $\beta_\parallel$ for drawn crystalline polymers with $\beta_{c\parallel}$ indicates that macroscopic $\beta_\parallel$ may considerably exceed

$\beta_{c_\parallel}$. Therefore, at least two negative contributions to $\beta_\parallel$ must exist. One of them is really determined by the value of $\beta_{c_\parallel}$ for crystallites oriented along the orientation axis. The second contribution may be due to the shrinkage of oriented chains in the amorphous regions above T_g. Thus, the shrinkage of oriented crystalline polymers on heating seems to include two molecular mechanisms: the conformational elasticity of oriented amorphous chains and the shrinkage of oriented crystallites. In contrast to the isotropic state, in the well-drawn state all crystallites are oriented along the drawing axis and therefore crystallite axes are parallel to the draw direction. On heating they become shorter. This contribution seems to be negative both below and above T_g. The contribution of the amorphous regions is determined by the degree of crystallinity, the portion of the tie chains and the degree of their orientation. Thus, for understanding the feature of the thermal shrinkage of drawn crystalline polymers, it is very important to consider relationships typical for the stretched elastomers.

A typical dependence of $\beta_\parallel(\lambda)$ on the degree of extension which has been observed for NR and some other elastomers up to $\lambda = 2$ is shown in Fig. 3.8 [42–45]. The theoretical relation of β versus the degree of extension for typical polymer networks is given in Chap. 6 [see Eqs. (6.23) and (6.24)]. As can be seen from Fig. 3.8, at λ up to 2.0 the experimental data follow the theoretical prediction very well. At larger extensions, $\beta_\parallel$ is even larger than predicted by Eq. (6.23). In particular, for NR at $\lambda = 1.5$, $\beta_\parallel = -1.3 \times 10^{-3}\,\mathrm{K}^{-1}$, at $\lambda = 3$, $\beta_\parallel = -3.7 \times 10^{-3}\,\mathrm{K}^{-1}$ and at $\lambda = 4.5$, $\beta_\parallel = -4.2 \times 10^{-3}\,\mathrm{K}^{-1}$. At $\lambda > 3$, $\beta_\parallel$ depends only weakly on the degree of extension, and $\beta_\parallel \Rightarrow \mathrm{const}$ at large λ.

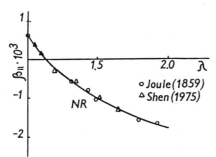

Fig. 3.8. Dependence of the linear thermal expansion coefficient of NR on extension ratio [43]

On stretching a crystallizable network, a portion of the chains are packed in crystallites their c-axes being oriented along the stretching direction. The thermal shrinkage of macromolecules in crystallites is about 2–2.5 orders of magnitude less than that of chains in the non-crystalline part. Therefore, if the degree of crystallinity increases on extension, the thermal shrinkage may decrease (Fig. 3.9). According to the two-phase model, a stretched semi-crystalline elastomer can be represented by a successive arrangement of crystallites and amorphous regions, which leads to

$$\beta_\parallel = \beta_{a_\parallel}(1 - w) + f_c\beta_{c_\parallel}w \tag{3.33}$$

where f_c is the function of orientation of the crystallites. Because $\beta_{a_\parallel} \gg \beta_{c_\parallel}$, then at $\lambda < 1$, $\beta_\parallel \cong \beta_{a_\parallel}$. Thus, a stretched crystallizable elastomer shrinks mainly because of the tendency of the stretched amorphous chains to increase their conformational entropy.

If a stretched elastomer comes back to its initial unstretched state at the same temperature, it shrinks to its initial length and the crystallites melt. In contrast, the stresses existing in the amorphous regions of oriented crystalline polymers are balanced by crystallites. Another very important difference from stretched crystallizable elastomers is that only some chains in the amorphous regions are the tie-molecules. Some possible ways of reaching the negative thermal expansivity with degree of drawing are shown in Fig. 3.9 [43]. One can suppose that at moderate degrees of cold-drawing of polymers with a low degree of crystallinity, a large portion of the chains in the amorphous regions are sufficiently stretched so that $\beta_\parallel \gg \beta_{c_\parallel}$ (Fig. 3.9 curve 2). On increasing the degree of drawing further, $\beta_\parallel$ may decrease, which leads to appearance of a minimum on the dependence on λ. Accepting the idea of rubber-elastic behavior of amorphous regions of crystalline polymers and neglecting the role of intermolecular interactions, $\beta_\parallel$ must be always considerably larger than $\beta_{c_\parallel}$; besides, $\beta_\parallel$ may approach $\beta_{c_\parallel}$ only from the region of large negative values. If the conformational mobility is strongly restricted by intermolecular interactions, a second way seems to be possible, namely if $\beta_\parallel < \beta_{c_\parallel}$ (Fig. 3.9, curve 3) and $\beta_\parallel$ approaches $\beta_{c_\parallel}$ from the region of small negative values. This is typical for polymers with a large degree of drawing.

Dilatometric studies have demonstrated negative thermal expansion for many drawn crystalline polymers [46, 47–53]. The results of these experimental studies may be summarized as follows. Cold-drawing of PE below T_m [49] and solid-state extrusion under elevated pressure [47, 48] lead to a monotonous decrease of the positive thermal expansion coefficient with increasing draw ratio. At a certain degree of drawing, dependent on temperature, $\beta_\parallel$ becomes negative with $\beta_\parallel < \beta_{c_\parallel}$ (Fig. 3.10). This is the second way of reaching negative expansion, which is also applied, for example, to POM ($w = 63\%$, $T_{dr} = 423$ K [50]). For less crystalline polymers such as LDPE with $w = 42\%$, one gets $\beta_\parallel = -5 \times 10^{-5}$ K^{-1} at $\lambda = 2$ and $\beta_\parallel = -20 \times 10^{-5}$ K^{-1} at $\lambda = 4.2$ [50, 46]. This data correspond to the first way (curve 2, Fig. 3.9). On the other hand, the same way was also found for polymers with a higher degree of crystallinity, e.g. HDPE with $w = 70\%$, for which at $20°$ C $\beta_\parallel = -2.4 \times 10^{-5}$ K$^{-1} > \beta_{c_\parallel}$ and at $\lambda = 8$ $\beta_\parallel$ is independent of λ [51–53] (cf also $\beta_\parallel$ for HDPE with $\lambda = 10$ and 20 in Table 5.4, Chap. 5). Thus, both ways of reaching the negative expansion indeed apply to oriented crystalline polymers. In what way the given negative value of $\beta_\parallel$ is reached depends on the conditions of orientation, first of all on temperature and degree of drawing. The degree of crystallinity is important, because for polymers with low w, a large amount of drawing cannot be achieved.

A very important feature follows from the comparison of $\beta_\parallel$ values of oriented crystalline polymers and stretched elastomers and especially of the temperature dependence of $\beta_\parallel$. If one estimates $\beta_{a_\parallel}$ using Eq. (6.23) and

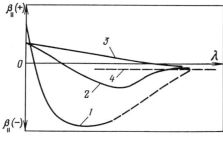

Fig. 3.9. Hypothetical dependence of the linear thermal expansion coefficient $\beta_\parallel$ on the degree of drawing [42]. *1 –* crystallizable stretched elastomer; *2,3* – crystalline drawn polymers; *4* – $\beta_{c\parallel}$

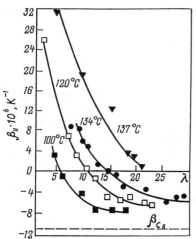

Fig. 3.10. The linear thermal expansion coefficient of HDPE as a function of degree of drawing at various temperatures [47, 48]

known values of $\beta_\parallel$, $\beta_{c\parallel}$ and w, then $\beta_{a\parallel} = -5.2 \times 10^{-5}\,\mathrm{k^{-1}}$ for the samples studied in [51–53]. This value is about two orders of magnitude lower than that for stretched elastomers. The same relation is typical for other high oriented polymers. Such large difference in the negative expansion of amorphous regions of crystalline polymers and stretched elastomers may be only due to the interchain interaction which prevents oriented chains from a free shrinkage. With increasing temperature, the interchain interaction and internal stresses decreases as a result of their relaxation. This means that $\beta_\parallel$ of oriented crystalline polymers must strongly increase with increasing temperature. According to Eq. (6.23), $\beta_\parallel$ is a weak function of temperature.

The temperature dependence of $\beta_\parallel$ shown in Fig. 3.11, typical for oriented crystalline polymers, reveals a very large increase of negative values of $\beta_\parallel$ with temperature. In a rather narrow temperature range (50–60 K) $\beta_\parallel$ increases several times. The temperature coefficient $d\beta_\parallel/dT$ does not depend much on polymer type, although the absolute value of $\beta_\parallel$ of LDPE is approximately ten times larger than that of other polymers. The following fact is also worth mentioning: For many oriented polymers such as POM with $\lambda = 6.5$, HDPE with $\lambda = 4.2$, PP with $\lambda = 5.5$, PA with $\lambda = 2.5$ the thermal expansion coefficient $\beta_\parallel$ is positive below T_g, increases in the region of glass transition and decreases monotonously on further heating and becomes neg-

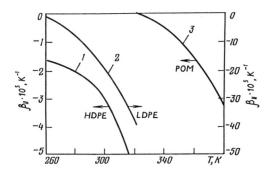

Fig. 3.11. Temperature dependence of $\beta_\parallel$ for drawn crystalline polymers. *1* – HDPE [51–53] (w = 70%, λ = 8–15); *2* – LDPE [49] (w = 43%, λ = 4.2); POM [49] (w = 63%, λ = 6.5)

ative at a temperature characteristic for a polymer and its thermomechanical history [50]. Such behavior is consistent with the idea that the intermolecular interaction in amorphous regions is the factor controlling the expansivity and thermomechanical properties of oriented crystalline polymers.

Summarizing the consideration of the thermal expansivity of oriented crystalline polymers, one may conclude that in two-phase polymers the behavior of their amorphous regions above T_g is controlled by conformational changes strongly restricted by interchain interaction.

3.5.2 Anisotropy of Thermal Expansion

Polymer crystallites exhibit a striking anisotropy of thermal expansion. A strong anisotropy of this property is also typical for solid drawn polymers and stretched elastomers. The linear thermal expansion decreases along the orientation axis $(\beta_\parallel)$ and increases in the direction perpendicular to the draw direction $(\beta_\perp)$ both for crystalline and glassy polymers. However, the anisotropy of oriented crystalline and glassy polymers may be considerably different at the comparable draw ratio. The ratio $\beta_\perp/\beta_\parallel$ is strongly dependent on the type of polymer. For glassy PS and PMMA at a draw ratio of $\lambda = 4$, the ratio is equal to 1.1 and 2.5, respectively, and for PVC which has a low degree of crystallinity, it is equal to 4.5 at a draw ratio of $\lambda = 3$ [54]. Wang, Choy and Porter [55, 56] studied the linear thermal expansion parallel and perpendicular to the extrusion direction of oriented PMMA with extrusion draw ratios $1 < \lambda < 4$ between 150 and 280 K and of atactic and isotactic PS with extrusion ratio $1 < \lambda < 15$ and $1 < \lambda < 7.5$, respectively. Experimental results for PMMA are shown in Fig. 3.12. The larger decrease of $\beta_\parallel$ occurs between $\lambda = 2$ and 3. For the most drawn sample with $\lambda = 4$, $\beta_\parallel$ has a weaker temperature dependence than $\beta_\perp$, which leads to a larger anisotropy $\beta_\perp/\beta_\parallel$ at higher temperature. Similar results have been also obtained for atactic and isotactic PS. Below $\lambda = 5$ the anisotropy is nearly the same for both PS, but at higher λ anisotropy for isotactic PS becomes much larger due to rapid rise in crystallinity.

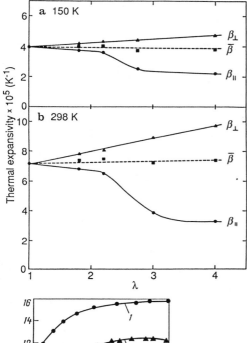

Fig. 3.12. Extrusion ratio dependence of $\beta_\perp$ and $\beta_\parallel$ for extruded PMMA [56] [$\bar{\beta} = 1/3(\beta_\parallel + 2\beta_\perp)$] is the average linear thermal expansion

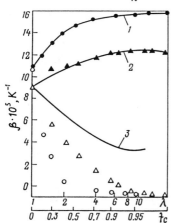

Fig. 3.13. Anisotropy of linear thermal expansion of drawn crystalline polymers as a function of draw ratio [57]. $\circ$ – ($\beta_\parallel$) and $\bullet$ – ($\beta_\perp$) HDPE (w = 80%, $T_{dr} = 353\,K$); $\triangle$ – ($\beta_\parallel$) and $\blacktriangle$ ($\beta_\perp$) PP (w = 60%, $T_{dr} = 398\,K$). *Solid curves: 1,2 –* $\beta_\perp$ from Eq. (3.34), *3 –* $\beta_\parallel$ from Eq. (3.35)

A typical dependence of $\beta_\parallel$ and $\beta_\perp$ on the draw ratio of crystalline polymers is shown in Fig. 3.13 [57].

Two different approaches have been used for the theoretical analysis of the anisotropy of thermal expansion of drawn crystalline polymers. One is based on the analysis of the effective expansion coefficients using the methods applied in the theory of thermoelasticity of microheterogeneous materials [50, 59]. These methods require the complete compliance tensors of both phases, and of the composite as well, which are usually not available. However, a more essential restriction of this approach is the assumption that in the composite material the properties of the components are constant and identical with their initial properties. Although for two-phase crystalline polymers this suggestion is not valid because of the existence of the amorphous regions, some attempts have been made to roughly estimate the expansivity of crystalline polymers [50]. Expressions for the thermal expansivities perpendicular and parallel to the draw axis, have the following form

$$\beta_\perp = w\beta_c + (1 - w)\beta_a + \frac{1}{3}wf_c(\lambda)(\beta_{c\perp} - \beta_{c\parallel}) \tag{3.34}$$

$$\beta_\parallel = w\beta_c + (1 - w)\beta_a - \frac{2}{3}wf_c(\lambda)(\beta_{c\perp} - \beta_{c\parallel}) \tag{3.35}$$

In these equations $\beta_c = (1/3)(2\beta_{c\perp} + \beta_{c\parallel})$ is the average coefficient of linear thermal expansion of crystallites and $f_c(\lambda)$ is the distribution function of the crystallites axes. For the isotropic sample $f_c = 0$ and both Eqs. (3.34) and (3.35) reduce to

$$\beta = w\beta_c + (1 - w)\beta_a \tag{3.36}$$

From Eqs. (3.34) and (3.35), it is evident that the average linear thermal expansion coefficient, i.e. $\beta_{av} = 1/3(2\beta_\perp + \beta_\parallel)$ or $1/3\alpha$ is independent of draw ratio and equals to to the isotropic value β_{iso}. According to this approach, the dependence on λ of the linear thermal expansion is determined solely by the orientation of the crystallites along the draw direction. When all the crystallites have become completely oriented ($f_c = 1$), a saturation occurs and both $b_\perp$ and $\beta_\parallel$ become independent of λ.

This model seems to be incorrect because the linear thermal expansion for both components in the isotropic and oriented state is assumed to be the same. The consequences of this assumption are quite different for $\beta_\perp$ and $\beta_\parallel$. For $\beta_\perp$, it does not play any essential role, because in the isotropic and oriented state perpendicular to the draw axis the thermal expansion is determined solely by intermolecular interactions. For $\beta_\parallel$, this assumption may lead to a principle inconsistency. This conclusion is evident from comparison of the calculated and experimental λ-dependencies of $\beta_\perp$ and $\beta_\parallel$ for PE and PP according to Eqs. (3.34) and (3.35). For $\beta_\perp$, the agreement between the model calculation and the experiment is quite satisfactory for all draw ratios. On the other hand, Eq. (3. 35) does not describe the λ-dependence of $\beta_\parallel$ at all. This equation does not yield negative values of $\beta_\parallel$ even in case of limited orientation of crystallites ($f_c = 1$) because it is based on the suggestion that β_a is always positive and $|\beta_a| > |\beta_c| \gg |\beta_{c\parallel}|$.

The other approach concerns the analysis of models of oriented crystalline polymers [41, 42, 49]. The detailed consideration of these models is given in Chap. 5 in the connection with their thermomechanical analysis. The expression for $\beta_\parallel$ of these models are given there [Eqs. (5.31) and (5.36)]. According to this consideration for the most advanced Peterlin–Prevorsek model of drawn crystalline polymers, the linear thermal expansion coefficient along the drawing axis $\beta_\parallel$ may have negative values depending on the draw ratio and the ratio of inter- and intrafibrillar amorphous tie-molecules. The expression for $\beta_\perp$ in this model can be represented as

$$\beta_\perp = \frac{\beta_{c\perp}wE_{c\perp} + \beta_{AB\perp}E_{AB\perp}(1 - w)}{E_{c\perp}w + (1 - w)E_{AB\perp}}(1 - X) + \beta_{AI\perp}X \tag{3.37}$$

According to Eq. (3.37) $\beta_\perp$ must only be positive. Therefore, the analysis of the structural models of drawn crystalline polymers seems to predict the anisotropy of thermal expansion more correctly.

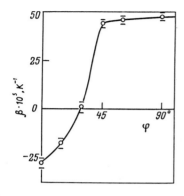

Fig. 3.14. Dependence of the linear coefficient of thermal expansion of LDPE on the angle to the drawing direction [42]. Values of β_φ were obtained from the heat of elastic stretching

Additionally, one can consider the thermal expansion in an arbitrary orientation direction. For films, the linear coefficient of thermal expansion at angle φ to the orientation axis is determined by

$$\beta_\varphi = \beta_\| \cos^2 \varphi + \beta_\perp \cos^2 (90 - \varphi) \tag{3.38}$$

Because $\beta_\|$ may be negative, there may exist such directions in drawn crystalline polymers along which the thermal expansion is zero. The dependence of β_φ on the angle for LDPE is shown in Fig. 3.14 [42]. Values of β_φ were obtained from the heat of elastic stretching (see Chap. 5). It is seen that at angles of less than 30° the linear thermal expansion coefficient is negative and above 30° positive. Hence, $\beta_{30} \cong 0$. Similar results were also obtained for drawn PA-6 and PET.

3.6 Thermal Expansion of Filled Polymers and Polymer–Matrix Composites

Polymers have much higher thermal expansion coefficients than the inorganic materials (glasses), metals and metallic alloys, therefore, by combining polymers with these materials one has the opportunity of reducing the high expansivity of polymers. This possibility is, of course, of great technological importance since low thermal expansion composite materials based on polymers can be produced. In some cases the composite polymer materials having a coefficient of thermal expansion only one fifth of the initial unfilled polymer can be obtained.

In general, the thermal expansion of polymer composites is a function of the thermal and mechanical properties of the components, the shape or other aspect of the particle geometry of reinforcing elements, the volume concentration of fillers, the temperature, polarity of polymer and filler etc. A general theory which takes into account all the factors is still absent. In the majority of cases it is impossible to calculate the thermal expansion coefficient of the isotropic filled polymers. On the other hand, the thermal expansion of anisotropic unidirectional or bidirectional composites with polymer matrixes can be estimated quite correctly using simple expressions.

In the simplest case, namely, in the absence of interphase interaction (adhesion), one may expect the coefficient of thermal expansion of a two-phase polymer composite having φ_f volume content of filler follow the simple relation of mixtures given by

$$\alpha_c = \alpha_p + \varphi_f(\alpha_f - \alpha_p) \tag{3.39}$$

where index p and f refers to polymer and filler, respectively. However, this simple relation describes in practice the thermal expansion behaviour of filled polymers very rarely. Normally, between polymer and filler microstresses resulting, first of all, from difference in the thermal expansivity of the filler and polymer exist. Due to these microstresses the thermal expansion behaviour of the composites does not follow Eq. (3.39). There are numerous attempts in literature to take into account this thermomechanical interaction between phases in composite materials and many formulas have been proposed for the calculation of the thermal expansion coefficient of composite materials. Many of these expressions were discussed by Holliday and Robinson [60, 61]. The majority of the formulas are rather complex. The analysis of these formulas given by Holliday and Robinson showed that they predicted a very different thermal expansion behaviour of composites.

Figure 3.15 shows the dependence of the thermal expansion coefficient of a composite with the thermal and mechanical parameters close to polyamide filled with the glassy filler upon the volume concentration of the filler calculated according to the formulas widely used. As can be seen from the figure the calculated dependencies 1, 3, 6, and 7, corresponding to the composites with spherical fillers, predict significantly the different thermal expansion behavior of the composites. The difference in the predicted values of α can reach 4–5 times. All the dependencies are between the curves 1 and 9 corresponding to the mixture relation (3.39) and the Turner relation. Holliday and Robinson compared a large number of experimental data for various filled polymer systems with the theoretical predictions. The summary of this comparison is shown in Fig. 3.16. The main conclusions which can be drawn from the analysis can be formulated as follows. The thermal expansion of typical polymers can be reduced considerably by filling with a corresponding filler. The largest effect is reached by using glass fibers and fabrics while the application of powdered fillers is less effective in reducing the thermal expansion coefficient of composites. The thermal behavior of PTFE is highly anomalous. Although in many cases a reasonable agreement between the theoretical predictions and experimental data can be reached, nevertheless, the general picture may be built only taking into account such important factors as the shape and other geometrical characteristics of the fillers and interface characteristics which usually are ignored. Some studies attempt to take into account such characteristics [62].

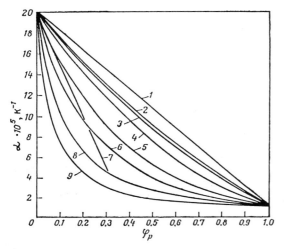

Fig. 3.15. The calculated dependence of the thermal expansion coefficient of a hypothetical composite material on the volume content of a filler. [60, 61]. *1* – according to Eq. (3.39), the Tummala formula; *2* – according to the Cribbs formula; *3* – according to the Kerner and Wang and the Kwei formulas; *4,5* – according to the Thomas formula with various parameters; *6* – according to the Blackburn and Kribbs (with another set of parameters); *7* – according to the Turner formula using the volume modulus of elasticity; *9* – according to Turner's formula using the Young modulus of elasticity

A number of recent studies concern the thermal expansion behavior of fiber-reinforced, polymer-matrix composite materials including epoxy, -polyimide and some other polymer matrices [63–69]. Typical unidirectionally and bidirectionally reinforced composite constructions involves the stacking of plies in a particular sequence followed by curing at elevated temperature and pressure to reach the properties required. Although these composite materials with a polymeric matrix are generally similar to those with other matrices (glass, metallic), however, they normally demonstrate very anisotropic thermal expansion behaviour due to the very large anisotropy of the thermal expansion of the basic reinforcements fibers. For example, graphite and Kevlar fibers which are most widely used for reinforcement of polymeric matrices are characterized by slightly negative thermal expansion coefficients in their axial directions and very large positive coefficients in the radial direction [68, 70]. Being combined with a homogeneous matrix having any given thermal expansion coefficient, the resultant composites should be highly anisotropic in their thermal expansion behavior.

Typical results for unidirectional and bidirectional reinforced Kevlar-epoxy composites are shown in Fig. 3.17. As can be seen, the expansion behavior is highly anisotropic in both types of composites. In unidirectional

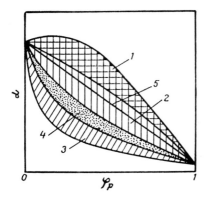

Fig. 3.16. The generalized dependence of the volume coefficient of thermal expansion of filled polymers on the volume concentration of filler [60, 61]. *1* – PTFE filled with inorganic powders; *2* – various thermoplastics filled with different inorganic powders; *3* – thermosets resins reinforced with the glass fabrics; *4* – thermoplastics filled with glass fibers; *5* – the curve corresponds to the mixture equation

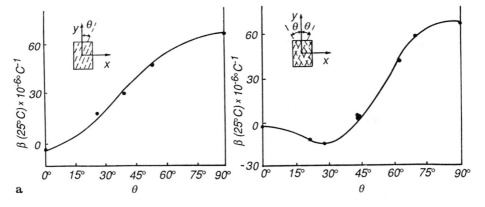

Fig. 3.17. The in-plane thermal expansion coefficient β_y, at 25° C for unidirectional (**a**), and bidirectional (**b**) composites reinforced in direction θ defined by θ in the figure. The *solid curves* are plots of the calculated values of β_y for a fiber volume fraction of 50% [67]

composites, β increased from a value of $-2.2 \times 10^{-6}°\text{C}^{-1}$ along the fiber axis to a value of $79.3 \times 10^{-6}°\text{C}^{-1}$ perpendicular to the fibers. For bidirectional composites, β measured parallel to the bisector of the angle 2θ between the principal fiber directions initially turned to be more negative with increasing 2θ, and then rapidly increased to highly positive values. The minimum calculated value happen to be -15.3×10^{-6} °C^{-1} for a fiber orientation of $\pm 32°$. Zero thermal expansion was predicted for fiber orientation of $\pm 43°$. The thermal expansion behavior of the Kevlar-epoxy composites could be predicted from the thermoelastic properties of the fiber and resin. Similar behaviour was also observed for graphite-epoxy composites [63] where it has been shown that the expansion coefficient of the unidirectional composites is described by a simple formula

$$\beta_0 = \beta_L \cos^2 \theta + \beta_T \sin^2 \theta \qquad (3.40)$$

where β_L and β_T denoted the expansion along and perpendicular to the fibers.

References

1. Gruneisen E (1926) In: Handbuch der Physik, vol 10, p1, Geifer H, Scheel K (ed), Springer Berlin
2. Novikova SI (1974) Thermal Expansion of Solids (in Russian), Nauka Moscow
3. Girifalco LA (1973) Statistical Physics of Materials, Wiley Inerscience New York
4. Barron THK (1955) Phil Mag *46:* 720
5. Barker RE Jr(1967) J Appl Phys *38:* 4234
6. Warfield RW (1974) Makromol Chem *175:* 3285
7. Shen M and Reese W (1975) In: Progress in Solid State Chemistry, vol 9, pp 241–268, McColdin, Samojai (ed), Pergamon Oxford
8. Shen M (1979) Polymer Eng Sci *19:* 995
9. Wada Y, Itani A, Nishi T and Nagai S (1969) J Polym Sci *part A-2 7:* 201
10. Pastine DJ (1968) J Chem Phys *49:* 3012
11. Wu CK, Jura G and Shen M (1972) J Appl Phys *43:* 4348
12. Broadhurst MG and Mopsik FI (1970) J Chem Phys *52:* 3634
13. Curro JG (1974) J Macromol Sci-Revs Macromol Chem *C11(2):* 321
14. Prigogine I, Trappeniers N and Mathot V (1953) Disc Faraday Soc *N15:* 93; Prigogine I, Bellemans A and Mathot V (1957) The Molecular Theory of Solutions, Wiley New York
15. Simha R (1980) J Macromol Sci Phys *18:* 377; Simha R and Somcynscy T (1969) Macromolecules *2:* 342; Olabisi O and Simha R (1975) Macromolecules *8:* 206, 211; Simha R (1979) J Polym Sci Polym Phys Ed *17:* 1929
16. Ainbinder SB, Tjunina EL and Tsirule KI (1981) Mechanics of compos mater *N3:* 387
17. Simha R, Roe JM and Nanda VS (1972) J Appl Phys *43:* 4312
18. Curro JG (1973) J Chem Phys *58:* 374
19. Gibbons TG (1974) J Chem Phys *60:* 1094
20. Shen M; Hansen WN and Romo PC (1969) J Chem Phys *51:* 425
21. Volodin VP and Gudymov SYu (1984) Solid State Phys (FTT) *26:* 1563
22. Devis GT, Eby RK and Colson JP (1970) J Appl Phys *41:* 4316
23. Kijima T, Koga K, Imada K and Takayanagi M (1975) Polym J *7:* 14
24. Engeln I, Meissner M and Pape UE (1985) Polymer *26:* 364
25. Perepechko II (1980) Low temperature properties of polymers, Pergamon Oxford
26. Kato E (1980) J Chem Phys *73:* 1020
27. Barron THK (1970) J Appl Phys *41:* 5044
28. Lifshitz IM (1952) J Exp Theor Phys (JETF) *22:* 475
29. Wakelin JH, Sutherland A and Beck LR (1960) J Polym Sci *42:* 278
30. Kobajashi J and Keller A (1970) Polymer *11:* 114
31. Baughman RH (1973) J Chem Phys *58:* 2976
32. Kan KN (1975) Theoretical Questions of Thermal Expansion of Polymers (in Russian), Izd LGU Leningrad

33. Chen FC, Choy CL and Young K (1980) J Polym Sci Polym Phys Ed *18:* 2312
34. Chen FC, Choy CL, Wong SP and Young K (1981) J Polym Sci Polym Phys Ed *19:* 971
35. Barron RM, Barron THK, Mummery PM and Sharkey M (1988) Can J Chem *66:* 718
36. Broadhurst MG and Mopsik FI (1971) J Chem Phys *54:* 4239
37. Dadobaev G and Slutzker AI (1981) Solid State Physics (FTT) *N7:* 1936; (1982) Vysokomol Soyed *A24:* 30; (1983) *A25:* 8
38. Frenkel YaI (1950) Introduction to the Theory of Metals, (in Russian) GITTL Moscow-Leningrad
39. Choy CL and Nakafuku C (1988) J Polym Sci Polym Phys Ed *26:* 921
40. White GK, Smith TF and Birch JA (1976) J Chem Phys *65:* 554
41. Godovsky YuK (1982) Colloid Polym Science *60:* 461
42. Godovsky YuK (1986) Adv Polym Sci *76:* 30
43. Godovsky YuK (1987) Progr Colloid Polym Science *75:* 70
44. Shen M and Croucher M (1975) J Macromol Sci *C12:* 287
45. Thile JL and Cohen RE (1980) Rubber Chem Technol *53:* 313
46. Godovsky YuK (1976) Thermophysical Methods of Polymers Characterization (in Russian), Khimiya Moscow
47. Mead WT, Desper CR and Porter RS (1979) J Polym Sci Polym Phys Ed *17:* 859
48. Capiati NJ and Porter RS (1977) J Polym Sci Polym Phys Ed *15:* 1427
49. Choy CL, Chen FC and Young K (1981) J Polym Sci Polym Phys Ed *19:* 395
50. Choy CL, Ito M and Porter RS (1983) J Polym Sci Polym Phys Ed *21:* 1427
51. Wolf F-P and Karl V-H (1980) Angew Makromol Chem *92:* 89
52. Wolf F-P and Karl V-H (1981) Colloid Polym Science *259:* 29
53. Wolf F-P and Karl V-H (1981) Makromol Chem *182:* 1787
54. Hellwege K-H, Hannig J and Knappe W (1963) Kolloid Z Z Polym *188:* 121
55. Wang L-H, Choy CL and Porter RS (1982) J Polym Sci Polym Phys Ed *20:* 633
56. Wang L-H, Choy CL and Porter RS (1983) J Polym Sci Polym Phys Ed *21:* 657
57. Choy CL, Chen FC and Ong EL (1979) Polymer *20:* 1191
58. Kardos JL et al (1979) Polym Eng Sci *19:* 1000
59. Shermergor TD (1977) Theory of Elasticity of Microheterogeneous Solids (in Russian), Nauka Moscow
60. Holliday L and Robinson JD (1973) J Mater Sci *8:* 301
61. Holliday L and Robinson JD (1977) In: Polymer Engineering Composites, chap 6, Richardson MOW (ed), Applied Science Publishers London
62. Pinheiro MFF and Rosenberg HM (1980) J Polym Sci Polym Phys Ed *18:* 217
63. Fahmy AH and Ragai AN (1970) J Appl Phys *41:* 5112
64. Rogers KF, Phillips LN, Kingston-Lee DM, Yates B, Overy MJ, Sergent JP and McCalla BA (1977) J Mater Sci *12:* 718
65. Yate B, Overy MJ, Sergent JP, McCalla BA, Kingston-Lee DM, Phillips LN and Rogers KF (1978) J Mater Sci *13:* 433
66. Ishikawa T, Koyama K and Kobayaski S (1978) J Composite Mater *12:* 153
67. Stife JR and Prevo KM (1979) J Composite Mater *13:* 264
68. Tompkins SS (1987) Intern J Thermophysics *8(1):* 119
69. Bowles DE and Tompkins SS (1989) J Composite Mater *23:* 270
70. Vyshvanjuk VI, Alypov VT and Vishnevskii ZN (1982) Mekhanika Composizionnych Materialov *N6:* 1102

4 Experimental Methods and Instrumentation

4.1 Heat Capacity

Two methods are usually used for the determination of the heat capacity of polymers over a wide temperature range: adiabatic calorimetry and differential scanning calorimetry (DSC).

4.1.1 Adiabatic Calorimetry [1–3]

Precision measurements of the heat capacity are normally carried out in adiabatic calorimeters. The principal construction of a typical adiabatic calorimeter used for the measurements of specific heat capacities consists of an electrical source of heat, which supplies an exactly known amount of heat into the sample under study, and a temperature measuring instrument, which determines precisely the change in temperature of the sample (Fig. 4.1). In order to minimize the heat losses, the calorimeter itself is surrounded by a radiation shield, the temperature of which is electrically controlled to be very close to that of the calorimeter. The smaller the temperature difference can be kept between calorimeter and shield, the smaller is the heat loss. Normally, calorimeter and jacket are located in an evacuated insulated container. Energy input to the calorimeter is chosen in such a way as to arrive at heating rates of less than $1\,\mathrm{K/min}$ after heating periods involving 1 to a few degrees with large equilibrium intervals between the heating periods. The energy input is usually corrected to the heat leakage and water value of the calorimeter itself.

An average heat capacity for the heating interval for the sample of known mass m is

$$\bar{C}_\mathrm{p} = \frac{\Delta Q}{m\,\Delta T} \qquad (4.1)$$

where $\bar{C}_\mathrm{p}$ is the mean value of the heat capacity for the given temperapture rise ΔT, ΔQ the corrected energy input. The overall precision of typical adiabatic calorimeters is 0.1–0.5%.

Depending on the temperature interval of operation, the adiabatic calorimeters used for the measurements of the heat capacity of polymers can be subdivided into two groups: low temperature calorimeters, which operate below $300\,\mathrm{K}$ and high temperature ones working in the temperature region 250–$600\,\mathrm{K}$. Heat capacity measurements below about $10\,\mathrm{K}$ need special cryostats and low temperature thermometry.

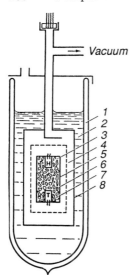

Fig. 4.1. Schematic diagram of the Nernst adiabatic calorimeter. *1* – Dewar vessel; *2* – liquid gas; *3* – heater; *4* – adiabatic shield; *5* – calorimeter; *6* – sample; *7* – temperature sensor; *8* – vaccum vessel

The heat capacity measurements of polymers by means of the adiabatic calorimeters of both these groups involve some difficulties closely related to the low thermal conductivity of polymers, their low density, which lead to a very unfavorable ratio of the heat capacity of the sample and the calorimeter and, finally, the possibility of a slow temperature drift occurring resulting from the slow relaxational processes in polymers. For the high temperature measurements a very important aspect is the oxidation and degradation of polymers. The precise data on heat capacities of most polymers especially at low temperatures have been obtained by means of adiabatic calorimetry. Adiabatic calorimeters for heat capacity measurements of polymers have been described in many papers. The references to these publications can be found in [4–5].

Adiabatic calorimetry is the most precise method of direct measurement of heat capacity of polymers, it has, nevertheless, some disadvantages. First of all, their operation is very slow because of the need to reach equilibrium after each step, which is usually chosen between one and a few degrees. Therefore, such measurements are rather time-consuming. As a rule, every adiabatic calorimeter needs large samples (from several grams up to 20–50 grams). In addition each instrument is as a rule custom built, which is also time-and money-consuming. Finally, polymers are not perfectly suitable for adiabatic calorimetry because of their metastability and sensitivity to thermal pretreatment. These problems stimulated the development of dynamic calorimetry which is widely used nowadays for the thermal analysis of polymers including the heat capacity measurements. This area of thermal analysis of polymers has enlarged enormously lately, therefore, we will restrict our consideration only to the main principles of heat capacity measurements and typical instrumentation.

4.1.2 Differential Scanning Calorimetry (DSC) [3, 6–9]

In contrast to adiabatic calorimetry, measurements of heat capacity in all DSC instruments are carried out in dynamical regimes, normally at a constant heating rate. All DSC instruments are of differential type, which is reflected by the situation that many of these instruments are closely related to classical DTA principle. Excellent consideration of the modern state of DSC and its application in physics and physical chemistry of polymers was given by Richardson [7, 8], Bershtein and Egorov [9] and in the book edited by E. Turi [10].

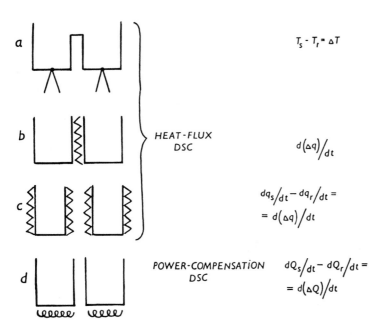

Fig. 4.2. Sensor configuration in various DSC types [7, 8]. s – sample; r – reference; Δ – difference; dq/dt – spontaneous thermal flux; $\wedge$ – thermocouple; – thermopile; eeee – heater

Figure 4.2 shows sensor configuration for most widely used DSC instruments [7, 8]. The first three types belong to heat-flux DSC, which have a common source of heat for both sample and reference cell. The last configuration belongs to power-compensation DSC. Although the term "differential scanning calorimeter" was originally introduced to describe this particular power-compensation type of instrument, however, nowadays this term is used to emphasize the quantitative mode of operation but gives no indication concerning the principle of operation. Four types of DSC are commercially available: Du Pont and Stanton Redcraft produce DSC based on the Fig. 4.2a principle; a five-junction thermopile (Fig. 4.2b) is used in the DSC 20 and

DSC 30 Mettler TA 3000 system; a Calvet-type DSC[1] (Fig. 4.2c) marketed by Setaram (DSC-111), and AN SSSR (DAK-1-IA); and power-compensation DSC is produced by the Perkin-Elmer Corporation (DSC-2 and DSC-7) and SKB "Biopribor" (DSM-2M and DSM-3). The DSC-2 Perkin-Elmer seems to be the most widely used instrument. A schematic diagram of the working block of the DSC-2 and DSC-7 is shown in Fig. 4.3.

The main operating characteristics of all the instruments are fairly standard: operating temperatures from 100–150 K to 700-1000 K; heating and cooling rates from a few tenths of a degree to a hundred degrees; sample weight – from some milligrams up to some tens of milligrams. Richardson [7] underlined that "the current status of DSC is such that true heat capacity curves are produced almost in parallel with the actual measurement and users are free to concentrate on the interpretation of results without major concern for the technique itself". Although the accuracy of the heat capacity determination in DSC experiments is significantly worse than in adiabatic calorimetry, being 1–3%, nevertheless, the DSC approach has replaced practically all classical adiabatic calorimetry in studies of heat capacity of polymers above about 100 K.

According to the theory of DSC methods, in order to determine heat capacity by means of DSC it is necessary to carry out three "runs": 1) with an empty sample cell; 2) with a calibrant; 3) with the sample. On the basis of these three runs one can arrive at the following relation for heat capacity of sample

$$C_{ps} = \frac{(S_s - S_e)m_c C_{pc}}{(S_c - S_e)m_s} \tag{4.2}$$

where S is the signal on heating and subscripts s, c, e refer to sample, calibrant and empty cell, respectively. This is the basic DSC heat capacity equation. The accuracy of the evaluation of C_p according to this equation depends upon the conditions of obtaining the experimental curves and their treatment. Typical computerized treatment, which is normal in modern DSC instruments, usually gives C_p values with an accuracy of about 1%.

Application DSC for the determination of the thermal conductivity of polymers is considered in the next section.

4.2 Thermal Conductivity and Diffusivity

Most polymers are poor thermal conductors possessing a low thermal conductivity and diffusivity. Experimental methods for measuring the thermal conductivity can be separated into two groups [6, 11–13]. To the first group

---

[1] An instrument based on a similar principle was designed by the author of this book at the beginning of 1960s [Godovsky YK, Barsky YP (1965) Plast Massy N7: 57], later it was reproduced at some institutions in the USSR and is still in use

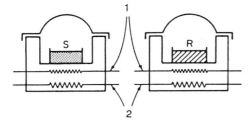

Fig. 4.3. DSC Perkin Elmer measuring block. *1* – sensor; *2* – microheater; *S* – sample; *R* – reference

belong those methods which are based on the principles of steady state measurements, and to the second one – unsteady state. The theoretical basis of all experimental methods is the main equation of thermal conductivity which is related to the time and space temperature changes resulting from the heat fluxes [14, 15]

$$C_p\rho\frac{\partial T}{\partial t} = \chi\left(\frac{\partial^2 T}{\partial x^2} + \frac{\partial^2 T}{\partial y^2} + \frac{\partial^2 T}{\partial z^2}\right). \qquad (4.3)$$

In the absolute steady state heat flow method at one end of the specimen, which typically has the shape of a rod, constant electrical power is applied, while the other end is in contact with a heat sink. In such a system a static temperature distribution can be eventually reached and in the case of the one-dimensional heat flow the thermal conductivity χ can be determined from the temperature drop between two points along the specimen.

In general, Eq. (4.3) for thermal conductivity has an infinite number of solutions, which corresponds to the large variety of methods for measuring χ and a. The change of the temperature field in the space as a function of time has an initial completely unsteady period, when the initial conditions play an important role, and more regular stages, where the initial conditions are unimportant [14, 15]. The border between these stages is characterized by the Fourier number: $F_0 = at/\delta^2$ (where δ is a characteristic dimension). For the completely unsteady stages $F_0 < 0.5$, and for the regular one $F_0 > 0.5$. There are three kinds of the regular regimes. In the case of the first kind of regular regime, a sample is heated at constant ambient temperature, while in the second kind of regular regime, heating proceeds at a constant rate. In the later case, the ambient temperature changes linearly with time which implies a constant heat flux effect on a sample. This regime is often called monotonic. Finally, a pulse thermal regime can be also used for determining the thermal conductivity and diffusivity.

The thermal diffusivity can be measured directly only in the unsteady state methods. It can be easily obtained from dynamic experiments, in which the temperature is recorded as a function of time at one or several points in a heated specimen. Thermal power can be supplied in the form of pulses or as some periodic function of time. Although the thermal conductivity χ and diffusivity a are closely related by a simple relation

$$\chi = a\,\rho\,C_p \qquad (4.4)$$

the methods of their measurements differ principally. For determining of χ it is necessary to measure the absolute or comparative heat flux, while for determining a, it is enough to carry out only the temperature measurements.

4.2.1 Steady State Methods

In the devices based on the steady state principle the temperature distribution in the sample studied does not depend on time. In this case Eq. (4.3) for the samples of simple geometrical shape and in which the heat flux is one-dimensional can be transformed to a simple form [14, 15]

$$\chi = \frac{Q\,d}{A\,\varDelta T} \tag{4.5}$$

where Q is the time rate heat flow, d is the specimen thickness, $\varDelta T$ is the temperature gradient across the sample, and A is the surface area. It is suggested that χ is independent of temperature which allows the use of only small temperature gradients (some degrees).

One of the most frequently used methods of this type is the guarded hot plate, which is normally regarded as the most accurate for the measurements of χ for low thermal conductivity materials. The general features of the apparatus which was widely used for measurements of χ of polymers in the temperature range $-180 + 100°$C are shown in Fig. 4.4 [16]. Normally the test specimen consists of two identical slabs 5–10 mm thickness which are placed on either side of a flat heater. Cooling units are placed against the nonheated face of the samples. The assembly may include the guard heaters or guard rings which prevent the heat losses from the ends of the system. The temperature gradient is measured by means of differential thermocouples which are fixed on the surfaces of the heater and cooling blocks. The absolute accuracy of the measurements is on the level of ±2–3%.

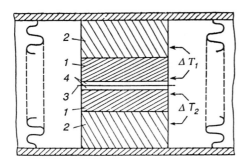

Fig. 4.4. Schematic diagram of the apparatus for heat conductivity measurements (without hot guard) [16]. *1* – samples; *2* – copper blocks; *3* – heater; *4* – insulation

Figure 4.5 shows an apparatus for measuring polymer thermal conductivities in steady-state experiments up to a pressure of 50 MPa and within the 80–350°C temperature range [17]. A cylindrical heating element enclosed in the sample with a common axis is emmersed in a pressurized fluid for applying the hydrostatic pressure to the sample. Isothermal conditions in the measuring cell can be maintained by a temperature-controlled oven. Thermal conductivity is deduced from steady-state measurements. Edge effects are taken into account by means of numerical simulation.

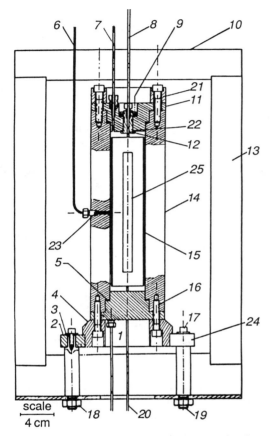

Fig. 4.5. High pressure conductivity meter, average temperature [17]: *1* – stopper; *2* – base; *3* – washer; *4* – lower clamp; *5* – gasket; *6* – thermocouple; *7* – purging tube; *8* – heating wire; *9* – nut; *10* – oven lid; *11* – bottom; *12* – support gasket; *13* – oven wall; *14* – cylinder; *15* – sample; *16,17* – bolts; *18* – nut; *19* – leg support; *20* – thermocouple; *21* – upper clamp; *22,23* – gaskets; *24* – nut; *25* – heating element

Another absolute technique is the hot wire method [13, 18]. This technique consists of the embedding in an electrically insulating material of a metal wire that is heated by a known constant power. The thermal conductivity of the material can be deduced from the temperature rise of the wire. The method can be used in two main variants: the cross wire and parallel wire scheme shown in Fig. 4.6 [18]. Normally the test assembly includes 2 or 3 standard specimens. The cross wire technique is more suitable for materials with $\chi \leq 2\,\mathrm{W\,m^{-1}K^{-1}}$. The temperature rise ΔT of the wire during the steady heating over a time interval $t_2 - t_1$ is given by

$$\Delta T = \frac{Q}{2\pi\chi} \ln\left[\frac{t_2 - t_c}{t_1 - t_c}\right] \tag{4.6}$$

where t_c is a constant dependent on contact resistances and the position of the temperature sensor. Hence, the knowledge of the applied power per

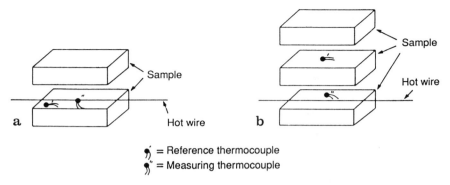

= Reference thermocouple
= Measuring thermocouple

Fig. 4.6. Schematic diagrams of sample arrangements for the hot wire technique [18]

unit length of the wire Q allows us to calculate χ by plotting ΔT versus $\ln[(t_2 - t_c)/(t_1 - t_c)]$ and analysing the slope $Q/4\pi\chi$. The limitation inherent in this procedure results from the difficulty in determining the initial temperature.

Various experimental equipments for determining the thermal conductivity of solid polymers and polymer melts by means of the absolute steady-state method are described in [19–36].

4.2.2 Unsteady-State Methods

Normally, direct measurements of the thermal conductivity are rather complicated because of the necessity of detecting the heat flux. The temperature measurements which are needed for determination of the thermal diffusivity, are much simpler. In the so-called modified Ångstrom methods [36–38], which are often used for the measurements of the thermal diffusivity, the thermal power is supplied to the sample in the form of a sinusoidal function of time and the thermal diffusivity can be determined by the measurements of the ratio of the temperature at two points separated by a certain length, the period of the thermal wave, and the phase difference. When a sinusoidal change of the temperature of angular frequency ω is applied to one end of a sample, which is at the steady temperature,

$$T(0, t) = T_0 + T_1 \cos \omega t \qquad (4.7)$$

and the other end of the sample is at surrounding temperature, then

$$T(x, t) = T_0 \exp(-m_0 x) + T_1 \exp(-m_1 x) \cos(\omega t - \beta x - \psi) \qquad (4.8)$$

where T_0, T_1 and ψ are constants. m_0 corresponds to the reduction of the steady temperature per unit length, and m_1 – the reduction of the periodic component of the temperature per unit length, is the phase shift per unit length, all these values being dependent on the thermal diffusivity of the material and on the heat loss. The diffusivity and heat loss increase with rising frequency. Normally, the experiments are performed at one frequency only, therefore, both the reduction of and the phase shift of the temperature wave are determined simultaneously. The thermal diffusivity of the sample can be determined then as

$$a = \frac{\omega}{2\beta m_1} = \frac{\omega}{2\beta \ln k}$$ (4.9)

where k is the amplitude decrement between two points x_1 and x_2 along the sample.

Among the purely unsteady methods the impulse method with a layer heat source is often used [39–41]. The scheme of the measurement is similar to that shown in Fig. 4.4, however, the calculations are based on the analysis of the initial unsteady period. All three thermophysical parameters, namely C_p, χ and a, can be obtained simultaneously during experiments using the measured temperature difference between the heating element and any fixed point in the sample as well as by the measurement of the temperature changes of the heating element. Typical dimensions of the samples used – 35 × 35 × 60 mm. Duration of the experiment – 5–7 min.

In the devices, which are based on the regular regimes of the first kind [11, 12, 42], the changes of the temperature distribution as a function of time in the sample placed into the media with constant temperature T_c are measured. Under the condition of an intensive heat exchange on the surface of the sample the thermal diffusivity can be obtained by

$$a = m K_{sh}$$ (4.10)

where m is an experimentally determined parameter which depends upon the shape and dimension of the sample and its thermophysical characteristics and K_{sh} is the coefficient of the shape. Such devices have been described in [43–46]. To obtain the thermal conductivity by this method it is necessary to carry out additional measurements of the heat exchange on the surface of the sample. The accuracy of the measurements of χ is 4–8%.

Among the unsteady-state methods which are very popular are the methods based on the regular regime of the second kind in which the measurement of a and χ are carried out at a constant heating rate. The measurement of the thermal diffusivity is based on the determination of the temperature at two points of the sample, separated by a constant length. The temperature difference between two points x_1 and x_2 in the one-dimensional case is inversely proportional to the thermal diffusivity

$$\Delta T = \frac{v}{2K_{sh}a}(x_2^2 - x_1^2)$$ (4.11)

where v is the heating rate. Such a method was used in [47]. To obtain the thermal conductivity it is necessary to determine the heat flux coming through the sample during the measurements. It can be done by means of determining the heating rate of some reference rod with a known heat capacity. This approach has been used in the tempeture range 300–650 K with an accuracy about 5% [48–50].

A very useful method of determining the thermal diffusivity is the heat pulse or flash method [51, 52]. In the flash radiometry method the front face of a small disk- or film-shaped specimen is subjected to a very short pulse of radiant energy coming from a laser or xenon flash lamp with an irradiation time of the order of one millisecond or less. The resulting temperature rise of the back surface of the sample is recorded as a function of time. Thermal diffusivity values are computed from the temperature rise versus time data. The simplest way consists of using the formula

$$a = \frac{L^2}{\pi^2 t_c} \tag{4.12}$$

where L is the thickness of the sample and t_c is the thermal diffusion time constant. The flash methods turned out to be ideal for studying the thermal diffusivity of oriented polymer films.

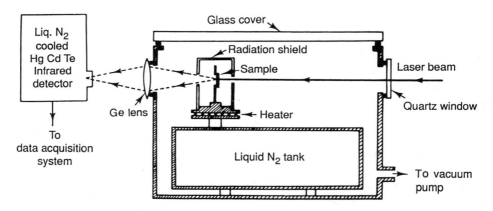

Fig. 4.7. Schematic diagram of the apparatus for flash radiometry measurement [54]

Choy and coworkers [53, 54] have applied this method to isotropic and oriented polymers. They developed two experimental setups: one for relatively thick samples and the other for thin films. A schematic diagram of the apparatus for flash radiometry measurements on thin films is shown in Fig. 4.7. The disk-shaped sample of diameter 4–12 mm and thickness 10–250 μm is fixed in the sample holders which is connected with a heater and a coolant bath. A laser beam passes through the quartz window and arrives at the sample. A germanium lens on the back surface serves to focus the emitted thermal radiation onto the infrared detector. A Nd/Yag laser provides excitation pulses of 10 ns width at any one of the four wavelengths: 266, 353, 532 and 1060 nm. The ultraviolet beam at 266 nm is especially suitable for polymers due to its strong absorption by many polymers. The pulse energy, which is needed to give a temperature rise of about 1 K, is below 1 mJ. The HgCdTe detector, operating in photoconductive mode, possesses a high sensitivity to 0–13 μm infrared radiation, a peak sensitivity being at about 11 μm, and has a rise time of 0.1 μs. Normally the experiment is carried out by averaging from 20 to 200 laser shots. For collecting and fitting the data to the theoretical expression a microprocessor is used.

The flash radiometry measurements can be performed in the temperature range between 150 and 500 K. For measurements in the temperature range 150–300 K cooling with liquid N_2 is used and the sample chamber is evacuated to 10^{-3} Torr to prevent condensation of moisture. The temperature of the sample holder is automatically controlled to 0.05 K. The radiation and air conduction effects for thin films can normally be neglected. Typical behavior of the flash radiometry signal for two films of different thicknesses is shown in Fig. 4.8. The diffusivity a can be deduced from the exponential decay time constant t_c by using Eq. (4.12). The thermal conductivity can then be determined using Eq. (4.4) and the relevant values of C_p and ρ which are available in the literature.

A useful approach to the determination of the thermal conductivity of polymers consists of the application of DSC technique [9, 18, 55–57]. A number of studies have shown that modifications of DSC can give reasonably good χ data. Three methods of operation have been suggested. In the first method two cylinders of identical size and shape, of measured and reference, are used (Fig. 4.9a). In the steady-state conditions using the recorded DSC output and the temperature difference ΔT across each cylinders, the thermal conductivity χ of the sample can be determined

$$\chi = \chi_r = \frac{\Delta H}{(S/d)\Delta T} \qquad (4.13)$$

where S and d is the surface area and thickness of the cylinders, respectively and χ_r is the thermal conductivity of the reference cylinder. It is possible to carry out the measurements of χ using only one cylinder of unknown χ, suggesting that the heat loss in the reference cell is fixed.

In the second method of operation a small sample is sandwiched between the DSC sample holder thermocouple and a copper rod, the opposite end of which is cooled by a large heat sink (Fig. 4.9b). A thermocouple embedded into the copper rod allows one to measure the temperature on the upper side of the sample. At a series of fixed temperatures the unit is equilibrated and the DSC output together with the temperatures at the lower and upper sides of the sample are measured. Similar experiments are also conducted with a standard sample of known χ_s after which χ of the unknown sample can be determined.

Both methods described possess a number of disadvantages such as comparatively large sample sizes, impossibility of studying the thermal conductivity in the molten state. The third method suggested is especially suitable for measurements of the thermal conductivity of very small samples even in the molten state, it needs little time and no instrument modification (Fig. 4.9c) [9, 18, 57]. The basic characteristic of any DSC is the thermal

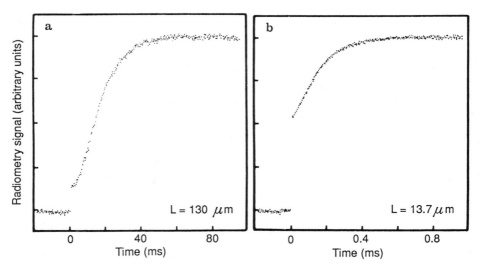

Fig. 4.8. Flash radiometry signal for polyimide film of thickness (**a**) $130\,\mu$m and (**b**) $13.7\,\mu$m [54]

resistance $R = R_0 + R_s$ where R_0 is the thermal resistance of the device resulting from the separate placement of the heater and sample and R_s is the thermal resistance which includes the thermal resistance of the sample itself and the resistance of the thermal contacts between the sample and sample pan. For small metallic samples, which are first melted in a pan, the thermal contact between the pan and the sample is close to ideal, therefore, $R_s \cong 0$. On the other hand, R_s for rather thick polymer samples of poor thermal conductivity, is the dominating factor in the total thermal resistance R. Hence, the main idea of determining the thermal conductivity of polymers with this method consists of placing a thin layer of calibrant metal (In, Ga, Wood alloy etc.) of 50–$100\,\mu$m of thickness, which completely covers the surface of the polymer sample, and studying the endothermal peak of the melting of the metal. Typical endothermal peaks of melting of Ga placed on the surface of PMMA and graphite plates of various thickness is shown in Fig. 4.10. It is seen that the slope of the peaks decreases with increasing thickness, which corresponds to the increase of R_s. The total thermoresistance $R = \operatorname{ctg} \varphi$. The thermal conductivity of any plates covered with metals can be determined from the equation

$$\chi = \frac{d}{(\operatorname{ctg} \varphi - \operatorname{ctg} \varphi_0)S} \tag{4.14}$$

where $\operatorname{ctg} \varphi_0$ is the slope of the peak corresponding to melting of metal without any sample, d and S is the thickness and surface area of the sample, respectively. The values of the thermal conductivity of PMMA, PS, PVC, LDPE estimated using this approach are in good accord with values obtained with traditional methods. Hence, this method is a very promising approach to thermal conductivity measurements.

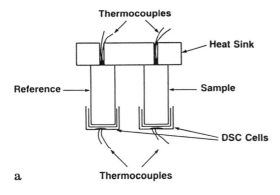

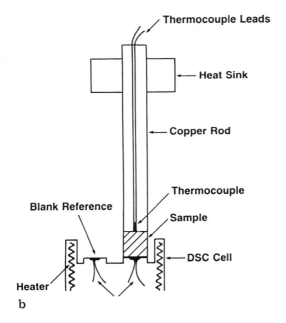

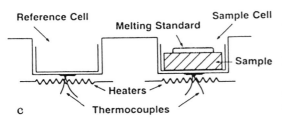

Fig. 4.9. (a) General features of a modified Perkin Elmer DSC method; (b) general features of modified Du-Pont DSC method; (c) sample arrangement for measuring thermal conductivity in a standard DSC unit [18]

Anisotropy of the thermal conductivity and diffusivity of sample of large sizes has been examined by means of the steady-state methods [29–32, 58–60]. Solid oriented samples are cut into separate rectangular parts from which plates are built, the large surfaces of which are either perpendicular or parallel to the orientation direction. If samples are in the rubbery state, they are

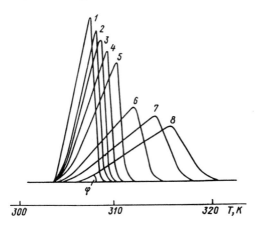

Fig. 4.10. Endothermal peaks of melting of Ga (m = 10 μg) without any sample (*1*) and placed on the surface of a graphite sample (*2–5*) with thicknesses of 0.25, 0.50, 0.75, and 1.0 mm, respectively, and of a PMMA sample (*6–8*) with thicknesses of 0.82, 1.34 and 1.90 mm, respectively; the surface of the samples S = 18 mm² [9]

stretched to a predetermined extension ratio and then are fixed in a metallic frame. The frame with the fixed sample is placed to an instrument.

Anisotropy of oriented polymer films may be estimated by the very elegant method of de Senarmont, which is based on the analysis of the figures of the melting of thin covers, placed on the surface of the sample, while applying a point source of heat to the sample. The shape of the molten figure allows one to estimate the anisotropy of the thermal conductivity and diffusivity. If there is no anisotropy of the thermophysical properties of the film, the molten figure should be a circle, however, if such anisotropy exists, the molten figure is, as a rule, an ellipse. Solution of the corresponding two-dimensional problem of the thermal conductivity [40, 61] gives

$$\frac{\chi_x}{\chi_y} = \frac{a_x}{a_y} = \frac{b_x^2}{b_y^2} \tag{4.15}$$

where b_x and b_y are the small and large semi-axis of the ellipse.

Such a method of estimation of the anisotropy of the thermal conductivity and diffusivity of drawn polymers was used by Kargin and Slonimsky with coworkers [62–64], who used thin paraffin layers, and Müller and Kilian with coworkers [61, 65], who used special thermoindicators.

4.3 Thermal Expansion

A group of methods which deal with the measurements of the temperature dependence of dimensional changes of samples is referred to as dilatometry. The linear dilatometry technique is rather simple. The determination of coefficient of linear thermal expansion, as performed in a classical quartz-tube dilatometer, is based on the following equation

$$\beta = \frac{\Delta L}{L_0 \Delta T} \qquad (4.16)$$

where ΔL is the change in the length of the sample as a result of temperature change, L_0 is the length of the test sample at any initial temperature and ΔT is the temperature difference over which the change in length of the specimen is measured. The resulting value of β is, of course, an average value for the temperature range covered. The actual or differential value of β is a more valuable characteristic of the thermal expansion especially for polymers which have a rather large coefficient of thermal expansion. The differential value of β can be obtained using the simple formula

$$\beta(T) = \frac{1}{L_0} \frac{dL}{dT} \, . \qquad (4.17)$$

The linear dilatometers can be divided into two groups: for absolute measurements of linear thermal expansion and for comparative measurements, in which a reference sample with a known coefficient of thermal expansion is used. A large number of linear dilatometers of both types are described in a monograph by Amatuni [66].

A number of linear dilatometers for polymers were constructed by Bartenev and coworkers [67–70]. Figure 4.11 shows the schematic diagram of one of them. The cylindrical sample, placed in a tube, is connected by a rider to the dial gauge which is used for reading the displacements. A special automatic cooling and heating system allows cooling and heating of the system at a small constant rate. In order to achieve good contact between the sample and the rider the sample is stressed with a small amount of tension. The accuracy of the measurements is 2–5%.

Some special construction of linear dilatometers for thin polymer films and fibers are described in [71, 72].

Now a lot of commercial equipment for thermomechanical analysis is available. In typical thermomechanical instruments, the length of the sample is monitored while the sample is subjected to a temperature program. In the course of the analysis a predetermined force can be applied to the sample with the measuring sensor. This force can either be constant or variable. If the dimension of the sample is measured under negligible force the linear thermal expansion can be obtained. On the other hand, measurements under force allow one to estimate temperature transitions, force of shrinkage for stretched films and fibers etc. One of the most powerful thermomechanical

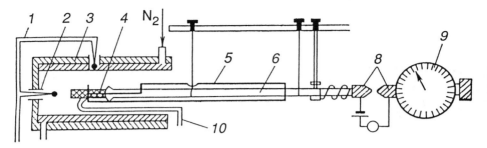

Fig. 4.11. Linear dilatometer [67]. *1* – regulated thermocouple; *2* – working chamber; *3* – cooling camera; *4* – sample; *5* – tube; *6* – rider; *7* – support; *8* – electrical contact; *9* – dial gauge; *10* – calibrated thermocouple

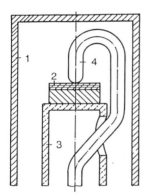

Fig. 4.12. Measuring principle of Mettler TMA 40 [73]. *1* – metal furnace assembly with electrical heating; *2* – sample; *3* – sample holder (adjustable in height according to sample thickness); *4* – measuring probe with a rounded end

analysers, the TMA 40, a diagram of which is shown in Fig. 4.12, has been developed by Mettler [73].

References

1. Kagan DN (1984) In: Compendium of thermophysical property measurement methods, vol 1, chap 12, Maglic KD, Gazairliyan A, Peletsky VE (ed), Academic New York
2. Hemminger W and Höhne G (1984) Calorimetry, Verlag Chemie Weinheim
3. Wunderlich B (1990) Thermal Analysis, Academic New York
4. Dole M (1960) Fortschr Hochpolym Forsch *2*: 221
5. Wunderlich B and Baur H (1970) Fortschr Hochpolym Forsch *7*: 151
6. Godovsky YuK (1976) Thermophysical Methods of Polymers Characterization (in Russian), Khimija Moscow
7. Richardson MJ (1984) Application of Differential Scanning Calorimetry to the Measurement of Specific Heat, In: Compendium of Thermophysical Property Measurement Methods, vol 1, chap 17, Maglic KD, Gezairliyan A, Peletsky VE (ed), Plenum New York
8. Richardson MJ (1989) Thermal Analysis, In: Comprehensive polymer science, vol 1, chap 36, Booth C, Price C (ed), Pergamon Oxford
9. Bershtein VA and Egorov VM (1990) Differential Scanning Calorimetry in Physical Chemistry of Polymers (in Russian), Khimija Leningrad

10. Turi EA (ed) (1981) Thermal Characterization of Polymer Materials, Academic New York
11. Chudnovski AF (1962) Thermophysical Characteristics of Dispersion Materials (in Russian), Fizmatgiz Moscow
12. Kondratyev GM (1957) Thermal measurements (in Russian), Mashgiz Moscow
13. Maglic KD, Gazairliyan A, Peletsky VE (ed) (1984) Thermal Conductivity Measurement Methods, In: Compendium of Thermophysical Property Measurement Methods, vol 1, part I, Plenum New York
14. Carslaw HS and Jager JC (1976) Conduction of Heat in Solids, Clarendon Press Oxford
15. Lykov AV (1967) Theory of Thermal Conductivity (in Russian), Vysshaya Shkola Moscow
16. Knappe W (1971) Adv Polymer Sci *7:* 477
17. Gobbe C, Bazin M, Gounot J and Dehay G (1988) J Poly Sci Polym Phys Ed *28:* 857
18. Khanna YP, Taylor TJ and Chomin G (1988) Poly Eng Sci *28:* 1034
19. Schallamach A (1941) Proc Phys Soc *53:* 214
20. Holzmüller W and Munx M (1958) Kolloid Z *159:* 25
21. Zanemonets NA and Fogel VO (1959) Kautchuk i Rezina *N2:* 21
22. Knappe E (1960) Z Angew Phys *12:* 508
23. Holzmüller W and Lorenz J (1961) Plaste u Kautschuk *8:* 351
24. Eiermann K and Werner K (1962) Z Angew Phys *14:* 484
25. Kline DE (1961) J Polym Sci *50:* 441
26. Sheldon RP and Lane K (1965) Polymer *6:* 77
27. Shoulberg RH and Shetter JA (1962) J Appl Polym Sci *6:* 532
28. Anderson AC, Reese W and Wheathley JC (1963) Rev Sci Instr *34:* 1386
29. Pasquino AD and Pilswarth MN (1964) J Polym Sci Part B *2:* 253
30. Anderson DR (1966) Chem Rev *66:* 677
31. Lohe P (1965) Koll Z Z Polym *203:* 115
32. Kolouch RJ and Brown RG (1968) J Appl Phys *39:* 3999
33. Erk S, Keller A and Poltz H (1937) Phys Z *38:* 394
34. Tautz H (1959) Exper Techn Phys *7:* 1
35. Tauth H (1961) Koll Z Z Polym *174:* 128
36. Abeles B, Cody CD and Beeres DS (1960) J Appl Phys *31:* 1585
37. Vandersande JW, and Pohl RO (1980) Rev Sci Instr *51:* 1694
38. Phyllipov LP, in [13] Temperature Wave Technique, chap 9
39. Vertinskaya AB and Novichenok LN (1960) Inzh Fiz Zh (IFZH) *3:* 65
40. Novichenok LN and Shul'man ZP (1971) Thermophysical properties of polymers (in Russian), Nauka i Technika Minsk
41. Steere RC (1966) J Appl Phys *37:* 3338
42. Volokhov GM and Kasperovich AS, in [13] Monotonic Heating Regime Methods for the Measurement of Thermal Diffusivity, chap 11
43. Zamoluev VK (1960) Plast Massy *N8:* 46
44. Gengrinovich BI and Fogel VO (1957) Kautchuk i Rezina *N9:* 27
45. Kirichenko YuA, Oleinik BN and Chadovich TZ (1964) Inzh Fiz Zh (IFZH) *7:* 70
46. Hattori M (1965) Koll Z Z Polym *202:* 11
47. Shoulberg RH (1963) J Appl Polym Sci *7:* 1597
48. Eiermann K, Hellwege K-H and Knappe W (1961) Koll Z Z Polym *174:* 134
49. Platunov ES (1961) Izv Vuzov Priborostroyenie *N4:* 90
50. Kurepin VV and Platunov ES (1961) Izv Vuzov Priborostroyenie *N4:* 119
51. Parker WJ, Jenkins RJ, Butler CP and Abbott GL (1961) J Appl Phys *32:* 1679
52. Taylor RE and Maglic KD, in [13] Pulse Method for Thermal Diffusivity Measurement, Chap. 8

53. Chen FC, Poon YM and Choy CL (1977) Polymer *18:* 129
54. Choy CL, Leung WP and Ng YK (1987) J Poly Sci Polym Phys Ed *25:* 1779
55. Brennan WP, Miller B and Whitwell JC (1968) J Appl Polym Sci *12:* 1800
56. Chiu J and Fair PG (1979) Thermochim Acta *34:* 267
57. Hakvoort G, Van Reijen L and Aarsten A (1985) Thermochim Acta *93:* 317
58. Eiermann K (1961) Kunststoffe *51:* 512
59. Hennig J (1964) Koll Z Z Polym *196:* 136
60. Hennig J and Knappe W (1964) J Polym Sci Part C *N6:* 167
61. Hellmuth W, Kilian H-G and Müller FH (1967) Koll Z Z Polym *218:* 10
62. Kargin VA, Slonimsky GL and Lipatov YuS (1955) DAN SSSR *104:* 96
63. Slonimsky GL and Dikareva TA (1965) Vysokomol Soyed *7:* 1276
64. Slonimsky GL, Dikareva TA and Pavlov VI (1966) Vysokomol Soyed *8:* 1717
65. Kilian H-G and Pietralla M (1978) Polymer *19:* 664
66. Amatuni AN (1972) Methods and equipments for determination of linear thermal expansivity of various materials (in Russian), Izdatelstvo Standartov Moscow
67. Bartenev GM, Gorbatkina YuA and Lukyanov IA (1963) Plast Massy *N1:* 56
68. Bartenev GM and Lukyanov IA (1955) Zh Fiz Khim (ZFKh)*29:* 1468
69. Bartenev GM and Gorbatkina YuA (1959) Vysokomol Soyed *1:* 769
70. Bartenev GM and Garzman VI (1962) Zavodskaya Laboratoriya *28:* 245
71. Sidorovich AV and Kuvshinsky EV (1959) Zavodskaya Laboratoriya *25:* 1124
72. Kaimin IF (1966) Plast Massy *N9:* 62
73. Mettler (1984) Collected Applications, Titration and Thermal Analysis

Part Two:
Thermal Behavior of Polymers under Mechanical Deformation and Fracture

5 Thermomechanics of Glassy and Crystalline Polymers

Conventionally, the deformation of solids is treated within the framework of elasticity theory in terms of stresses and strains, i.e. in purely mechanical terms. Although experimental determination of stress-strain relationship provides important information concerning the deformation process, it is quite obvious that this conjugated variable pair is only one of several pairs which can be used to describe the response of a material on deformation. The necessity of using the thermodynamic instead of mechanical approach became evident when investigators began to consider deformation in terms of thermodynamic potentials, such as internal energy, free energy etc., rather than in terms of potential energy. The first law of thermodynamics shows that energy is conserved in all deformation processess either reversible or irreversible. Therefore, the mechanical response of any material reflects exactly that amount of energy which accompanied the deformation process as heat and/or changes of internal energy. What this means to an experimentalist is that the temperature or thermal variations brought about by adiabatic or isothermal processes have to be measured simultaneously with stresses and strains. The thermal effects and temperature variations accompanying deformation of solids are usually rather small and it is their correct measurement which constitutes the major difficulty in passing from mechanical to a thermodynamic approach. But the thermal effects in condensed materials must carry profound information about the molecular changes occurring due to deformation. Physically, therefore, the thermodynamic approach must be preferred. A briliant example of applying the thermodynamic approach to molecular processes in deformation is the thermodynamic analysis of rubber elasticity, which will be described in Chap. 6.

Experimental investigations of the thermodynamics of reversible deformations of glassy and crystalline polymers has attracted much less attention compared to rubberlike materials. Often such experiments are devoted to the test of the Kelvin relation between stress and temperature. It is well known that this and other similar relations can be obtained in the classical theory of thermoelasticity. An important step in the thermomechanical study of glassy and crystalline polymers was taken in the late 1950s when deformation calorimetry was first developed and applied by Müller and collaborators. This technique has now been greatly improved (see Chap. 9) and important progress has been recently achieved in investigating calorimetric behavior of glassy and semicrystalline polymers under deformation. In this chapter an overview of the thermomechanical behavior of glassy and crystalline polymers is presented with a special focus on relationships following from the thermomechanical equations of state for various deformation modes and on the elucidation of molecular mechanisms of macroscopic deformation and structural changes accompanying the reversible deformation of solid polymers.

5.1 Phenomenological Aspects of the Thermomechanics of Elastic Materials

According to the first law of thermodynamics the change in the internal energy U of an isotropic condensed material subjected to several forces of various types can be represented by [1, 2]

$$dU = dQ + \sum_{i=1}^{n} \xi_i dx_i \qquad (5.1)$$

where dQ is the element of heat absorbed, ξ_i is the generalized force conjugated with the generalized coordinate x_i and n is the number of generalized forces. Experimental thermomechanical studies of solid polymers are normally carried out under the action of two generalized forces, one of which is the external pressure P. In this case for a reversible quasistatic process Eq. (5.1) yields

$$dU = TdS - PdV + \xi dx \qquad (5.2)$$

where V is the volume of the system, T the temperature and S the entropy. Using Eq. (5.2) and the standard definition of the associated free energy F, the enthalpy H and the free enthalpy G we arrive at

$$dF = -SdT - PdV + \xi dx \qquad (5.3)$$

$$dH = TdS + VdP + \xi dx \qquad (5.4)$$

$$dG = -SdT + VdP + \xi dx . \qquad (5.5)$$

The thermodynamic potentials introduced differ from those in use in the thermodynamics of gases and liquids in the additional ξdx term. For solids, it is convenient to introduce four additional thermodynamic potentials [3–6]

$$U^* = U - \xi x ; \quad dU^* = TdS - PdV - x\, d\xi \qquad (5.6)$$

$$F^* = U^* - TS ; \quad dF^* = -SdT - PdV - x\, d\xi \qquad (5.7)$$

$$H^* = U^* - \xi x ; \quad dH^* = TdS + VdP - x\, d\xi \qquad (5.8)$$

$$G^* = H^* - TS ; \quad dG^* = -SdT + VdP - x\, d\xi . \qquad (5.9)$$

Using these thermodynamic potentials introduced, 24 Maxwell relations containing certain partial derivatives can be obtained [5, 6]. These relations together with the corresponding specific heats $C_{y,z} = T(dS/dT)_{y,z}$ (where y represents either V or P, and z represents either ξ or x) permit us to describe phenomenological relationships between the deformation (or stress) in solid materials and the accompanying thermal and temperature effects. In particular, the thermal effect of isobaric-isothermal extension of a solid rod by the force f can be derived from Eq. (5.9)

$$(\partial Q/\partial f)_{P,T} = T(\partial S/\partial f)_{P,T} = T(\partial L/\partial T)_{P,f} = T\beta_{P,f}L \qquad (5.10)$$

or in the integral form

$$Q_{P,T} = T\beta_{P,f}Lf \qquad (5.11)$$

where L is the length of the rod and $\beta_{P,f} = 1/L(\partial L/\partial T)_{P,f}$ is the linear thermal expansion coefficient at constant P and f. According to Eq. (5.11) the extension of a rod with a positive $\beta_{P,f}$ is accompanied by absorption of heat. Such behavior is very characteristic of the majority of solids.

If the force is applied instantaneously, the deformation of the rod is adiabatic and is accompanied by a relative temperature change. The temperature change can be derived from Eq. (5.8):

$$(\partial T/\partial f)_{P,S} = -(\partial L/\partial S)_{P,f} = -\beta_{P,f}LT/C_{P,f} \qquad (5.12)$$

or in the integrated form

$$\Delta T_{P,S} = -\beta_{P,f}LTf/C_{P,f} . \qquad (5.13)$$

Equation (5.13) is an expression well known for the thermoelastic or Kelvin (Thompson) effect. It shows that solid with a positive $\beta_{P,f}$ gets cooler with adiabatic extension and warmer under adiabatic compression.

A very important problem in the thermodynamics of deformation of solids is the relationship between heat and work. From Eqs. (5.2) and (5.4) by integration, the internal energy and enthalpy can be derived. Similar to other condensed systems, the enthalpy differs from the internal energy at atmospheric pressure only negligibly, since the internal pressure in condensed systems $P_i \gg P$. Therefore, the work against the atmospheric pressure can be neglected in comparison with the term $\xi_i x_i$. Hence it follows that

$$\Delta U \approx H = T\Delta S + W \qquad (5.14)$$

where $W = \int f_i d\xi$.

Using Eq. (5.14) one can arrive at a convenient form for characterizing the elastic system by introducing the characteristic ratio [6–8]

$$\eta = Q/W = \Delta U/W - 1 = \omega - 1 . \qquad (5.15)$$

Two idealized limiting models can be discussed:

1. The material with ideal elastic energy ($\eta = 0$, $\omega = 1$), in which the reversible work W is totally conserved as the internal energy U. Deformation of such a system is not accompanied by thermal or temperature effects. The classical theory of elasticity treats deformation of elastic systems from this point of view.
2. The ideal entropy elastic system ($\eta = -1$, $\omega = 0$) in which the work exchanged is totally transformed into a change in entropy. A well-known example of such systems is the ideal gas.

Deformation of real condensed systems can be accompanied by both the internal energy and entropy (or temperature) change, but phenomenological thermodynamics cannot answer the question concerning the molecular nature of the changes. This is the consequence of the fact that the phenomenological approach does not allow to determine the change in the potential energy of the particles of a condensed system under deformation and, therefore, does not permit us to determine the change of its temperature or entropy. The answer can be obtained only by analysis of appropriate thermomechanical equations of state. Let us therefore consider a concrete elastic system characterized by the corresponding thermomechanical equation of state. We shall see that the characteristic ratios η and ω introduced above allow us to describe elastic systems in a particularly convenient way with reference to one of the limiting models.

5.2 Linear Thermomechanics of Quasi-Isotropic Hookean Solids

5.2.1 Uniform (Volume) Dilation and Compression

The thermomechanical equation of state of such a solid body may be obtained by combining Hook's law and the law of thermal expansion [6, 7]. Hence, for uniform dilation we arrive at (as a first approximation)

$$\sigma = K \left[\frac{V}{V_0} (1 - \alpha T) - 1 \right] \tag{5.16}$$

where σ is the stress, K the isothermal bulk modulus, V_0 the volume in the unstrained state (at initial temperature T_0), V the volume in the strained state and α the volume thermal expansion coefficient. Material constants K and α are assumed to be independent of temperature in the linear theory. As a first approximation one can also neglect the second order terms, such as αT^2. The transition from dilation to compression in Eq. (5.16) may be carried out by the replacement of σ by $-P$.

The analysis of Eq. (5.16) has led to the conclusion [6] that the strain-energy function W in the mode of uniform deformation is parabolic with a minimum potential energy in the unstrained state

$$W = Ke^2/2 \qquad (5.17)$$

where $e = (V - V')/V'$, in which V' is the volume in the unstrained state. On the other hand, the heat of deformation is the linear function of deformation

$$Q = T\Delta S = \alpha T\sigma = \alpha TKe . \qquad (5.18)$$

The characteristic parameter η of the Hookean solid can be found immediately

$$\eta = 2\alpha T/e . \qquad (5.19)$$

Hence, we arrive at the conclusion that only in the limit $\alpha \to 0$ is the Hookean solid the ideal energy-elastic one ($\eta = 0$) and the uniform deformation of a real system is accompanied by thermal effects. Equation (5.19) also shows that the dependence of the parameter η (as well as ω) on strain is a hyperbolic one and α, the phenomenological coefficient of thermal expansion in the unstrained state is determined solely by the heat to work and the internal energy to work ratios. From Eqs. (5.17) and (5.18) the internal energy change of Hookean material can be obtained from

$$\Delta U = Ke^2/2 + \alpha TKe . \qquad (5.20)$$

The relationship between the heat, work and internal energy change for a linear thermoelastic Hookean solid in uniform deformation at constant temperature is shown in Fig. 5.1. It can be easily demonstrated that for a Hookean solid, a thermomechanical inversion of the internal energy ($\Delta U = 0$) must occur at the deformation

$$e_{inv} = -2\alpha T . \qquad (5.21)$$

We see from Eq. (5.21) that the internal energy inversion occurs on compression of the solid with a positive thermal expansivity and on dilatation with the negative one. Occurence of the thermomechanical internal energy inversion in Hookean solids is a result of a different dependence of the work and heat on strain (Fig. 5.1). The work depends on K which represents essentially the harmonic character of the binding forces, while heat depends also on α which is associated with the anharmonicity. Table 5.1 tabulates the main parameters of the thermomechanical internal energy inversion for typical solids.

Recently Lyon and Farris [9] following the linear theory of thermoelasticity [10] arrived at the same conclusions in a more general form.

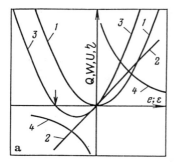 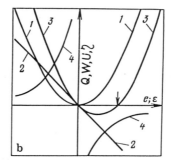

Fig. 5.1a,b. Mechanical work W (*1*), elastic heat Q (*2*), internal energy change ΔU (*3*) and heat work ratio Q/W (*4*) as a function of strain e (uniform deformation) or ε (unidirectional deformation) for quasi-isotropic Hookean solids [6, 7]. **a** – positive α and β; **b** – negative α and β. The *arrows* indicate inversion points (see text)

Table 5.1. Parameters of the internal energy inversion under deformation of condensed systems (T = 300 K)*

Materials	Uniform Extension (Compression)				Uniaxial Extension (Compression)			
	α $\times 10^5$ K	K_T GPa	e_{inv} %	U_{min} J/cm³	β $\times 10^5$ K	E GPa	ε_{inv} %	U_{min} J/cm³
Metals	4	50	−2	−0.5	2	50	−1	−0.1
Inorganic glasses	2	30	−1	−0.05	1	50	−0.5 < −0.1	
Polymers:								
Glasslike	20	3	−10	−0.5	7	4	−2	−0.01
Crystalline	30	3	−15	−1.0	10	3	−5	−0.1
Rubberlike	70	2	−40	−4	–	–	–	–
Organic Liquids	80	1	−50	−3	–	–	–	–
Water	20 (T > 4C)	2	−10	a−0.5	–	–	–	–
	−6.1	–	+4	−0.05	–	–	–	–
AgI	−0.4	40	+0.1	−0.01	−0.13	40	+0.1	< −0.001

* Typical values of α, β, E and K_T for given class of materials

5.2.2 Simple Elongation and Compression

The thermomechanical equation of state of an isotropic Hookean solid rod subjected to force f along the rod axis can be obtained analogously to that used for the uniform deformation [2, 6, 7]

$$\sigma = f/A_0 = E \left[\frac{L}{L_0}(1 - \beta_{P,f}T) - 1 \right] \tag{5.22}$$

where E is Young modulus, A_0 the cross-sectional area and $\beta_{P,f}$ the linear thermal expansion coefficient under constant P and f. Besides $\varepsilon = (L - L')/L'$ with $L' = L_0(1 + \beta_{P,f}T)$. For most solids, one can neglect the difference between $\beta_{P,f}$ ($\alpha_{P,f}/3$ for an isotropic solid) and the coefficient of thermal expansion at constant P is usually used. Therefore, we may use β and α without subscripts. Assuming, as in the previouse case, that E and β are independent of temperature one can arrive at relations similar to Eqs. (5.17–21). To do this, it is necessary to replace in Eq. (5.16) the volume deformation e by ε, the modulus K by E and α by β (see Fig. 5.1)

$$W = E\varepsilon^2/2 \tag{5.17'}$$

$$Q = \beta TE\varepsilon \tag{5.18'}$$

$$\eta = 2\beta T/\varepsilon \tag{5.19'}$$

$$\varepsilon_{inv} = -2\beta T . \tag{5.20'}$$

It is interesting to note that

$$\varepsilon_{min} = -\beta T \tag{5.23}$$

$$\Delta U_{min} = -E(\beta T)^2/2 . \tag{5.24}$$

Since the values β and α for solid polymers are usually in the range of 10^{-4}–10^{-5} K, the thermomechanical inversion of the internal energy must occur at the deformation of a few percent (see Table 5.1).

The above analysis demonstrates that the thermal and temperature effects resulting from the reversible uniform and unaxial deformations are determined by the coefficients of thermal expansion. As demonstrated in Chap. 3, the thermal contraction along the chain axis seems to be a general phenomenon in crystalline polymers. As a result of this conclusion we are immediately faced with the question what the consequences are for the thermomechanical behavior of crystalline polymers with negative thermal expansion along the c-axis of the crystalline lattices. For bulk unoriented polymers, it seems to have no effect because the value of the negative thermal expansion coefficient is much smaller than that of two other positive coefficients perpendicular to the chain axis, which leads to the positive volume thermal expansion. Because of the statistical arrangement of crystallites in unoriented crystalline polymers and, consequently, the chain axes, it is clear that the thermal expansion in macroscopic samples along any directions must be a positive one. The second reason is the presence of amorphous regions in crystalline polymers. In contrast to crystalline lattices, the anisotropy of the thermal expansion in the amorphous regions seems to be absent. These arguments allow us to suppose that the thermal shrinkage of crystalline macromolecules will have no consequences for macroscopic thermal properties of undrawn crystalline polymers. However, the situation may change dramatically in oriented crystalline polymers due to the orientation of both crystallites and amorphous regions. Hence, in drawn crystalline polymers one can expect to observe unusual thermomechanical behavior.

5.2.3 Shear Torsion

According to the theory of elasticity, the pure shear and torsion are accompanied only by a change in shape of an elastic solid but its volume remains unchanged. The equation of state of the Hookean solid for pure shear can be represented in the following form [6, 11]

$$\tau = G_0(1 - mT)\gamma_s \tag{5.25}$$

where τ is the shear stress, γ_s the shear deformation, G_0 is the shear modulus and m a small positive constant. Equation (5.25) yields the following expressions for work W and heat Q

$$W = \tau\gamma_s/2 = G_0(1 - mT)\gamma_s^2/2 \tag{5.26}$$

$$Q = TG_0 m\gamma_s^2/2 . \tag{5.27}$$

Comparing W and Q, we find $\eta = mT/(1 - mT) \ll 1$. Hence, we see that the heat to work ratio for shear is independent of deformation. It means that the thermal effects resulting from pure shear can be ignored (as a first approximation) compared to the deformation modes described above. This is also true for torsion.

This result stresses the fact that the heat effects can be detected only at such deformation modes of quasi-isotropic Hookean solids that are accompanied by a volume change. Thus, the thermal and temperature effects are the consequence of the changes of the vibrational entropy which in turn is a result of the volume change. It is very important to emphasize that the internal energy and entropy changes resulting from the deformation of solids are closely related and their values are quite comparable. This conclusion undermines the widely accepted belief that the elasticity of solids is entirely energetic. In contrast to this opinion we see that in solids the energy and entropy variations are interrelated and normally comparable in magnitude.

5.3 The Thermoelastic Effect in Glassy and Crystalline Polymers

The magnitude of the temperature change as a result of the reversible application of stress or hydrostatic pressure under adiabatic conditions is predicted by the Kelvin (Thompson) Equation (5.13). Such thermoelastic studies are performed primarily under uniaxial tension or compression, limiting stresses to small values because of the reversibility requirement. Haward and collaborators [12–14] reported quantitative studies of the thermoelastic effect in PMMA, PS, PC and some epoxy resins using uniaxial stresses varied between 2 and 50 MPa over a temperature range 220–350 K. Typical results are shown in Fig. 5.2. ΔT gives, in general, a good straight line fit as a function of stress for both tensile and compression results. Similar results were also obtained in some other studies [15]. As a result of these investigations it was concluded that all the materials investigated obey the Thompson equation. Using the thermoelastic measurements Haward and collaborators estimated values of the linear thermal expansion coefficient β, and, what is more important, the value of the Gruneisen constant γ_T. By assuming that $B_s/C_p = B_T/C_v$ they obtained the relationship

$$\gamma_T = 3(dT/d\sigma)B_s/T = 3(dT/d\sigma)(B_T/T)C_v/C_p \qquad (5.28)$$

which allows the direct evaluation of γ_T from the measurements of $dT/d\sigma$ and a separate measurement of B_T. The values of γ_T obtained for PS, PMMA and PC agree well with the literature data. A similar approach for estimating the Gruneisen constant γ_T has been also suggested recently [16, 17].

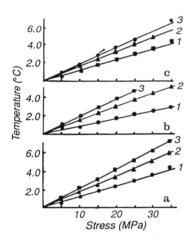

Fig. 5.2. Thermoelastic temperature changes as a function of stress [13]. **a** – PS: *1* - 222 K; *2* – 298 K; *3* – 333 K. **b** – PMMA: *1* - 222 K; *2* – 298 K; *3* – 333 K; **c** – PC: *1* – 222 K; *2* – 303 K; *3* – 358 K

A more complex thermoelastic behavior has been observed for crystalline polymers [17–19]. It has been established that, unlike the amorphous glassy polymers, in typical crystalline polymers such as HDPE, LDPE, PP and polypenten the adiabatic temperature changes follow the equation

$$\Delta T_s = -TV/C_p \left[\beta\sigma - (\partial E_s/\partial T)\sigma^2/2E_s^2\right] \qquad (5.29)$$

which takes into account the temperature dependence of the modulus of elasticity. As it has been assumed above in the Thompson equation E considers as temperature independent.

In general, all these studies performed under rapid uniaxial extension or compression are limited to rather small stresses because of the limitations of plastic deformation. Recently, Rodriguez and Filisko [20–22] have measured temperature changes as a result of large rapid compressive deformations using a specially-modified capillary rheometer and an experimental setup for measurements under hydrostatic pressure. The measurements were performed for PMMA and HDPE. Figure 5.3 illustrates the temperature changes as a function of stress for these polymers at different temperatures. A linear dependence of the temperature changes with stresses applied is very characteristic behavior even at rather high compressive deformations. Thermoelastic coefficients $(\partial T/\partial \sigma)$ are summarized in Table 5.2. First of all, HDPE shows higher values than PMMA. Secondly, a linear dependence of the coefficients on temperature is observed for both polymers. Finally, the predicted values for the thermoelastic coefficient agree rather well only for PMMA at 303 and 323 K. However, relatively large differences are observed at 343 and 373 K. For HDPE there is disagreement between the predicted and the experimental values.

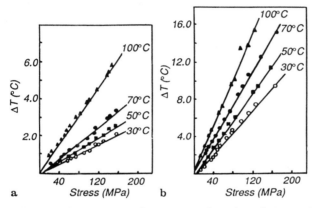

Fig. 5.3. Temperature changes as a function of stress at different temperatures for PMMA (**a**) and HDPE (**b**) [21]

Table 5.2. Thermoelastic coefficients $(\partial T/\partial \sigma)$ values for PMMA and HDPE [21]

Temperature	PMMA		HDPE	
K	Experiment °C/MPa	Predicted* °C/MPa	Experiment °C/MPa	Predicted* °C/MPa
303	0.0141	0.0141	0.0613	0.0480
323	0.0174	0.0161	0.0758	0.0528
343	0.0231	0.0178	0.0911	0.0605
373	0.0366	0.0217	0.1232	0.0617

* From the equation $(\partial T/\partial \sigma) = \beta T/\rho C_p$

Temperature changes as a result of large rapid hydrostatic pressure changes were measured for PMMA at various temperatures from ambient to 368 K and for various pressure increments up to 200 MPa [20–22]. It has been established that the ratio of the temperature change to the pressure change can be represented either by a quadratic equation or through the empirical equation $\Delta T = a(\Delta P)^b$, where a and b are constants, $\Delta T = T - T_0$

and $\Delta P = P - P_0$, where T_0 is the reference temperature and P_0 is the atmospheric pressure. In this case the agreement between the thermoelastic coefficients $(\partial T/\partial P)_s$ obtained from experimental data and predicted by the Thompson equation is considerably better. The thermoelastic coefficients were also used to estimate the thermodynamic Gruneisen parameter γ_T for PMMA. The Gruneisen parameters were found to be a function of pressure and temperature.

Low-temperature thermoelastic effects in PMMA and vitreous silica in the temperature range 0.5–10 K were studied by Wright and Phillips [23]. Both ramp and sinusoidal strains were used. The temperature changes induced were used to estimate the Gruneisen parameter which agreed well with values derived from thermal expansion in silica above 4 K and in PMMA above 2 K.

5.4 Thermomechanics of the Undrawn Glassy and Crystalline Polymers

For investigation of the reversible thermomechanical effect in undrawn glassy and crystalline polymers, calorimetric measurements of heat effects resulting from the stretching of films and fibers were used. These measurements were carried out on typical glassy polymers such as PMMA, PS, PC, amorphous PET and cured epoxy resins and typical crystalline polymers such as polyolefins, polyesters, polyamides. Quantitative analysis of the thermomechanical effect can be in principle carried out using Eqs. (5.16)–(5.24). The first question which arises immediately is: to what extent are the relationships obtained for the thermomechanics of typical solids in the analysis of the simple equations of state in agreement with the real thermal phenomena accompanying the reversible deformations of glassy and crystalline polymers. A priori, one could expect them to give a close estimate for glasslike polymers, since the assumptions used in deriving the equations of state are no doubt valid for glassy polymers. On the other hand, it is well known that crystalline polymers are as a rule two-phase systems consisting of both crystalline and amorphous phases and therefore, their thermomechanical behavior is more complex than that of glassy polymers, particularly if the amorphous component is above its glass transition temperature. In this case, a two-phase crystalline polymer can be considered as a network in which crystallites are rigid filler particles and act as multifunctional crosslinks. The reversible deformation is localized in the amorphous regions. It has been suggested that the elastic properties of such two-phase polymers should be related to the conformational changes in the amorphous region. According to this widely accepted approach the elastic properties of two-phase crystalline polymers can be analysed using the theory of rubber elasticity. The chains in the amorphous phase are supposed to be in a highly strained state even in the absence of external macroscopic stress, so that their elasticity has to be considered in terms of inverse Langevin statistics (see Chap. 6). The change in the enthalpy on deformation of amorphous regions resulting from the intermolecular changes is supposed to be small and can be neglected. According to this approach, the free energy of deformation of two-phase crystalline polymers above T_g is purely intrachain. However, as it will be demonstrated in this section this approach is incorrect because the changes of interchain interactions in amorphous regions occur during deformation accompanied by volume change so that the enthalpy changes considerably. The volume changes correspond to that accompanied deformation of solids and not of elastomers. All these facts mean that the free energy of deformation of semicrystalline polymers has an intermolecular origin. Keeping this idea in mind, let us consider the experimental data concerning the thermoelastic behavior of undrawn solid polymers.

The results for two typical glassy polymers shown in Fig. 5.4 demonstrate that the polymers behave classically over the reversible stress range, i.e. heat effects Q resulting from elastic elongation were a linear function of stress. The linear dependencies have been obtained regardless of small stress relaxation (time) effects which occur after a quick straining of the sample by several percent. At small strains that do not exceed 1% the deformation was fully reversible for such glassy polymers as PMMA and PS without any hysteresis effects.

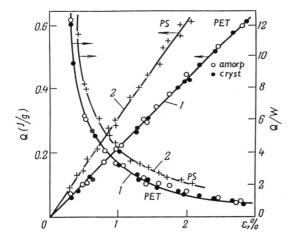

Fig. 5.4. Heat Q and heat to work ratio Q/W as a function of deformation at 20°C [6, 7]. *1* – PET (○ – amorphous, ● – crystalline, w = 50%). *2* – PS. *Solid curves* were obtained by equations $Q = \beta TE\varepsilon$ and $Q/W = 2\beta T/\varepsilon$ with the following values of parameters: PS – $\beta = 6.8 \times 10^{-5}\mathrm{K}^{-1}$, E = 2.0 GPa; PET – $\beta = 5.4 \times 10^{-5}\mathrm{K}^{-1}$, E = 1.8 GPa

The linear thermal expansion coefficients calculated from the measurements are in exellent agreement with literature data obtained by the conventional method, as is seen from Table 5.3. The characteristic heat to work ratio depends hyperbolically on strain which is also in an exellent agreement with the prediction following from the thermomechanical analysis. Hence, the results of thermomechanical measurements similar to thermoelastic data demonstrate that typical undrawn polymeric glasses are well described by the classical thermomechanical theory.

As seen from Fig. 5.4, the crystallinity does not change the value and sign of thermal effects and heat to work ratio in PET in comparison with a completely amorphous glassy sample. Hence, the thermomechanical behavior of crystalline polymers at temperatures below T_g is fully identical with that of a completely amorphous glassy polymer.

Table 5.3. Linear thermal expansion coefficients for undrawn glassy and crystalline polymers and metallic materials calculated from deformation calorimetry data (room temperature)

Polymers and Metallic Materials (Foils and Wires)	Calorimetry $\beta \times 10^5 \deg^{-1}$	Dilatometry $\beta \times 10^5 \deg^{-1}$	References
PS	6.8	7.0	7, 24, 25
PET	5.4	5.0	7, 24, 25
PC	4.2	3.9	26
LDPE*	24.6	23.7	7, 25
LLDPE*	10.4	10.0–22.0	26
HDPE*	14.6	10.0–15.0	7, 25
UHMWPE*	6.6	7.2	26
PP*	11.2	8.45	7, 25
Chronifer 1808 (Chrom-nickel alloy)	1.4–1.6	1.37	27, 28
Invar (Nickel-iron alloy)	0.1	0.1	27
Nickel	1.31	1.31	29
Niobium	0.7	0.625	29
Paladium	1.1	1.17	29
Aluminium	1.9	1.9–2.4	26
Copper	1.9	1.7	26

* Depends on degree of crystallinity; LLDPE – linear low density polyethylene; UHMWPE – ultra high molecular weight polyethylene

Similar to glassy polymers and crystalline polymers with glassy amorphous regions, for crystalline polymers above their T_g, absorbtion of heat is observed on stretching and the return to the initial strain-free state was accompanied by evolution of heat. Such thermal behavior is qualitatively inconsistent with the suggestion that chains in the amorphous regions are in a highly strained state since the stretching of such chains must be accompanied by liberation of heat. The observed effect is of opposite sign. The thermomechanical behavior of typical semicrystalline polymers above T_g is shown in Figs. 5.5 and 5.6. As is seen from the Q-ε dependence, linear behavior is observed only at small initial deformations. Further extension is accompanied by a considerable slow down in the increase of heat, which is a result of the

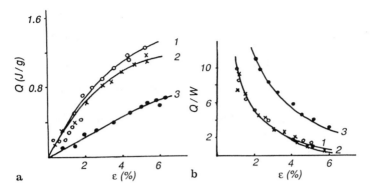

Fig. 5.5. Heat Q (a) and heat to work ratio Q/W (b) as a function of strain at room temperature [24, 30]. *1* – PP; *2* – HDPE; *3* – LDPE

appearence of inelastic deformation. The values of the coefficients of thermal expansion calculated from the initial heat effects agree well with dilatometric results (Table 5.3). Data for various PE demonstrate the influence of the degree of crystallinity on the value of thermal effects and thermal expansion coefficients.

Most undrawn crystalline polymers possess spherulite morphology with a radial arrangement of fibrils which are complex aggregates of crystallites and amorphous regions. One can conclude that the presence of crystallites prevents the amorphous chains from deforming exclusively due to rotational rearrangement and the deformation is accompanied by volume changes. For undrawn PE and PP with degrees of crystallinity of 60–70%, the Poisson's ratio is approximately 0.35 at room temperature [6, 7]. This value is typical for solids and it corresponds to the increase of the volume during deformation. Thus, from these studies of thermomechanical effects in isotropic crystalline polymers, one can finally conclude that the volume elasticity of their amorphous regions is very characteristic of such polymers. This volume elasticity seems to be a consequence of a statistical arrangement of crystallites and amorphous regions and their covalent linkage. Therefore, the thermomechanical behavior of isotropic crystalline polymers contradicts the idea that these polymers can be considered as classical semicrystalline networks and that their elastic properties are determined only by the conformational changes of highly stretched amorphous parts of chains [31, 32]. This picture becomes especially evident after considering the thermomechanical behavior of drawn crystalline polymers (see below).

For undrawn crystalline polymers during the deformation mode extension-stress relaxation-contraction stress relaxation occurs after extension to a predetermined strain. Because the extension during this strain cycle was carried out in the ballistic mode (see Chap. 9) using this data it is possible to estimate the heat accompanying the stress relaxation. The results obtained after such a treatment are shown in Fig. 5.7. The stress relaxation in undrawn polymers is accompanied by liberation of heat, the value of heat being a linear

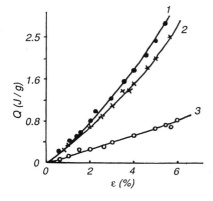

Fig. 5.6. Internal energy change as a function of strain at room temperature [24]. *1* – PP; *2* – HDPE; *3* – LDPE

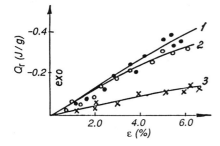

Fig. 5.7. Heat resulting from stress relaxation as a function of strain for undrawn PP (*1*), HDPE (*2*) and LDPE (*3*) [24, 30]. Duration of stress relaxation – 10 min; room temperature

function of strain. A comparison with the mechanical characteristic of the stress relaxation $-\Delta\sigma$ showed that the decrease of internal energy during stress relaxation corresponds to the dissipated heat. This means that no energy storage occurs during the stress relaxation.

5.5 Thermomechanics of Drawn Polymers

5.5.1 Drawn Amorphous Polymers

The most thorough deformation calorimetric investigation of drawn glassy polymers was carried out by Bonart et al. [34] on PET-films. The effect of the degree of cold-drawing of PET-films on the value and sign of thermal effects at small and fully reversible stretching was studied. Typical results are shown in Fig. 5.8. When the degree of cold drawing increases the endo-effect Q resulting from the stretching to the same strain decreases. With degrees of cold drawing larger than $\lambda = 4$ the heat effect changes its sign and stretching of the film is accompanied by liberation of heat. It means that due to cold drawing, the linear thermal expansion coefficient changes its sign from positive to negative. In spite of this change, the dependence of heat to work ratio η on strain is hyperbolic as predicted by the theory (Fig. 5.1). It is worth mentioning that the samples of a higher degree of cold drawing than 4 possess some crystallinity, because the cold drawing occurred near the glass transition. However, as shown in Chap. 7, cold drawing of PET-films at room

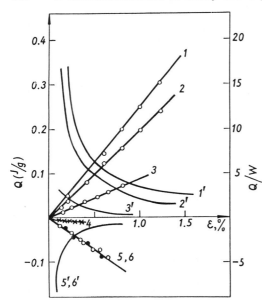

Fig. 5.8. Dependence of elastic heat Q (*1–5*) and heat to work ratio Q/W (*1'–5'*) on strain for PET of various degrees of drawing [34]. Degree of drawing: *1, 1'* – undrawn; *2, 2'* – $\lambda = 2$; *3, 3'* – $\lambda = 3$; *4* – $\lambda = 4$; *5, 5'* – $\lambda = 5$; *6, 6'* – $\lambda = 6$

temperature which leads only to the appeerence of so-called amorphous c-texture without any crystallization, also causes a change in sign [7]. Hence, although stretching of well-oriented PET-films at room temperature (which is well below glass transition) is accompanied by liberation of heat, the strain dependence of η as well as the elastic inversion of the internal energy correspond to the thermomechanical behavior of the solid with a negative thermal expansion along the stretching axis and is not characteristic of elastomers. The parameters of internal energy inversion and the values of β obtained with the aid of Eqs. (5.23) and (5.24) using values of ε_{inv} are listed in Table 5.4.

To explain the observed internal energy inversion in PET a thermodynamic model was proposed by Bonart et al. [34] in which the stretching of a two-phase crystalline polymer is simulated by an assembly of a parallel-connected perfect spring and a piston moving in a cylinder filled with an ideal gas. The model suggests that the spring simulates the behavior of crystallites and the gas of the amorphous regions. According to this thermodynamical model, the liberation of heat during stretching of drawn PET-film and inversion of internal energy is due to compression of amorphous regions in looped fibrils. Although the model permits us to explain formally the behavior of U, however, such simulation cannot be considered to have a sound physical meaning. In particular, the amorphous regions of PET are glassy at the experimental temperature (20°C) and there are no reasonable grounds for identifying their behavior with that of ideal gases. Besides, this approach is quite unable to account for the entire range of the thermoelastic phenomena in solid polymers and other solids because its underlying assumption is that the systems contain two types of regions with ideal energy and entropy elasticity. For these reasons this approach seems to be not very stimulating.

Table 5.4. Parameters of internal energy inversion on stretching drawn and compression undrawn polymers [25]

Polymer	Characteristics of samples[a]	ε_{inv} %	ε_{min} %	ΔU_{min} J/cm^3	$\beta_{\parallel} \times 10^5$ K^{-1}	E GPa	Poisson's ratio
Extension of drawn polymers							
LDPE	$\lambda = 4$; $T_{dr} = 20°C$	10.5	5.3	−0.43	−19.40	0.27	–
HDPE	$\lambda = 10$; $T_{dr} = 95°C$	4.5	2.2	−0.40	−7.50	1.70	0.56–0.63
HDPE	$\lambda = 20$; $T_{dr} = 95°C$	2.2	1.0	−0.45	−3.65	8.0	–
PP	$\lambda = 8$; $T_{dr} = 20°C$	3.4	1.8	−0.39	−6.30	2.35	0.55
PP	$\lambda = 30$; $T_{dr} = 140°C$	2.2	1.0	−0.22	−3.70	3.8	–
PA-6	Commercial film	5.0	2.4	−0.15	−8.20	0.55	0.50–0.55
PET	$\lambda = 5.4$; $T_{dr} = 20°C$ $\rho = 1.338 \, g/cm^3$	0.5	0.2	−0.012	−0.80	5.0	0.58
PET [34]	$\lambda = 6.0$; $T_{dr} = 85°C$ $\rho = 1.380 \, g/cm^3$	0.27	0.13	−0.015	−0.50[b]	14.0	–
Compression of undrawn polymers[c]							
LDPE	$w = 0.40$	−14.4	−7.2	−0.25	24.60	0.1	0.35
HDPE	$w = 0.55$	−8.5	−4.2	−0.54	14.60	0.6	–
PP	$w = 0.64$	−6.2	−3.2	−0.32	11.20	0.62	0.35
PS	amorphous	−4.0	−2.0	−0.38	6.80	2.0	0.33
PET	amorphous	−3.2	−1.5	−0.22	5.40	1.8	–
PET	$w = 0.40$	−3.2	−1.5	−0.22	5.40	1.9	–

[a] w – degree of crystallinity: T_{dr} – drawing temperature
[b] Values of ε_{inv}, ε_{min} and ΔU_{min} were computed according to expressions given in 5.2.2
[c] β and E obtained in extension and compression (small strain) experiments

Our analysis of the thermomechanics of solids did not use any specific structural models and, therefore, allows us to predict the thermomechanical behavior of any solids. It shows, in particular, that the inversion of internal energy is unrelated with two-phase structure of solids but is a rather typical characteristic of homogeneous solids. The only condition for the inversion is a negative coefficient of thermal expansion. Sufficiently oriented PET films do have a negative thermal expansion along the axis of drawing.

5.5.2 Drawn Crystalline Polymers

Drawn crystalline polymers have turned out to be extremely interesting thermomechanical materials. In an early study [27], it was found that cold drawing of PA-6 leads to a change in sign of the thermal effect accompanying reversible stretching of a cold-drawn sample i.e. absorbtion of heat of an undrawn sample changes to evolution of heat on stretching of a cold-drawn sample. Further comprehensive studies [7, 17, 30, 35] showed that the change in sign is observed in many crystalline polymers. Typical results are shown in Fig. 5.9. The origin of the sign change is determined by the nature of the negative thermal expansion of drawn crystalline polymers (see Chap. 3).

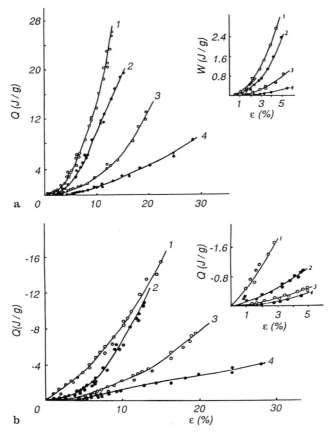

Fig. 5.9a,b. Mechanical work (W) and heat of elastic elongation (Q) as a function of strain for drawn crystalline polymers at room temperature [30, 35]. *1* – HDPE ($\lambda = 10$); *2* – PP ($\lambda = 8$); *3* – PA-6 (commercial film); *4* – LDPE ($\lambda = 4$)

For drawn crystalline polymers a considerable stress relaxation occurs after extension. The heat effects resulting from the stress relaxation are shown in Fig. 5.10. In contrast to the undrawn crystalline polymers, in this case only PA-6 and LDPE stress relaxation is accompanied by the liberation of heat. In two other polymers, HDPE and PP, stress relaxation is accompanied by absorption of heat. At the same time the relative stress relaxation $\Delta\sigma/\sigma$ for HDPE and PP is considerably higher. All these results seem to indicate that stress relaxation in these polymers is connected not only with conformational transformations in the amorphous regions of microfibrils but also with some structural transformations in these regions. The difference in the thermomechanical behavior of HDPE and PP in comparison with LDPE and PA-6 one can attribute to the difference in degree of crystallinity and structure of their amorphous regions.

Hence, the reversible deformation of drawn crystalline polymers may be accompanied by the considerable thixotropy effects. These effects are pro-

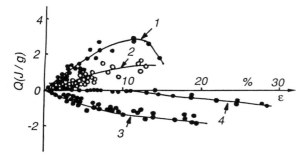

Fig. 5.10. Heat resulting from stress relaxation as a function of strain for drawn HDPE (*1*), PP (*2*), PA-6 (*3*), LDPE (*4*) [24, 30]. Duration of stress relaxation – 10 min, room temperature

nounced especially obviously in thermal effects resulting from contraction of the extended samples to the initial stress-free state. Figure 5.11 illustrates typical results for PP and HDPE. Absorption of heat during contraction has been observed only after small deformations while after larger extensions the sign of the heat effect during contraction changes very sharply. This inversion seems to indicate that some changes in the molecular mechanism of deformation occurs. The thixotropic effects depend also upon the time during which the sample was under stress.

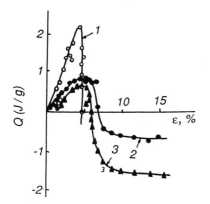

Fig. 5.11. Heat effect resulting from the elastic contraction as a function of strain at room temperature [30, 35]. *1* – HDPE ($\lambda = 20$, relaxation time under stress – 6 min); *2* – HDPE ($\lambda = 10$, relaxation time – 6 min); *3* – HDPE ($\lambda = 10$, relaxation time – 1 hour)

The inversion of the internal energy on elastic extension is a very important feature of the thermomechanical behavior for drawn crystalline polymers. Such a behavior is predicted by Eqs. (5.20) and (5.24) for solids having negative thermal expansivity. Such an inversion of the internal energy is really observed in drawn crystalline polymers. Figure 5.12 illustrates the inversion behavior for various drawn polymers and in Table 5.4 the main inversion parameters are tabulated. Although a similar inversion of the internal energy may also occur upon extension in elastomers (see Chap. 6) the sign of the inversion and its molecular origin is quite different. Indeed, for drawn crystalline polymers, ΔU at first decreases and only then begins to increase. In elastomers, such a change of ΔU cannot be observed in principle. It is important to point out that the inversion of the internal energy in crystalline polymers is determined only by the value of $\beta_{\parallel}$, while the intrachain conformational energy changes do not play any role in this case.

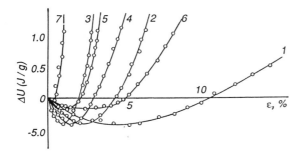

Fig. 5.12. Internal energy changes ΔU as a function of strain for drawn crystalline polymers [6, 25]. *1* – LDPE ($\lambda = 4$); *2* – HDPE ($\lambda = 10$); *3* – HDPE ($\lambda = 20$); *4* – PP ($\lambda = 8$); *5* – PP ($\lambda = 30$); *6* – PA-6 (commercial film); *7* – PET ($\lambda = 5.4$)

According to Eq. (5.20) the compression of isotropic solid polymers having positive thermal expansion must be accompanied by the internal energy inversion. ΔU inversion at compression has been estimated [6] to occur at strains of 5–15%. On compression, irreversible plastic deformations occur which prevent a correct experimental determination of ΔU. With inversion parameters, ΔU for isotropic polymers under compression may be calculated with formulas for inversion at elongation using the values of β and E obtained both for stretching films and compression of small cylindrical samples (Table 5.4). Finally, it must be pointed out that evolution of heat resulting from reversible stretching is not connected with any crystallization process in amorphous regions and is a thermodynamic consequence of their negative thermal expansivity.

Above T_g, moderately drawn crystalline polymers have an elasticity modulus of the order of a few GPa and are capable of being reversibly deformed by 15 to 25%. Since this deformation is concentrated in the amorphous regions, the true strain in these regions at typical values of crystallinity (30 to 70%) may be several times higher than these values. These reversible deformations of amorphous regions with the relatively small elasticity modulus must, of course, be due to stretching of the otherwise coiled conformations. Such transitions in drawn crystalline polymers (HDPE, PP, PA-6) have been registered by IR spectroscopy [36, 37]. However, in contrast to elastomers, these *gauche/trans*-transitions in amorphous regions of microfibrils occur simultaneously with variations of volume and intermolecular energy, which determine the elasticity modulus since the *cis-trans*-transitions alone cannot assure rather high values of elasticity modulus if the intermolecular interactions remain practically constant. This is true not only for weakly stretched elastomers but also for highly stretched macromolecules such as the portions of chains in the amorphous regions. For constant intermolecular interaction, the thermomechanics of strongly stretched oriented chains (the elasticity modulus may in this case be two to three orders of magnitude higher than that of unstretched elastomers) must be controlled by the intramolecular entropy and energy variations brought about by *cis-trans*-transitions.

The available data on the thermomechanical behavior of two-phase crystalline polymers prompt the conclusion that, in contrast to the classical elastomers, in crystalline polymers it is the intermolecular interaction which controls the thermomechanics of deformation both below and above the glass transition temperature of the amorphous regions. This peculiar thermomechanical behavior of crystalline polymers led to the conclusion [6, 24, 35], that, despite the localization of reversible deformations in amorphous regions, the submicroscopic size of crystallites and amorphous regions, as well as the molecular bonding of the amorphous and crystalline regions by means of the tie molecules create, during deformation of this microheterogeneous system, the conditions, in which it is principally impossible for the amorphous regions to be deformed only on account of shape variation (conformational elasticity) and they undergo complex deformations involving volume variation. These volume variations are typical of solid polymers rather than of elastomers and, what is even more essential, the free energy of deformation is related to these volume variations. In other words, the work of deformation is spent to overcome the volume elasticity of the amorphous regions. Although conformational changes (the *cis-trans*-transitions) also occur in this case, making considerable reversible deformations geometrically possible, the thermodymanics of such deformations is controlled by the intermolecular interactions between chains in the amorphous regions. These complex volume deformations of amorphous regions including elements of both conformational and volume elasticity are responsible for the unique thermomechanical properties of crystalline polymers.

The question of the elasticity modulus and other properties of amorphous regions of crystalline polymers at temperatures above T_g is one of the most complex in the analysis of crystalline polymers. So far there has been no reliable method for calculating the elasticity modulus of amorphous regions and all estimates have been, to a larger or lesser degree, based on comparison with elastomers. In view of the preceding, it becomes obvious that the conventional approach to the analysis of elasticity modulus of crystalline polymers based on treating them as crystallite-reinforced elastomers is oversimplified.

Now we are in a position to consider the relation between morphology of drawn crystalline polymers and their thermomechanical behavior. Numerous studies of the structure and properties of drawn crystalline polymers have led to the microfibrilar models of fibrous morphology [36, 38, 39]. According to these models the long and thin microfibrils are the basic elements of the fibrous structure. The microfibrils consist of alternating crystallites and amorphous regions. The axial connection between the crystallites is accomplished by intrafibrilar tie-molecules inside each microfibril. The adjacent microfibrils may be interconnected by interfibrillar tie-molecules. Following Takayanagi's approach [40] of calculating the modulus of elasticity of two-phase polymers, let us analyse the thermomechanical behavior of various structural models of drawn crystalline polymers. These models are schematically shown in Fig. 5.13.

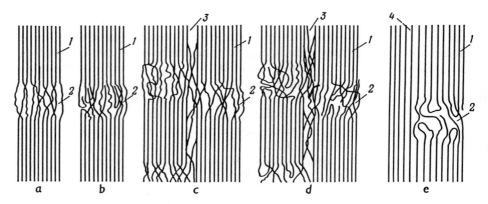

Fig. 5.13. Schematic diagrams showing the stuctural models of drawn crystalline polymers [36, 40, 41]. a – Hess–Statton–Hearle model; b – Hosemann–Bonart model; c – Hess–Statton–Hearle microfibrils with interfibrillar amorphous chains; d – Peterlin–Prevorsek model; e – "statistical crystal bridge" model. 1 – crystallite; 2 – amorphous intrafibrillar region; 3 – amorphous interfibrillar region; 4 – crystal bridge

If crystallites alternate with amorphous regions where all the molecules leaving the preceding crystallite extend through to the succeeding one, the situation is what is called the Hess-Statton-Hearl model (HSH). Microfibrils of this type consist of only alternating crystalline and amorphous regions. The thermomechanical behavior of such model is defined by the following equations:

$$E_{\parallel} = \frac{E_{AB}E_{CR}}{(1-w)E_{CR} + w\,E_{AB}} \tag{5.30}$$

$$\beta_{\parallel} = \beta_{AB}(1-w) + \beta_{CR}w \tag{5.31}$$

$$Q = T\beta_{\parallel}E_{\parallel}\varepsilon = \frac{T[\beta_{AB}(1-w) + \beta_{CR}w]E_{AB}E_{CR}}{(1-w)E_{CR} + w\,E_{AB}}\,\varepsilon \tag{5.32}$$

$$W = \frac{E_{\parallel}\varepsilon^2}{2} = \frac{E_{AB}E_{CR}}{2[E_{CR}(1-w) + E_{AB}w]}\,\varepsilon^2 \tag{5.33}$$

$$\frac{Q}{W} = \frac{2T[\beta_{AB}(1-w) + \beta_{CR}w]}{\varepsilon} \tag{5.34}$$

where the subscript "AB" relates to the amorphous intrafibrillar regions. Stretching of such a model at above T_g must always involve liberation of heat $Q < 0$, since β_{CR} and β_{AB} are both negative. Usually, $E_{CR} \gg E_{AB}$ and $\beta_{AB} \gg \beta_{CR}$, therefore one may neglect the contribution of crystallites into the thermomechanical properties in tension. The macrodeformation of an HSH microfibril must coincide with the long period variation in deformation, and the amorphous region density must increase.

In the case when only some of the macromolecules extend from one crystallite through the amorphous region to another and a partial returns to the same one, the model is the Hosemann-Bonart (HB) one. The thermo-mechanics of reversible deformation of an HB microfibril will be given by the same Eqs. (5.30)–(5.34) but the sign of $\beta_\parallel$ and, consequently Q, must be either positive or negative in this case. It is determined by the percentage of the tie-chains, and one may assume that if that is small or they are unsufficiently stretched, $\beta > 0$, and $Q > 0$. Although the tie-chains must have a negative coefficient of thermal expansion, the integral coefficient of the layer may still be positive. The relationship may, apparently, vary, but what is particularly important is that, unlike the HSH model, an HB model with a few tie-molecules may have a positive thermal effect in stretching. Where the number of tie-chains is small, the Poisson ratio $\left(\mu = -\frac{\varepsilon_\perp}{\varepsilon_\parallel}\right)$ is apparently less than 0.5, so the density of this layer must decrease in stretching. The macrodeformation of an HB microfibril must correspond to the deformation of the long period in elongation.

The investigations of structure and properties of drawn crystalline polymers have shown [38, 40] that the interfibrillar chains play an important role in their deformation behavior and strength. The models considered below comprise HSH and HB microfibrils combined into a macrofibril by interfibrillar chains that form amorphous interlayers between microfibrils. The interfibrillar chains are assumed to be strongly stretched, and hence $\beta_{AI} < 0$. The thermomechanics of such combined models are described by the following relations:

$$E_\parallel = E_{AI}X + \frac{(1-X)^2 E_{AB} E_{CR}}{(1-w-X)E_{CR} + w\,E_{AB}} \tag{5.35}$$

$$\beta_\parallel = \frac{\beta_{AI}X\,E_{AI}}{E_{AI}X + \dfrac{(1-X)^2 E_{AB} E_{CR}}{(1-w-X)E_{CR} + w\,E_{AB}}} + \frac{(1-X)E_{AB} E_{CR}}{E_{AI}X + \dfrac{(1-X)^2 E_{AB} E_{CR}}{(1-w-X)E_{CR} + w\,E_{AB}}}$$
$$\times \left[\frac{\beta_{AB}(1-w) + \beta_{CR}(w-X)}{(w-X)E_{AB} + (1-w)E_{CR}}\right] \tag{5.36}$$

$$Q = T\left\{\beta_{AI}X\,E_{AI} + (1-X)E_{AI}E_{CR}\left[\frac{\beta_{AB}(1-w) + \beta_{CR}(w-X)}{(w-X)E_{AB} + (1-w)E_{CR}}\right]\right\}\cdot\varepsilon \tag{5.37}$$

$$W = \left[E_{AI}X + \frac{(1-X)^2 E_{AB} E_{CR}}{(1-w-X)E_{CR} + w\,E_{AB}}\right]\cdot\frac{\varepsilon^2}{2} \tag{5.38}$$

$$\frac{Q}{W} = \left\{\frac{\beta_{AI}X\,E_{AI}}{E_{AI}X + \dfrac{(1-X)^2 E_{AB} E_{CR}}{(1-w-X)E_{CR} + w\,E_{AB}}} + \frac{(1-X)E_{AB} E_{CR}}{E_{AI}X + \dfrac{(1-X)^2 E_{AB} E_{CR}}{(1-w-X)E_{CR} + w\,E_{AB}}}\right.$$
$$\left.\cdot\left[\frac{\beta_{AB}(1-w) + \beta_{CR}(w-X)}{(w-X)E_{AB} + (1-w)E_{CR}}\right]\right\}\frac{2T}{\varepsilon} \tag{5.39}$$

where X is volume fraction of interfibrillar amorphous regions (subscript AI).

The HSH model with interfibrillar layers is characterized only by heat liberation under stretching. Thus, the thermomechanical behavior of a simple HSH microfibril and an HSH microfibril with interfibrillar layers is qualitatively the same and the difference may only be quantitative. Both inter- and intrafibrillar regions must be densified under tension. A major difference of the behavior of microfibrils with interfibrillar amorpous regions seems to be the non-coincidence of the macrodeformation with the deformation of the long period.

In the HB model with interfibrillar amorphous regions, the sign of heat on stretching depends on the ratio between β_{AI} and β_{AB} and may be a function of elongation even at small strains. The constriction during stretching may in this case be larger than elongation and this will mean $\mu > 0.5$. This is indicative of volume decrease in stretching of such a microheterogeneous system even though according to the classical theory of elasticity the isothermal Poisson ratio of a homogeneous solid cannot exceed 0.5 [42], for heterogeneous systems with anisotropy of properties this limitation may not hold [43]. Direct microscopic measurements of Poisson's ratios of a number of drawn crystalline polymers (HDPE, PP, PET, PA-6) [30] (Table 5.4) and comparison of them with those determined from measured elasticity modulus [44] shows that in such polymers the value of μ may be higher than 0.5, which shows that the proposed deformation mechanism is quite probable. The last one of the models discussed herein is presently the most widely accepted and experimentally supported model of a drawn crystalline polymer (the Peterlin-Prevorsek (PP) model) [38, 39].

The fact that macrodeformation does not coincide with deformation of the long period in specimens with a large content of interfibrillar amorphous regions seems to be due to reversible slip of microfibrils relative each other [7] and, therefore, must bring about an interesting thermomechanical effect. The slip of microfibrils must result in an additioanl heat release, since the work of internal friction must, at least partly, be converted to heat. On contraction, the microfibrils must return to their original position, i.e. slip back, which must again involve heat liberation. One may assume, therefore, that when such polymers are stretching, the beginning of slipping must be manifested by a sharp increase of the exo-effect and, further, the sign of the thermal effect of contraction may be reversed from endo to exo.

The above treatment is formally similar to the model analysis of elasticity moduli of two-phase systems proposed by Takayanagi [40] and currently used for the calculation of elasticity moduli of crystalline polymers. However, although Takayanagi's phenomenological models do describe the mechanical properties of two-phase systems closely enough, they fail to provide any information about the microstructure of crystalline polymers and their behavior on deformation. The treatment considered above indicates that the thermomechanical analysis may yield such information. Therefore, our next step will be a thorough analysis of the thermomechanical behavior of drawn crystalline polymers with their morphology and those morphological transformations which accompanied reversible deformation of drawn crystalline polymers.

As can be seen from Fig. 5.9 and Table 5.4 all the polymers studied release heat in stretching and have inversion of internal energy. None of the polymers investigated were observed to absorb heat during stretching along the orientation axis. It would seem to be reasonable to conclude from these results that the HSH-model is the one most closely describing the structure of drawn polymers. However, since no X-ray investigations of the long period were carried out simultaneously it is impossible to draw any definite conclusions about the absence or presence of interfibrillar amorphous regions from the thermomechanical data alone. On the other hand, analysis of stretching and contraction heat shows (Fig. 5.14) their variations to be consistent with the the slippage of microfibrils with interfibrillar amorphous regions in thermal effects which was discussed above. These results may be regarded as an indirect demonstration of the presence of interfibrillar amorphous regions in drawn PE and PP. The sharpness of the reversal of sign of contraction heat is strongly dependent on the draw ratio and this is not unexpected if one considers that length differences among interfibrillar chains are smaller when the drawing is greater. The Poisson ratios (Table 5.4) also support this conclusion. But the absence of an endo-effect in stretching is not an unambiguous indication of a large number of tie-molecules in the intrafibrillar interlayers, because it may well be that these effects are overwhelmed by the exo-effects of deformation of the interfibrillar chains. X-ray analysis of drawn PE showed [36, 40] that low-molecular-weight specimens have almost no interfibrillar amorphous regions, whereas in high-molecular-specimens the interlayers may be large. Accordingly, one would expect drawn specimens of low-molecular-weight PE whose structure apparently corresponds to the HB smooth microfibril model, to absorb heat on stretching, while the high-molecular-weight specimens, whose structure corresponds rather to the PP model, to liberate heat under tension.

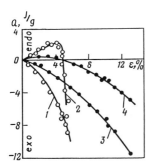

Fig. 5.14. Heat of elastic elongation (*1, 3*) and contraction (*2, 4*) as a function of strain for HDPE (*1, 2*) with degree of cold-drawing $\lambda = 20$ and for PP (*3, 4*) with degree of cold-drawing $\lambda = 8$ [7]

The correlation of the thermomechanical behavior with molecular parameters (MW, MWD and degree of branching) and orientation parameters (temperature and draw ratio) has been established for PE [45–49]. The study included a calorimetric investigation of thermomechanics combined with an

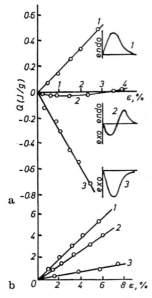

Fig. 5.15. Heat of elastic elongation (**a**) and deformation of long period (**b**) versus macroscopic deformation of PE-1 (*1*), PE-2 (*2*) and PE-3 (*3*). Characteristics of the samples are shown in Table 5.5. The form of corresponding heat effects is also shown [45, 46]

X-ray analysis and the analysis of molecular mobility with the aid of a molecular probe. This complex study showed that the sign and values of the thermal effects on stretching indeed depend on the molecular weight and strain as shown in Fig. 5.15. Combined analysis of the thermomechanical behavior and structural changes showed that the heat evolution seems to be due to a compression of the interfibrillar amorphous molecules and the heat absorption to expansion of the intrafibrillar regions containing a small number of tie-molecules. The resultant thermal effect depends on the relative ratio of the compressed and expanded amorphous regions. Based on the thermal effect and long-period variations in deformation as shown in Fig. 5.15, a method was proposed for the quantitative evaluation of the volume fraction occupied in PE by interfibrillar tie-molecules. The method involves summing up the contributions of the thermal effects of intra-and inter-fibrillar amorphous regions. The structural and thermomechanical characteristics of PE analysed with this method are tabulated in Table 5.5. As can be seen from the results, low-molecular-weight specimens are practically free of interfibrillar amorphous regions. An increase in molecular weight causes a monotonous increase of the fraction of interfibrillar regions for comparable temperatures and degrees of drawing. Furthermore, their fraction increases with the degree of drawing at a given temperature and with decreasing draw temperature. These experimental findings are quite consistent with theoretical estimates [41] and X-ray structural measurements [50]. Thus, the thermomechanical behavior of drawn PE can be described well in terms of the PP model.

Table 5.5. Structure and thermophysical characteristics of drawn PE [7]

Specimens and its thermal history	w %	l nm	L nm	X_{AI} %	$1 - (w - X_{AI})$ %	E GPa	$\beta_\parallel \times 10^5$ K^{-1}	Thermal effect of initial elongation
PE-1 $\bar{M}_\eta = 4.2 \times 10^4$; $\bar{M}_\eta/\bar{M}_n = 3.0$; 0.1 CH$_3$/100 C $\lambda = 6$	72	21	26	0	28	4.4	1.1	*endo*
PE-2 ($\bar{M}_\eta = 3 \times 10^5$; $\bar{M}_\eta/\bar{M}_n = 1.95$ 0.1 CH$_3$/100 C) $\lambda = 6$	63	23	28	12	25	1.43	−0.5	*exo*
Isometric annealing	79	35	40	10	11	1.43	−	*endo*
Free annealing	78	32	38	8	14	0.63	−	*endo*
Annealing under pressure of 0.7 GPa at 264°C	98	100	−	6	0	0.38	−	*endo*
$\lambda = 15$	68	25	28	23	9	5.00	−1.1	*exo*
Isometric annealing	81	27	28	17	2	4.90	−	*exo*
Free annealing	81	39	41	15	4	3.40	−	*endo-exo*
PE-3 ($\bar{M}_\eta = 7 \times 10^6$ $\bar{M}_\eta/\bar{M}_n = 2.5$; 0.1 CH$_3$/100 C) $\lambda = 5.5$	44	19	29	27	29	0.70	−10.0	*exo*

Symbols: λ – draw ratio; l – longitudinal dimension of crystals according to WAXS; L – long period according to SAXS; X_{AI} – volume fraction of interfibrillar amorphous regions; $1 - (w + X_{AI})$ – volume fraction of intrafibrillar amorphous regions. The degree of crystallinity of a specimen was determined from calorimetric measurements of fusion heat. In all cases the cold-drawing temperature was 108°C. Annealing temperature was 132°C under atmospheric pressure; annealing time 30 min

Annealing free or fixed drawn crystalline polymers is accompanied by considerable changes in their supermolecular structure and, as a consequence, in their mechanical properties [36, 38, 40]. The changes are most radical in highly stretched interfibrillar chains and tie-chains in the intrafibrillar amorphous interlayers. Such abrupt changes can affect the thermomechanical behavior. The data represented in Fig. 5.16 show that the changes are most tangible in highly drawn specimens after annealing in the free state, which causes reversion of the sign of the extension heat. The variations observed under all annealing conditions may be explained due to desorientation and loosening of highly stretched interfibrillar molecules. As a result of these changes, the contribution of the endothermal effect may become predomi-

nant during the initial stretching. However, with increasing deformation the
endo-effect turns to *exo*. Free state annealing considerably reduces the elasticity modulus. Although in isometric annealing the elasticity modulus remains
practically unchanged, the thermal effect of stretching and respectively $\beta_\parallel$
vary considerably and in some cases sign inversion may follow. Hence, the
thermomechanical phenomena occurring in deformation of annealed samples
reflect closely the changes brought out during annealing.

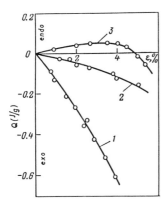

Fig. 5.16. Effect of annealing of HDPE ($\lambda = 15$)
on the heat of elastic extension [7]. *1* – initial unan-
nealed sample; *2* – after isometric annealing at 132°C
for 30 min; *3* – after free annealing at 132°C for 30
min

High pressure annealing of drawn PE results in formation of the so-called
extended-chain crystals [51, 52]. The crystallite dimension along the extension axis is over 100 nm and the density is 0.996–0.997 g/cm³, the heat of
fusion $Q_f = 285$ J/g corresponding to a crystallinity of 96–98%. Such specimens usually have a laminar structure in which the crystalline layers of
large longitudinal and transverse dimensions alternate with unordered layers where only a few chains are tie-chains. As seen from the Table 5.5, such
specimens are characterized by a small elasticity modulus and endothermal
effect on stretching. These results show that large crystallites and a high
degree of crystallinity are not sufficient for assuring high mechanical properties of drawn crystalline polymers. The thermomechanical reason for this
is that the work of deformation is mostly spent for overcoming the weak intermolecular interactions in the unordered regions whereas the deformation
of valence angles and bonds of the transconformers of crystallites remains
largely unrealized.

Thus, the thermomechanical behavior of annealed drawn polymers can
also be accounted for in terms of structural changes predicted by PP microfibrillar model. It may be supposed that the model will be valid for those
crystalline polymers, which in an undrawn state possess lamellar spherulites
since their transformation to a fibrillar structure usually involves formation
of lamellar microfibrils with a limited number of tie-chains.

Using cold-drawing of glassy polymers it is sometimes possible to obtain
an amorphous texture of such crystallizable polymers as PET [53] where
most chains are in *trans*-conformational state. The chains are quite regu-

larly shifted along the texture axis, as they are in crystalline polymers, but the azimuthal rotation and spacing of the chains are strongly distorted. The coefficient of thermal expansion of such a texture must obviously be negative. As it follows from Table 5.4 (specimen 7) the elasticity modulus of an amorphous texture obtained by drawing at 20°C is 2.5 times higher than that of a non-drawn specimen, and the texture has a negative value of $\beta_{\parallel}$. This is consistent with the conception that such a texture contains a large number of *trans*-conformations. The data of Bonart et al. [34] (specimen 8) show that as the temperature of drawing is raised to 85°C, the modulus E is sharply increased and the coefficient of thermal expansion decreased but remains negative. According to the density data, the texture is a partly crystalline one. The increase of the modulus is related to the increased orientation of *trans*-isomers during crystallization. According to the IR-data, the total number of the *gauche*-isomers is below 10% ([34], Fig. 11), which suggests that there cannot be a large number of folds in the amorphous regions and allows such a texture to be regarded as an example of drawn systems in which the intrafibrillar amorphous regions predominantly contain the tie-chains in the stretched conformations. Crystallization of the amorphous texture may generate boundaries between individual microfibrils and cause interfibrillar amorphous regions to appear. A brief isometric annealing at elevated temperatures (10 min, 175°C) somewhat reduces the elasticity modulus and brings the coefficient to zero, which is apparently due to the *trans-gauche* transition in the amorphous regions and appearence of a certain number of folds.

Unlike PET, PBT cannot be obtained in the completely amorphous state even by rapid quenching. Therefore, the specimens drawn at room temperature are semicrystalline and their stretching both at room temperature and at elevated temperatures (120–140°C) involves heat absorbtion. As a result one may conclude that a microfibrillar structure with a small number of tie-chains in the intrafibrillar amorphous regions with a small percentage of interfibrillar amorphous chains is the dominant one in drawn PBT. It is worth mentioning that the drawn PBT specimens exhibit the unusual phenomenon of polymorphism in deformation, which manifests itself in the change of the lattice type at a certain stress [54]. This phase transition occurs as a result of variation of the lattice spacing along the macromolecuar axis "c" due to the ultimate elongation of the chains under mechanical stress. In calorimetric experiments this transition manifests itself as sharp slope change of the elastic heat versus elastic force relationship. The unusual thermomechanical behavior is shown in Fig. 5.17. A considerable hysteresis of heat (as well as stress) occurs during contraction of the sample due to re-transition into the initial crystalline structure.

The above analysis of the thermomechanical and structural data was concerned with drawn crystalline polymers usually having an elasticity modulus of only a few percent of the theoretical modulus of the crystal lattice. However, at least in one case – the highly drawn PP [55] – the elasticity modulus was as high as 25–30% of that of the crystal lattice. Nevertheless, it is possible to account unequivocally for the thermomechanical properties

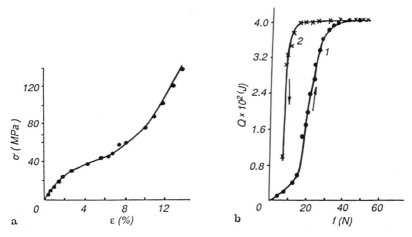

Fig. 5.17. Stress-strain curve (**a**) and elastic heat Q (**b**) as a function of elastic force f for drawn PBT film. *1* – elongation; *2* – contraction

and structural changes occurring in original and annealed samples during reversible deformation in terms of the PP model.

In conclusion, we will consider in brief the thermomechanical properties of superdrawn polymers. Articles of this kind have been obtained in recent years [41] from PE, PP and POM polymers by colddrawing, hydrostatic extrusion in the solid state and by controlled crystallization of a flow of polymer melt or solution. From the standpoint of thermomechanics, two characteristics of such superdrawn polymers have great importance: the very high elasticity modulus which may be as high as $E_\parallel > 0.5$ Ecr at $-150°$C and $(0.3-0.35)$ Ecr at $20°$C and the negative coefficient $\beta_\parallel$. At the same time the super-drawn polymers are characterized by morphological and structural features typical of standard well-drawn polymers [56], viz. the clearly outlined fibrillar structure and high concentration of lamellar crystallites. A morphological feature of superdrawn polymers which is most essential is that they contain crystalline regions whose length is much greater than the long period.

A "statistical crystal bridge" model (Fig. 5.13) has been proposed [57, 58] for describing the properties of super-drawn polymers. Transition from the usual modulus to a high modulus is a continuous process determined by the fraction of the tie-molecules in the amorphous regions and their degree of orientation. Neither the PP nor the "statistical crystal bridge" models presume the existence of very large crystals.

The thermomechanical data concerning the thermomechanical behavior of super-drawn polymers are shown in Fig. 5.18. Despite the absence in super-drawn crystalline polymers of extended chain crystals, their properties are quite comparable with those of crystal lattices. One can suggest that the ultimate properties are a result of the existence of perfectly oriented chains in intra- and interfibrillar amorphous regions, which are a kind of "shunt" assuring transmission of stress not via a weak intermolecular interaction but

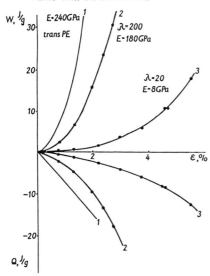

Fig. 5.18. Elastic work W and heat Q as a function of strain. 1 – crystalline $trans$-PE (calculated according to $Q = \beta TE\varepsilon$ with E $= 240\,$GPa, $\beta = -1.3 \times 10^{-5}\deg^{-1}$); 2 – HDPE ($\lambda = 200$, E $= 180\,$GPa); 3 – HDPE ($\lambda = 20$, E $= 8\,$GPa)

through the chain skeleton. If the intercrystalline intra- and interfibrillar bridges are really crystalline in the conventional sense of the word, the super-drawn polymers must be expected to have an excellent dimensional stability up to the melting point due to the small value of $\beta_{\|}$. This is indeed observed in super-drawn PE in a temperature range from -150 to $+50°$C [60, 61], but above $50°$C the negative coefficient begins to rise rapidly. This has been attributed to high-temperature annealing ($120°$C). However, this fact may be explained in a different way if one assumes that the inter-crystallite bridges are not truly crystalline but are made up of ultimately oriented chain portions. In contrast to the dimensional variations with the temperature range of $-150°$ to $+50°$C, those occurring above $50°$C must be irreversible – and this is so indeed [59]. Similar effects are also observed in high-modulus PP fibers [55, 61].

As shown in Chap. 3, drawn polymers posess a strong anisotropy of thermal expansivity. Using Eq. (3.38) one can obtain the linear coefficient of thermal expansivity β_φ at any angle to the draw direction. The corresponding modulus of elasticity E_φ can be calculated according to the following expression [62]

$$E_\varphi = E_{\|}/(\cos^4 \varphi + b\sin^2 2\varphi + c\sin^4 \varphi) \qquad (5.40)$$

where $c = E_{\|}/E_\perp$, $b = E_\perp/E_{45°} - [(1 + c)/4]$. Figure 5.19 shows that dependencies of the elastic heat on elastic force follow Eq. (5.11) in any direction. Similar behavior has been observed for drawn PET and PA-6 [24, 30]. Dependencies of heat to work ratios upon strain for all these drawn polymers support the idea that in any direction to the draw axis the thermomechanical behavior is controlled by the intermolecular interaction [7, 25, 30].

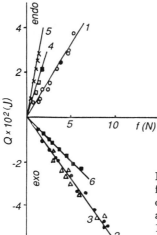

Fig. 5.19. Elastic heat Q as a function of elastic force f for LDPE [24, 30]. *1* – isotropic; *2* – drawn; *3* – redrawn; *4* – at an angle 45° to the draw direction; *5* – at an angle 90° to the draw direction; *6* – at an angle 15° to the draw direction

5.6 Microphase-Separated Block Copolymers with a Solid Matrix

The first loading of microphase-separated block copolymers with a rather high content of the hard phase often produces an initial rapid rise in stress (Chap. 6). Thermomechanical investigations [63, 64] have shown that such behavior is a result of solid-like deformation of the stress-supported rigid phase. Figure 5.20 demonstrates the thermomechanical characteristics for various block copolymers as a function of strain. For a comparison, data for LDPE are also included. These dependencies are really typical for solid polymers and they support the idea indeed that the initial rapid rise in stress is a consequence of the deformation of the stress-supporting rigid phase.

The continuity of the phase seems to depend strongly on the rigid block content and its morphology. Two ways are especially useful for changing the morphology of block copolymers: casting of films from various solvents and processing technique. Figure 5.21 shows the dependence of β determined from the heat effects in the initial elastic deformation of films and modulus of elasticity E on the solubility parameters of the solvent used. These results agree rather well with a general concept of the role of the solvent selectivity in forming a network of rigid glassy domains [65]. It has been well established that the thermodynamic quality of the solvent determines the composition of the coexisting phases and the possibility of formation of a continuous network.

Finally SBS block copolymers with "single crystal" morphology (Chap. 6) showed differing thermomechanical behavior along the cylinder's axis, and perpendicular to the extrusion direction.

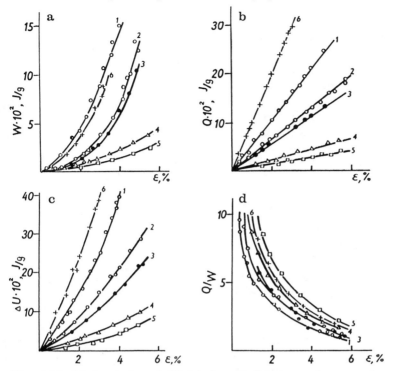

Fig. 5.20. Mechanical work W (**a**), heat Q (**b**), internal energy changes ΔU (**c**), and heat to work ratio Q/W (**d**) as a function of strain for thermoelastoplastics: polyurethane (*1*), polyester-polyether (*2*), SBS (Solprene 406) (*3*), SBS (Solprene 411) (*4*), SBS (DST-30) (*5*). Data for LDPE (*6*) are also shown [63]

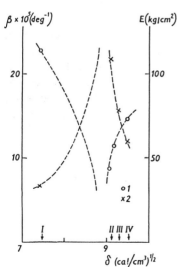

Fig. 5.21. Effect of the solvent solubility parameter δ on the linear coefficient of thermal expansion β (*1*) and modulus of elasticity E (*2*) of films of linear SBS thermoelastoplastics with 28.3% PS obtained from solutions. The solvents are indicated on the absciss axis: *I* – *n*-heptane, *II* – tetrahydrofurane, *III* – benzene, *IV* – chlorobenzene [63]

5.7 Filled Solid Polymers

Fine solid fillers markedly affect not only the physical and mechanical properties of elastomers but of solid polymers as well. A study of the interaction of a glassy plasticized PVC-matrix with reinforcing (silica) and inactive (chalk) fillers by deformation calorimetry has shown that filling of PVC with silica is accompanied by a considerable increase of W, Q and ΔU on stretching [66]. The same amount of the inactive filler does not practically affect the thermomechanical properties of PVC. Therefore, fine silica promotes a structuring of the PVC-matrix and the change in the internal energy plays a decisive role in reinforcement of the glassy PVC.

5.8 Biopolymers

Using deformation calorimetry, Ebert et al. [67–69] studied the change in the internal energy of a number of solid poly-α-aminoacids, such as poly-L-alanin and its copolymers with S-cystien, poly-L-lysin and poly-γ-methyl-L-glutamate. Most of these studies were devoted to the determination of irreversible changes of internal energy as a result of conformational changes accompanying irreversible deformation of these poly-α-amino acids. Up to 1% stretching, reversible energy–elastic behavior is predominant: stretching of the films and fibers of poly-α-amino acids is accompanied by *endo*-effects and contraction by *exo*-effects. The linear thermal expansion coefficients obtained from calorimetric measurements are in a good agreement with the data obtained by linear dilatometry. One can conclude that the general thermomechanical behavior of solid poly-α-amino acids is quite consistent with the thermomechanics of other solid polymers.

Godovsky et al. [70] investigated the thermomechanics of a water-gelatin system. Gelatin which is obtained from nonenzymatic denaturation of collagen can be considered as a copolymer. The thermomechanical behavior of water-gelatin systems strongly depends on the amount of water. If the water content is less than 25%, the samples exhibit typical glass like behavior at room temperature. They are reversibly strained by only 1–2% and exhibit heat absorbtion. At 25% of sorbed water, glass-to-rubber transition occurs. The internal energy of films containing 50–72% water decreases on stretching, which is connected with a negative energy contribution in the rubberlike water-gelatin systems. The thermomechanical heat inversion was not observed in these systems.

References

1. Landau LD and Lifshitz EM (1976) Statistical physics (in Russian), Nauka Moscow
2. Rummer YuB and Ryvkin MSh (1977) Thermodynamics, Statistical Physics and Kinetics (in Russian), Nauka Moscow
3. Callen HB (1960) Thermodynamics, Wiley New York
4. Reiss H (1965) Methods of Thermodynamics, Blaisdell New York
5. Sharda SC and Tschoegl NW (1974) Macromolecules 7: 882
6. Godovsky YuK (1982) Thermal Physics of Polymers (in Russian), Khimiya Moscow
7. Godovsky YuK (1982) Colloid Polym Sci 260: 461
8. Kilian H-G (1982) Colloid Polym Sci 260: 895
9. Lyon RE and Farris RJ (1987) Polymer 28: 1127
10. Sneddon IN (1974) The linear theory of thermoelasticity, Springer Berlin Heidelberg New York
11. Nadai A (1963) Theory of Flow and Fracture of Solids, McGraw-Hill New York
12. Trainor A and Haward RN (1974) J Mater Sci 9: 1243
13. Gilmor IW, Trainor A and Haward RN (1978) J Polym Sci Polym Phys Ed 16: 1277
14. Gilmor IW, Trainor A and Haward RN (1978) J Polym Sci Polym Phys Ed 16: 1291
15. Butyagin PYu, Garanin VV and Kuznetzov AR (1974) Vysokomol Soedin A16: 327
16. Ossi PM, Bottani CE and Rossitto F (1978) J Phys C Solid State Phys 11: 4921
17. Volodin VP and Gudimov SYu (1984) Solid State Phys (FTT) 26: 1563
18. Anisimov SP, Volodin VP, Orlovsky IYu and Fedorov YuN (1978) Solid State Phys (FTT) 20: 77
19. Fedorov YuN (1983) Vysokomol Soedin B25: 912
20. Rodriguez EL and Filisko FE (1982) J Appl Phys 53: 6536 (1982)
21. Rodriguez EL and Filisko FE (1986) Polym Eng Sci 26: 1060
22. Rodriguez EL (1988) J Polym Sci Polym Phys Ed 26: 459
23. Wright OB and Phillips WA (1984) Phil Mag B50: 63
24. Godovsky YuK (1976) Thermophysical Methods of Polymers Characterization (in Russian), Khimiya Moscow
25. Godovsky YuK (1986) Adv Polym Sci 76: 31
26. Adams GW and Farris RJ (1988) J Polym Sci Polym Phys Ed 26: 433
27. Müller FH (1969) Thermodynamics of deformation. Calorimetric investigation of deformation processes. In: Rheology, vol 5, p 417, Academic New York
28. Goritz D (1986) J Polym Sci Polym Phys Ed 24: 1839
29. Andrianova GP, Arutyunov BA and Popov YuV (1978) J Polym Sci Polym Phys Ed 16: 1139
30. Godovsky YuK (1972) Doctoral thesis, Karpov Institute of Physical Chemistry Moscow
31. Krigbaum WR, Roe R-J and Smith KJ Jr (1964) Polymer 5: 533
32. Heise B, Kilian H-G and Pietralla M (1977) Progr Colloid Polym Sci 62: 16
33. Lohse DJ and Gaylord RJ (1978) Polym Eng Sci 18: 512
34. Morbitzer L, Hentze G and Bonart R (1967) Kolloid Z Z Polym 216/217: 137
35. Godovsky YuK, Slonimsky GL, Papkov VS and Dikareva TA (1970) Mechan polym N5: 785
36. Marikhin VA and Myasnikova LP (1977) Supermolecular Structure of Polymers (in Russian), Khimiya Leningrad
37. Pakhomov PM, Shermatov M, Korsukov VE et al (1976) Vysokomol Soedin A18: 132

38. Peterlin A (1987) Colloid Polym Sci *265:* 357
39. Prevorsek DC, Tirpak GA and Heged PJ (1974) J Macromol Sci Phys *B9:* 733
40. Ward IM (ed) (1975) Structure and properties of oriented polymers, Wiley New York
41. Ciferri A and Ward IM (ed) (1979) Ultra-High Modulus Polymers, Applied Science Publishers London
42. Landau LD and Lifshitz EM (1974) Theory of Elasticity, Nauka Moscow
43. Rabinovich AL (1970) Introduction to Mechanics of Polymer Composites, Nauka Moscow
44. Ward IM (1972) Mechanical Properties of Solid Polymers, Wiley Interscience London
45. Chvalun SN et al (1978) Vysokomol Soedin *B20:* 672
46. Chvalun SN et al (1980) Vysokomol Soedin *B22:* 359
47. Chvalun SN et al (1981) Vysokomol Soedin *A23:* 1381
48. Chvalun SN et al (1981) III Internat Symp Man-Made Fibers, Kalinin, Preprints, vol 1, pp 116–121
49. Karpova SG et al (1983) Vysokomol Soedin *A25:* 2435
50. Marikhin VA, Myasnikova LP and Viktorova NL (1976) Vysokomol Soedin *A18:* 1302
51. Zubov YuA, Selikhova VI and Konstantinopolskaya MB (1972) Vysokomol Soedin *A14:* 2090
52. Zubov YuA, Ozerin AN and Bakeev NF (1975) Dokl Akad Nauk SSSR *221:* 121
53. Kazaryan LG and Tsvankin DYa (1965) Vysokomol Soedin *A7:* 80
54. Brereton MG, Davis GR, Jakaways R et al (1978) Polymer *19:* 17
55. Godovsky YuK et al (1981) III Internat Symp Man-Made Fibers, Kalinin, Preprints, vol 1 (suppl), pp 66–72
56. Capaccio G (1981) Colloid Polym Sci *259:* 23
57. Gibson AG, Davies GR and Ward IM (1978) Polymer *19:* 683
58. Ward IM (1979) Vysokomol Soedin *A21:* 2553
59. Mead WT and Porter RS (1976) J Appl Phys *47:* 4278
60. Gibson AG and Ward IM (1979) J Mater Sci *14:* 1838
61. Zubov YuA, Bakeev NF, Kabanov VA et al (1981) III Internat Symp Man-Made Fibers, Kalinin, Preprints, vol 1 (suppl), p 53
62. Risyuk BD and Nosov MP (1978) Mechanical Anisotropy of Polymers (in Russian), Naukova Dumka Kiew
63. Godovsky YuK (1984) Makromol Chem Suppl *6:* 117
64. Godovsky YuK et al (1984) Paper presented at Internat Symp Rubber-84, Moscow, vol A2, paper A48
65. Noshay A and McGrath JE (1977) Block Copolymers. Overview and Critical Survey, Academic New York
66. Godovsky YuK, Bessonova NP and Guzeev VV (1983) Mechan polym *N4:* 605
67. Ebert G, Knispel G and Müller FH (1980) Colloid Polym Sci *258:* 495
68. Ebert G et al 1980 Progr Colloid Polym Sci *67:* 175
69. Ebert G, Maeda A and Müller FH (1982) Colloid Polym Sci *260:* 404
70. Godovsky YuK, Malzeva II and Slonimsky GL (1971) Vysokomol Soedin *A13:* 2768

6 Thermomechanics of Molecular Networks and Rubberlike Materials

On simple elongation, rubberlike materials are capable of undergoing very large reversible elastic deformations. The modulus of elasticity, which unlike solids is strongly dependent on deformation, is some order of magnitude lower than the bulk modulus. Unlike the solids, the modulus of elasticity of the deformed networks and rubberlike materials is proportional to the absolute temperature (excluding very initial deformations). A very striking feature of the thermomechanical behavior of elastomers is a strong dependence of the linear thermal expansivity on deformation: the initial positive thermal expansion decreases drastically with deformation and in the vicinity of 0–10% deformation the expansion becomes negative and at moderate deformations it reaches the value typical for gases. All these facts demonstrate that the thermomechanical behavior of rubberlike materials differs in principle from that of solids. High elastic deformations are characteristic only for those deformation modes which are connected with the elasticity of the form. Since the volume compressibility of rubberlike materials is very small (the same order as for liquids) the thermodynamics of their uniform (volume) deformation is similar to the uniform deformation of solids and liquids.

6.1 Thermomechanics of Molecular Networks (Theory)

6.1.1 Thermomechanics of Gaussian Networks

The classical theory of rubber elasticity for a Gaussian polymer network which took into account not only the change of conformational entropy of elastically active chains in the network but also the change of the conformational energy, led to the following equation of state for simple elongation or compression [1, 2]

$$f = vkTL_0^{-1}(\langle r^2 \rangle / \langle r^2 \rangle_0) \left(\lambda - \frac{V}{V_0 \lambda^2} \right) \qquad (6.1)$$

or alternatively

$$f = vkTL_i^{-1}(\langle r^2 \rangle_i / \langle r^2 \rangle_0)(\alpha_* - \alpha_*^{-2}) \qquad (6.2)$$

where L_0 and V_0 are length and volume of the network at zero force, zero pressure and temperature T; L and V are the corresponding quantities at force f, pressure P and temperature T; $\lambda = L/L_0$ is the elongation (compression), ν is the number of elastically active chains in the network, k the Boltzmann constant, $\langle r^2 \rangle$ the mean square end-to-end distance of the network chains in volume V_0, $\langle r^2 \rangle_0$ that of the corresponding free chains, $\langle r^2 \rangle_i$ the mean square end-to-end distance of a network in the undistorted state of volume V and $\alpha_* = L/L_i$, L_i the length of the undistorted sample at volume V. The value of α_* differs negligibly from λ. Below we will use the equation of state only in the form of Eq. (6.1).

According to the molecular theory of rubber elasticity, undeformed polymeric chains of elastic networks adopt random conformations or spatial arrangemens in the bulk amorphous state. The stress resulting from the deformation of such networks originates within the elastically active chains and not from interactions between them. It means that the stress exhibited by a strained network is assumed to be entirely intramolecular in origin and intermolecular interactions play no role in deformations (at constant volume and composition).

Making use of the equation of state in Eq. (6.1), the work, heat and internal energy change at V, T = const. can be derived [6]

$$W_{V,T} = \frac{C}{2} \frac{(\lambda - 1)}{\lambda} (\lambda^2 + \lambda - 2) \tag{6.3}$$

$$Q_{V,T} = T(\Delta S)_{V,T} = -\frac{C}{2} \left(1 - T \frac{d \ln \langle r^2 \rangle_0}{dT} \right) \frac{(\lambda - 1)}{\lambda} (\lambda^2 + \lambda - 2) \tag{6.4}$$

$$\Delta U_{V,T} = \frac{C}{2} T \frac{d \ln \langle r^2 \rangle_0}{dT} \frac{(\lambda - 1)}{\lambda} (\lambda^2 + \lambda - 2) . \tag{6.5}$$

In these equations $C = \nu k T L_0^{-1} \langle r^2 \rangle / \langle r^2 \rangle_0$.

The characteristic parameters

$$\eta_{V,T} = (Q/W)_{V,T} = -1 + T \frac{d \ln \langle r^2 \rangle_0}{dT} = (f_s/f) \tag{6.6}$$

$$\omega_{V,T} = (\Delta U/W)_{V,T} = T \frac{d \ln \langle r^2 \rangle_0}{dT} = (f_u/f) \tag{6.7}$$

are usually called the entropic and energetic components of the work (force). According to the Gaussian theory of rubber elasticity they should be constant and independent of deformation.

Equations (6.3)–(6.7) demonstrate very obviously the basic thermomechanical idea of the statistical theory of rubber elasticity. On deformation of the chains, some part of the work is connected with the intramolecular energy change resulting from the transition of the chains from one spatial conformation to another. The sign and value of the energy change are dependent on the chemical structure of the macromolecules. The parameter $d \ln \langle r^2 \rangle_0 / dT$ may be both positive and negative and, conseqently, the simple deformation of polymer networks may be accompanied by both increasing and decreasing of the internal energy. For the chains with free rotational states, d $\ln\langle r^2\rangle_0/dT = 0$, $\eta = -1$ which corresponds to the ideal entropy-elastic model. Hence, the factor d $\ln\langle r^2\rangle_0/dT$ may be considered as a parameter of nonideality of polymer chains [3]. Unlike gases, which can behave ideally by decreasing the pressure, this type of non-ideality cannot be removed by diluting since it is intramolecular in origin.

Although Eqs. (6.4) and (6.5) are very simple it is necessary to carry out the measurements at constant volume. This experiment is difficult in practice. Deformation of rubbers is usually accompanied by a change in volume $\Delta V/V \approx 10^{-4}$. This volume change can be neglected for determining the mechanical work. However, even small changes in volume during deformation are extremely important for determining the vibrational entropy and the internal energy changes resulting from the volume change. Thus, even a small change of the volume can strongly distort the intramolecular effects.

Achievement of the constant volume condition requires applying hydrostatic pressure during the measurements [4, 5]. Usually the thermomechanical experiments are carried out at the constant pressure condition. The expressions for W, Q, ΔU, η and ω under P, T = const. are

$$W_{P,T} = \frac{C\,(\lambda - 1)}{2}\frac{}{\lambda}(\lambda^2 + \lambda - 2) \tag{6.8}$$

$$Q_{P,T} = -\frac{C}{2}\left[\left(1 - T\frac{d\ln\langle r^2\rangle_0}{dT}\right) - \frac{2\alpha T}{\lambda^2 + \lambda - 2}\right]\frac{(\lambda - 1)}{\lambda}(\lambda^2 + \lambda - 2) \tag{6.9}$$

$$U_{P,T} = \frac{C}{2}\left(T\frac{d\ln\langle r^2\rangle_0}{dT} + \frac{2\alpha T}{\lambda^2 + \lambda - 2}\right)\frac{(\lambda - 1)}{\lambda}(\lambda^2 + \lambda - 2) \tag{6.10}$$

$$\eta = \left(\frac{Q}{W}\right)_{P,T} = -1 + T\frac{d\ln\langle r^2\rangle_0}{dT} + \frac{2\alpha T}{\lambda^2 + \lambda - 2)} \tag{6.11}$$

$$\omega = \left(\frac{U}{W}\right)_{P,T} = T\frac{d\ln\langle r^2\rangle_0}{dT} + \frac{2\alpha T}{\lambda^2 + \lambda - 2} \cdot \tag{6.12}$$

These expressions demonstrate that the change of entropy and internal energy on deformation under these conditions is both intra- and intermolecular in origin. Intramolecular (conformational) changes, which are independent of deformation, are characterized by the temperature coefficient of the unperturbed dimensions of chains d $\ln\langle r^2\rangle_0/dT$. The intermolecular changes are characterized by the thermal expansivity α and are strongly dependent on deformation. The difference between the thermodynamic values under P, T = const. and V, T = const. is very important at small deformations since at $\lambda \to 1$ $2\alpha T/(\lambda^2 + \lambda - 2)$ tends to infinity. Comparing Eqs. (6.6), (6.7) and (6.11), (6.12), we arrive immediately at

$$\eta_{\Delta V} = \eta_{P,T} - \eta_{V,T} = \frac{2\alpha T}{\lambda^2 + \lambda - 2} \tag{6.13}$$

$$\omega_{\Delta V} = \omega_{P,T} - \omega_{V,T} = \frac{2\alpha T}{\lambda^2 + \lambda - 2} \tag{6.14}$$

These expressions characterize the relative change of the entropy and internal energy resulting from the volume change. Therefore, they must be identical with the corresponding expressions for solids. It can be easily proved by introducing the strain ε instead of λ into Eq. (6.13) and (6.14) and neglecting ε^2 terms

$$\left(\frac{Q}{W}\right)_{\Delta V} = \left(\frac{\Delta U}{W}\right)_{\Delta V} = \frac{2\alpha T}{3\varepsilon} = \frac{2\beta T}{\varepsilon} \qquad (6.15)$$

this expression is fully identical with the expression for η for simple elongation of solids [Eq. (5.19′)].

The absolute values of the intermolecular change of the internal energy and entropy associated with the volume dilatation may be written as

$$T(\Delta S)_{\Delta V} = (\Delta U)_\Delta = C\alpha T \frac{\lambda - 1}{\lambda} . \qquad (6.16)$$

We recognize from Eq. (6.16) that the change of internal energy resulting from the volume change of a Gaussian network is exactly balanced by the equivalent change of entropy and, thus, this volume change gives no contribution to the free energy of deformation [6, 7]. Provided α and the isothermal compressibility κ are independent of strain, by using the thermodynamic equality $(dU/dV)p, T = \alpha KT$, one can obtain the equation for the volume change

$$\frac{\Delta V}{V_0} = \frac{C}{K} \frac{\lambda - 1}{\lambda} = C\kappa \frac{(\lambda - 1)}{\lambda} . \qquad (6.17)$$

All these expressions demonstrate the distinct between the mechanism of elasticity of solids and elastomers on simple deformation (under the conditions of constant pressure and temperature). The elasticity of a solid is a result of the resistance to a change in its volume, and all the thermodynamic properties are connected with the volume change. The elasticity of an elastomeric body is a result of the resistance to change of its shape and the work is spent only on the conformational change of chains. A small volume change accompanying the change in shape is an attendent effect, since the resulting energy change is fully compensated by the entropy change.

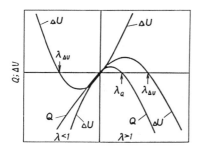

Fig. 6.1. Heat and internal energy changes as a function of deformation at $P, T = $ const. for a Gaussian polymer network [6, 55]. Q = heat; ΔU = internal energy (d $\ln\langle r^2\rangle_0/dT < 0$) and (d $\ln\langle r^2\rangle_0/dT > 0$). The *arrows* indicate inversion points (see text)

Equations (6.9) and (6.10) predict the inversions of heat and internal energy on deformation [6]. The inversion of heat must occur at (Fig. 6.1)

$$\lambda_Q = 1 + \frac{2}{3} \frac{\alpha T}{(1 - T\,d\,\ln\langle r^2\rangle_0/dT)} \;. \tag{6.18}$$

This inversion of heat is due to competition between the increase of the vibrational entropy connected with the volume change at deformation and the decrease of the conformational entropy. The deformation at which a maximum of heat is absorbed at elongation is given by

$$\lambda_Q^{\max} = 1 + \frac{1}{3} \frac{\alpha T}{3(1 - T\,d\,\ln\langle r^2\rangle_0/dT)} \tag{6.19}$$

and $Q^{\max}$ at this deformation is given by

$$Q^{\max} \approx \frac{C}{3} \frac{(\alpha T)^2}{(1 - T\,d\,\ln\langle r^2\rangle_0/dT)} \;. \tag{6.20}$$

For the internal energy inversion point, we arrive at

$$\lambda_U = 1 - \frac{2}{3} \frac{\alpha T}{T\,d\,\ln\langle r^2\rangle_0/dT} \;. \tag{6.21}$$

Unlike the thermomechanical inversion of heat, the inversion of internal energy is possible only for chains with $d\,\ln\langle r^2\rangle_0/d\Gamma \neq 0$. It is also evident from Eq. (6.21) that for $d\,\ln\langle r^2\rangle_0/dT < 0$ the inversion must occur at $\lambda > 1$ and vice versa. For values $\alpha = (6 - 10) \times 10^{-4}\,\mathrm{K}^{-1}$ and $d\,\ln\langle r^2\rangle_0/dT = (5-0.5) \times 10^{-3}\,\mathrm{K}^{-1}$, which are typical for polymeric networks, $\lambda_U = 1.3$–2.2 for extension and $\lambda_U > 0.5$ for compression. It must be emphasized that this thermomechanical inversion of internal energy is not connected with the stress-induced crystallization and arises from different signs of inter- and intrachain contributions to the internal energy. The extreme of the internal energy occurs at the deformation

$$\lambda_U^{ex} = 1 - \frac{1}{3} \frac{\alpha T}{T\,d\,\ln\langle r^2\rangle_0/dT} \;. \tag{6.22}$$

It is quite obvious that the thermomechanical inversions both of heat and internal energy must disappear under the condition of constant volume.

Finally, making use the equation of state (6.1) one can demonstrate that the linear thermal expansion coefficient depends on the deformation as follows (for isoenergetic chains) [8, 9]

$$\beta_\parallel(\lambda) = \frac{1}{T} \frac{\lambda^3 - 1}{\lambda^3 + 2} + \frac{\alpha}{\lambda^3 + 2} \tag{6.23}$$

$$\beta_\perp(\lambda) = \frac{1}{2T} \frac{\lambda^3 - 1}{\lambda^3 + 2} + \frac{\alpha}{2} \frac{\lambda^3 + 1}{\lambda^3 + 2} \;. \tag{6.24}$$

In the undeformed state, $\beta_\parallel = \beta_\perp = \alpha/3 > 0$. At large λ, one can neglect the terms with α and we arrive at

$$\beta_\parallel \approx \frac{1}{T} \frac{\lambda^3 - 1}{\lambda^3 + 1} \tag{6.25}$$

$$\beta_\perp \approx \frac{1}{2T} \frac{\lambda^3 - 1}{\lambda^3 + 2}.$$

(6.26)

These expressions show that a deformed polymer network is an extremely anisotropic body and possesses a negative thermal expansivity along the orientation axis of the order of the thermal expansivity of gases about two orders higher than that of macromolecules incorporated in a crystalline lattice (Chap. 3). In spite of the large anisotropy of the linear thermal expansivity, the volume coefficient of thermal expansion of a deformed network is the same as of the undeformed one. As one can see from Eqs. (6.23) and (6.24) $\beta_\parallel + 2\beta_\perp = \alpha$. Equation (6.23) shows also that the thermoelastic inversion of $\beta_\parallel$ must occur at $\lambda_{\mathrm{inv}}(\varepsilon_{\mathrm{inv}}) \approx 1 + (1/3)\alpha T$. It coincides with $\lambda_Q^{\max}$ for isoenergetic chains [see Eq. (6.19)].

Flory [1] showed that for Gaussian networks the connection between the temperature dependence of the force at constant volume and the temperature coefficient of molecular dimension $d \ln\langle r^2\rangle_0/dT$ holds for all types of distortions. According to Treloar [10], the equation of state for torsion of the Gaussian network is

$$M = \left(\frac{\pi\nu kT}{2V_0}\right) \left(\frac{\langle r^2\rangle}{\langle r^2\rangle_0}\right) \left(\frac{\theta}{L_0}\right) a_0^4$$

(6.27)

where M is the torque which is needed for twisting a cylinder of length L_0 by the angle θ; a_0 is the radius of an unstrained sample. From Eq. (6.27) we get

$$W = \int_0^\theta M \, d\theta \left(\frac{1}{4} \frac{\pi\nu kT}{V_0}\right) \left(\frac{\langle r^2\rangle}{\langle r^2\rangle_0}\right) \cdot \frac{a_0^4}{L_0} \theta^2$$

(6.28)

$$T(\Delta S)_{P,T,L} = W(-1 - \alpha T + T \, d \ln\langle r^2\rangle_0/dT$$

(6.29)

$$(\Delta U)_{P,T,L} = W(-\alpha T + T \, d \ln\langle r^2\rangle_0/dT)$$

(6.30)

$$T(\Delta S/W)_{P,T,L} = -1 - \alpha T + T \, d \ln\langle r^2\rangle_0/dT$$

(6.31)

$$(\Delta U/W)_{P,T,L} = -\alpha T + T \, d \ln\langle r^2\rangle_0/dT$$

(6.32)

$$T(\Delta S)_{V,T,L} = W(1 - T \, d \ln\langle r^2\rangle_0/dT)$$

(6.33)

$$(\Delta U)_{V,T,L} = WT \, d \ln\langle r^2\rangle_0/dT$$

(6.34)

$$T(\Delta S/W)_{V,T,L} = -1 + T \, d \ln\langle r^2\rangle_0/dT$$

(6.35)

$$(\Delta U/W)_{V,T,L} = T \, d \ln\langle r^2\rangle_0/dT.$$

(6.36)

We see that the relative change of entropy and internal energy at constant pressure is independent of the degree of twisting. This conclusion differs from that obtained for simple extension or compression. The entropic and energetic components in torsion are identical with the result for simple deformation. Equations (6.29) and (6.30) lead to the conclusion that there can be no thermomechanical inversions of heat and internal energy in torsion.

The change of entropy and internal energy, associated with the change of volume due to torsion is given by the equation

$$T(\Delta S)_{\Delta V} = (\Delta U)_{\Delta V} = -W\alpha T .$$

(6.37)

The corresponding change of volume is

$$\Delta V = -\frac{M\kappa\theta}{2} = -\left(\frac{1}{4}\frac{\eta\cdot\nu\cdot k\cdot T}{V_0}\right)\left(\frac{\langle r^2\rangle}{\langle r^2\rangle_0}\right)\frac{\theta^2}{L_0}a_0^4\kappa .$$

(6.38)

Hence, from Eqs. (6.37) and (6.38) one can conclude that the change in volume due to torsion does not contribute to the free energy of deformation, that torsion is accompanied by a decrease in volume and that this volume decrease is a parabolic function of the twisting angle.

In the classical statistical theory of rubber elasticity, it is suggested that the front-factor $\langle r^2\rangle/\langle r^2\rangle_0$ is independent of volume (and hence of pressure). Tobolsky and Shen [11, 12], however, have supposed that such a dependence occurs due to intermolecular forces. They have proposed a semiemperical equation of state taking into account this dependence

$$f = \frac{vkT}{L_0}\frac{\langle r^2\rangle}{\langle r^2\rangle_0}\left(\frac{V_0}{V}\right)^\gamma\left(\lambda - \frac{V}{V_0\lambda^2}\right)$$

(6.39)

where $\gamma = (\partial\ln\langle r^2\rangle_0/\partial\ln V)_{T,L}$ is a constant dependent on the chemical structure of the chains. If γ is taken to be zero, Eq. (6.39) is transformed to Eq. (6.1). From the equation of state, the energetic contribution of the work is readily derived as

$$\left(\frac{\Delta U}{W}\right)_{V,T} = \left(\frac{\Delta U}{W}\right)_{P,T} - \frac{2\alpha T}{\lambda^2 + \lambda - 2} - \gamma\alpha T .$$

(6.40)

Equation (6.40) differs from Eq. (6.12) by the term $(-\gamma\alpha T)$ and hence $(\Delta U/W)_{V,T}$ should be independent of extension ratio λ, since this term is not a function of deformation. Equation (6.39) leads also to the following expression for the strain-induced volume dilatation

$$\frac{\Delta V}{V_0} = C\kappa\frac{(\lambda - 1)}{\lambda}\left[1 + \frac{\gamma}{2}(\lambda^2 + \lambda - 2)\right] .$$

(6.41)

If γ is taken to be zero, Eq. (6.41) is identical with Eq. (6.17). The change of entropy and internal energy corresponding to the volume dilatation according to Eq. (6.41) are

$$(\Delta U)_{\Delta V} = T(\Delta S)_{\Delta V} = C\alpha T\frac{(\lambda - 1)}{\lambda}\left[1 + \frac{\gamma}{2}(\lambda^2 + \lambda - 2)\right] .$$

(6.42)

Recently, an attempt has been made to give a molecular interpretation of the parameter γ [13].

Schwarz [14] has modified the equation of state for the Gaussian polymer network by introducing a coefficient taking into account the influence of the packing of chains and their sizes on the mean square end-to-end distance of deformed chains.

In recent years the classical theory of rubber elasticity has been reconsidered significantly and a number of new models of rubber elasticity have been suggested [15–26]. Thermomechanics of the new models is considered in Appendix A.

6.1.2 Thermomechanics of Non-Gaussian Networks

Equations (6.1) and (6.2) are derived from the statistical theory based on the Gaussian statistics which describes the network behavior if the network is not deformed beyond the limit of the applicability of the Gaussian approximation [27, 28]. For long chains, this limit is close to 30% of the maximum chain extension. For values of r, which are comparable with r_{max}, the force–strain dependence is usually expressed using the inverse Langevin function [27, 28]

$$f = \frac{vkT}{3} p^{1/2} [\mathcal{L}^{-1}(\lambda p^{-1/2}) - \lambda^{-3/2} \mathcal{L}^{-1}(\lambda^{-1/2} p^{-1/2})] \qquad (6.43)$$

where $\mathcal{L}^{-1}$ is the inverse Langevin function, p is the number of statistical segments in a chain which determines the limit of chain extensibility $\lambda = p^{1/2}$. If $\lambda_{max} = \sqrt{p}$, the force f becomes infinite since $\mathcal{L}^{-1}(1) \to \infty$. According to estimations of Treloar [27] the difference in the value of force f corresponding to Eq. (6.1) and (6.43) is about 3% at $r = 0.3 \, r_{max}$ and about 8% at $r = 0.5 \, r_{max}$.

To elucidate the role of the limited chain extensibility of chains on the thermomechanical properties of non-Gaussian networks, Schwarz [29, 30] carried out theoretical considerations of the behavior of short-chain model networks consisting of chains with 4–14 bonds. As a result of calculations assuming no fluctuations of crosslinks, it was concluded that the energetic component resulting from simple elongation is equal to the theoretical value. The energy contribution depends on deformation, being a very weak linear function of the extension for the networks with positive energetic contribution and a strong one for the networks with a negative energetic component. This is a rather unexpected conclusion.

In other statistical theories of rubber elasticity (see, e.g. reviews [12, 28]) the Gaussian statistics are not valid even with small deformations and the intramolecular energy component is dependent on deformation.

The non-Gaussian theories of rubber elasticity have the disadvantage of containing parameters which generally can be determined only by experiment. Curro and Mark [31] proposed a new non-Gaussian theory of rubber elasticity based on rotational isomeric state simulations of network chain configurations. Specifically, Monte Carlo calculations were used to determine the distribution function of end-to-end dimension of the network chains. The utilization of these distribution functions instead of the Gaussian function yields a large decrease in the entropy of the network chains.

6.1.3 Phenomenological Equations of State

It is well known that the equation of state of Eq. (6.1) based on Gaussian statistics is only partly successful in representing experimental relationships of tension-extension and fails to fit the experiments over a wide range of strain modes [23, 38, 39]. The deviations from the Gaussian network behavior may have various sources and this was discussed by Dušek and Prins [28]. Therefore, phenomenological equations of state are often used. The most often used phenomenological equation of state for rubber elasticity is the Mooney–Rivlin equation [23, 38, 39]

$$\frac{f}{A_0} = \left(2C_1 + \frac{2C_2V_0}{V\lambda}\right)\left(\lambda - \frac{V}{V_0\lambda^2}\right)$$
(6.44)

where C_1 and C_2 are constants.

Shen [32] considered the thermoelastic behavior of the materials described by the Mooney–Rivlin equation of state and showed that the energetic component is given by

$$\left(\frac{f_U}{f}\right) = 1 - \frac{\partial \ln f}{\partial T} - \alpha T \left[\frac{1}{(\lambda^3 - 1)} + \frac{1}{\dfrac{C_1}{C_2}\lambda + 1}\right].$$
(6.45)

Equation (6.45) indicates that the energy contribution is a function of deformation and must decrease with an increasing extension ratio. The change in volume resulting from the deformation is given by

$$\Delta V/V_0 = 2\kappa(\lambda - 1)(C_1/\lambda + C_2).$$
(6.46)

Shen [32] also considered the thermoelastic behavior of another widely used phenomenological equation of state, the so-called Valanis–Landel equation. Valanis and Landel [33] postulated that the stored energy function W should be expressible as the sum of three independent functions of principle extension ratios. This hypothesis leads to the following equation of state

$$f = CA_0 \left[2\ln\lambda + \left(\frac{V}{V_0\lambda^3}\right)^{1/2}\ln\left(\frac{V}{V_0\lambda}\right)\right].$$
(6.47)

On the basis of Eq. (6.47), one can get the expression for the energetic contribution

$$\frac{f_U}{f} = \left(1 - \frac{\partial \ln C}{\partial \ln T} - \frac{\alpha T}{3}\right) - \frac{\alpha T}{3}\left[1 - \frac{2\lambda^{3/2} + 2}{(2\lambda^{3/2} + 1)\ln\lambda}\right].$$
(6.48)

Again, the energy contribution decreases with increasing deformation.

It is very important to stress that the decrease of the internal energy contribution with increasing extension ratio is due to a decrease of the intermolecular interaction with deformation, since the intramolecular contribution is independent of the deformation in full accord with the statistical theory. At very high strains, the λ-dependent part of f_U/f approaches a limiting value of -0.68 for the Mooney–Rivlin and -0.07 for the Valanis–Landel materials.

A comprehensive consideration of new phenomenological equations of state for rubber elasticity was carried out by Tschoegl et al. [34–38]. One of their equations of state is given by

$$f = \frac{2C}{n}\left(\frac{V}{V_0}\right)^{\gamma}\left(\lambda^{n-1} - \frac{(V/V_0)^{n/2}}{\lambda^{(n+2)/2}}\right) \tag{6.49}$$

where n is the parameter which characterizes the non-linear stress- strain behavior. If $n = 2$, this equation reduces into the Tobolsky–Shen equation [Eq. (6.39)] and if in addition $\gamma = 0$ it transforms into the equation of state of Eq. (6.1). It was supposed that the constant n is independent of temperature. In this case, the energy contribution corresponding to Eq. (6.49) should be independent of the deformation. Later, a modification of the theory was proposed which took into account the temperature dependence of the strain parameter n, and this modification led to a dependence of the energy contribution on extension.

A number of attempts were made to represent the stored energy function W by a power series containing more than two terms [39–42]. In particular. Chadwick [40] modified the expression for the stored energy function proposed by Ogden [39] and introduced the scalar response function for a compressible elastomer. The generalized form of the stored energy function W is represented by the following series

$$F = \sum_r (\mu_r/\mu\alpha_r)(\lambda_1^r + \lambda_2^r + \lambda_3^r - 3J^{1/3}\alpha_r) \tag{6.50}$$

where J is the ratio of the density of the deformed and undeformed materials and μ_r, α_r and μ are material constants, being the shear modulus. We see that this equation is still consistent with the Valanis–Landel hypothesis. Equation (6.50) leads to the following expression for the volume dilatation on extension

$$(\partial\ln V/\partial\lambda)_{P,T} = 1/2\kappa\Sigma(\alpha_r\mu_r\lambda^{-1/2}\alpha_r - 1)\,. \tag{6.51}$$

If $\alpha_r = 2$, $\mu_r = 2C_1$, $\alpha_r = -2$, and $\mu_r = -2C_2$, this equation reduces into Eq. (6.46) for the Mooney–Rivlin materials. A further development of this approach is given in [41, 42].

In a series of papers Kilian [43–45] proposed a new phenomenological approach to rubber elasticity and suggested a molecular network might be considered as a formelastic fluid the conformational abilities of which were adequately characterized by the model of a van der Waals conformational gas with weak interaction. The ideal network is treated as an ideal conformational gas. According to these assumptions, the van der Waals equation of state of real gases yields a thermomechanical equation of state for real networks under simple deformation

$$f = \frac{\nu kT}{L_0} \frac{\langle r^2 \rangle}{\langle r^2 \rangle_0} D(B - aD) \tag{6.52}$$

where $D = \lambda - \lambda^2$, $B = D_m/(D_m - D)$, $D_m = \lambda_m - \lambda_m^2$, $\lambda = L_m L_0$ is the strain related to the limited chain extensibility. The van der Waals parameter, a, takes into account weak interactions beetwen chains. It may be represented by

$$a = a_1(1 - a_2/\lambda^2) . \tag{6.53}$$

Thus this phenomenological equation of state includes the three parameters λ_m, a_1 and a_2 which can be obtained from the analysis of the shape of the stress–strain curve.

Starting from Eq. (6.52), Kilian [43] obtained the following expressions for the entropic and energetic components of the elastic force f in simple extension

$$\left(\frac{f_s}{f}\right)_{P,T} = \left(1 - T\frac{d\ln\langle r^2 \rangle_0}{dT}\right) - \beta T \left(\frac{3}{\lambda^3 - 1} + \frac{\bar{B} - a\bar{D} - \bar{a}D + (\partial a/\partial T)D/\beta}{B - aD}\right) \tag{6.54}$$

$$\left(\frac{f_H}{f}\right)_{P,T} = T\frac{d\ln\langle r^2 \rangle_0}{dT} + \beta T \left(\frac{3}{\lambda^3 - 1} + \frac{\bar{B} - a\bar{D} - \bar{a}D - (\partial a/\partial T)D/\beta}{B - aD}\right) \tag{6.55}$$

$$\bar{B} = \frac{1}{D_m - D}\left(\bar{D}_m - D_m\frac{\bar{D}_m - \bar{D}}{D_m - D}\right) ; \quad \bar{D}_m = \lambda_m + 2\lambda_m^{-2} ;$$

where $D = \lambda - 2\lambda^{-2}$;

$$\frac{\partial a}{\partial T} = \frac{\partial a_1}{\partial T}\left(1 - \frac{a_2}{\lambda^2}\right) - a_1\lambda^{-2}\frac{\partial a_2}{\partial T} , \quad \bar{a} = 2a_1 a_2\lambda^{-2} .$$

Integration of these equations leads to $W(\lambda)$, $Q(\lambda)$ and $\Delta U(\lambda)$. Kilian showed that the energetic component $f_U/f = (\Delta U/W)_{V,T} = d\ln\langle r^2 \rangle_0/dT$ is in full agreement with the statistical theory of rubber elasticity. Hence, if $d\ln\langle r^2 \rangle_0/dT$ and β are known, the full energy balance of deformation can be calculated using parameters λ_m, a_1, a_2 and $\partial a/\partial T$.

Making use of the expression

$$(\partial V/\partial L)_{P,T} = (\partial f/\partial T)_{T,L} \tag{6.56}$$

the change of volume with deformation is expressed by

$$\frac{\Delta V}{V_0} = \kappa_L L_0^{-2} \int f \left(\frac{3}{\lambda^3 - 1} + \frac{\bar{B} - a\bar{D} - \bar{a}D + \kappa^{-1}(\partial a_1/\partial P)D}{B - aD}\right) \tag{6.57}$$

with $\partial a/\partial P = (\partial a_1/\partial P)(1 - a_2/\lambda^2)$ and $\kappa_L = -(\partial \ln L_0/\partial P)_{T,L}$ being the linear isothermal compressibility in the unstrained state. These volume changes are dependent not only on the compressibility of the network but also on the van der Waals parameters and the pressure coefficient of the interaction parameter a.

The concept of the van der Waals conformational gas also yields a reduced equation of state [54]

$$f' = d \left(\frac{8t}{3d} - 3d \right) \tag{6.58}$$

where $f' = f/f_c$, $d = D/D_c$, $t = T/T_c$ with the critical coordinates $T_c = 8aD_m /27$, $D_c = D_m/3$ and $f'_c = aD_m/27$. The stability of the network under simple elongation is closely controlled by van der Waals parameters. Thus, one can conclude that this new phenomenological equation of state for rubber elasticity is a promising approach to the thermomechanical behavior of polymer networks and rubberlike materials.

Finally, it should be mentioned that some works on the phenomenological theory of rubber elasticity and thermoelasticity were published recently [46–48].

6.1.4 Thermoelasticity of Liquid Crystalline Networks

In recent years, the behavior of liquid crystalline polymers including elastomers has been a subject of considerable interest [49, 50]. It is known that small molecule liquid crystals turn into a macroscopic ordered state by external electric or magnetic fields. Similar behavior seems to occur for liquid-crystalline polymer networks under mechanical stresses or strains.

Jerry and Monnerie [51] have proposed a modified theory of rubber elasticity which includes anisotropic intermolecular interactions U_{12} (favoring the alignment of neighboring chain segments) in the form $U_{12} = \Sigma U_L(r_{12}P_L(\theta_1) P_L(\theta_2)$, where r_{12} is the intermolecular distance, θ_1 and θ_2 are the angles between the molecular axes and the symmetry axis of the medium, $P_L(\theta)$ are Legendre polynomials. Thus, such interactions are described in the mean field approximation by intermolecular potential having the same form as that usually used for the study of nematic phases. It is shown that the U_{12} interactions increase orientation of chain segments but do not significantly alter the stress–strain behavior of a network. Although some stress-optical coefficient data in the literature are consistent with the presence of nematic-like interactions, the birefringence measurements can be strongly perturbed by anisotropic internal field effects. The authors believe that linear dichroism and fluorescence polarization could bring more conclusive information.

Rusakov [52, 53] proposed a simple model of a nematic network in which the chains between crosslinks are approximated by persistent threads. Orientational intermolecular interactions are taken into account using the mean field approximation and the deformation behavior of the network is described in terms of the Gaussian statistical theory of rubber elasticity. Making use of the methods of statistical physics, the stress–strain equation of the network with its macroscopic orientation are obtained. The theory predicts a number of effects which should accompany deformation of nematic networks such as the temperature-induced orientational phase transition. The transition is affected by the intermolecular interaction, the rigidity of macromolecules and the degree of crosslinking of the network. The transition into the liquid crystalline state is accompanied by the appearence of internal stresses at constant strain or spontaneous elongation at constant force.

6.2 Thermomechanical Behavior of Molecular Networks

6.2.1 Entropy and Energy Effects with Small and Moderate Deformations

According to the theory of rubber elasticity, the elastic response of molecular networks is characterized by two mechanisms. The first one is connected with the deformation of the network, and the free energy change is determined by the conformational changes of elastically active network chains. In the early theories, the free energy change on deformation of polymeric networks was completely identified with the change of conformational entropy of chains. The molecular structure of the chains was fully ignored and the energy effects upon deformation of the network thought to arise only as a result of the compressibility (expansivity) characteristic for normal liquids and was connected with the volume change, i.e. due to changes of the intermolecular interactions upon deformation.

According to the current state of the theory, the deformation of polymeric network must be accompanied not only by the intrachain conformational entropy changes but intrachain energy changes which depend on the conformational energies of macromolecules. Therefore, reliable experimental determination of these intrachain energy changes and their interpretation by means of isomeric state theory is of fundamental importance for polymer physics.

Intrachain Energy Effects.   Numerous investigations concerning the intrachain effects in polymer networks and the determination of the contribution of the internal energy have been published over the last 20 years. The main results of these studies were summarized by Mark [2, 3]. The overwhelming majority of the studies were performed by measuring of the elastic force as a function of temperature under constant pressure and temperature. The intrachain effects were thus determined indirectly using corrections of the volume change upon deformation. In recent years, deformation calorimetry has been used widely for determining the intrachain energy contribution to the elasticity of polymeric networks.

The results for two most widely and thoroughly studied networks, namely NR and PDMS listed in Table 6.1, demonstrate that the reliable values of the intrachain energetic effects can be obtained both by thermoelastic and thermomechanical measurements at various deformation modes. It is very important, that these values are also independent of the condition of the experiments, i.e. whether the experiments were carried out at constant volume and temperature or constant pressure and temperature. Finally, the results show a full insensitivity of the energy contribution to the factors which can have an effect on the intermolecular interaction. Mark [54] and Shen and Croucher [11] who analysed the influence of such factors as degrees of crosslinking, crosslinking conditions, type of deformation, extent of deformation and swelling of the networks and have concluded that in the majority of the studies all these factors have no influence on the intrachain energy effects. Thus, the conclusions are in full accord with the basic postulate of the Gaussian theory of rubber elasticity – the free energy additivity principle.

Table 6.1. Energy contribution for NR and PDMS as obtained by various experimental methods

Method	Type of deformation	Experimental conditions (invariants)	$(\Delta U/W)_{V,T}$ $= f_u/f$ $= M_u/M$	Ref.
NR				
f-T	Extension	V, L	0.12 ∓ 0.02	4, 5
f-T	Extension	V, L	0.23	4, 5
f-T	Extension	P, L	0.18 ∓ 0.03	4, 5
f-T	Extension	P	0.17^*	2
f-T	Compression– Extension	P, L	0.18 ∓ 0.02	56
M-T	Torsion	P, L, θ	0.17 ∓ 0.02	57
Calorimetry	Torsion	P, T	0.20 ∓ 0.02	58
Calorimetry	Extension	P, T	0.18 ∓ 0.01	58, 59
Calorimetry	Extension	P, T	0.35^{**}	60
Calorimetry	Extension	P, T	0.28 ∓ 0.03	6, 55, 61
Calorimetry	Extension (heat invers.)	P, T	0.22 ∓ 0.03	6, 55
PDMS				
f-T	Extension	V, T	0.25 ∓ 0.01	5
f-T	Extension	P, L	0.27 ∓ 0.02	62
f-T	Extension	–	0.2^*	2
Calorimetry	Extension	P, T	0.30 ∓ 0.05	6, 55
Calorimetry	Extension (heat invers.)	P, T	0.25 ∓ 0.03	6, 55

* Mean value from the Table compiled by Mark [2];
** At 45°C

According to Eq. (6.6) the intrachain energy contribution must be independent of the deformation extent. The majority of the thermoelastic investigations confirm this theoretical conclusion [2, 6, 11]. However, in some studies dealing with the thermoelasticity of NR, EPR and some other networks, a dependence of f_u/f is found at small deformation ratios (see, e.g. [11, 35]), which is difficult to interpret since it occurs at $\lambda < 1.5$ at which Eq. (6.1) suitably predicts not only the intrachain energy effects, but also stress–strain behavior. Various attempts have been made to explain this dependence, but special studies carried out by Shen [11, 32] showed that this dependence seems to be a result of the sensitivity of measurements in the low-strain region and experimental error because of the correction term $(\lambda^3 - 1)^{-1}$. This term becomes very large in the region of small deformation. However, this correction does not apply to measurements at constant volume and then the results demonstrate the independence of f_u/f on the deformations with small strains [4, 5].

The problem of the dependence of intrachain effects on deformation in the region of low strain was studied by means of deformation calorimetry [6, 55, 61]. Typical results are shown in Fig. 6.2. For all networks studied in the regions of small strains $(Q/W)_{P,T}$ and $(\Delta U/W)_{P,T}$ increase sharply but intrachain changes of $(Q/W)_{V,T}$ and $(\Delta U/W)_{V,T}$ are virtually independent of strain including NR and EPR, for which, as mentioned above, a dependence of f_u/f on strain was observed in this region. A large scattering of the values of $(\Delta U/W)_{V,T}$ is seen in Fig. 6.2 at small strains. Since the term $(\lambda^2 + \lambda - 2)$ tends to be very small at large λ, it can be neglected and, hence, at large λ it is just $(\Delta U/W)_{V,T}$ which is measured.

Table 6.2. Comparison of the energy contributions obtained in thermoelastic (f_u/f) and thermomechanical $(\Delta U/W)_{V,T}$ measurements and values of d $\ln\langle r^2\rangle_0/dT$ calculated from the energy contributions and values of viscosity–temperature measurements on isolated chains

Polymer	$(\Delta U/W)_{V,T}$	f_u/f	d $\ln\langle r^2\rangle_0/dT \times 10^3$ from f_u/f and $(\Delta U/W)_{V,T}$	Viscosity – temperature
NR	0.21*	0.18*	0.66	–
PDMS	0.27*	0.24*	0.86*	0.71 [11], 0.66 [54]
PBR	0.12 [59]	0.11 [2, 54]	0.39	–
EPR	−0.42 [6, 55]	−0.45 [2]	−1.48	–
SBR	0 [6, 55]	−0.13 ($\mp$0.06) [2]	−0.21	–
NBR	0.06 [6, 55]	0.03 [2]	0.15	–
PCR	−0.10 [6, 55]	−0.10 [2]	−0.34	–
PE	–	−0.45 [2]	−1.53	−1.20 [3, 11]

* Mean value from Table 6.1

This conclusion permits comparison of the thermomechanical and thermoelastic results for various networks. The most reliable data are tabulated in Table 6.2. The temperature coefficients of the unperturbed dimensions of chains d $\ln\langle r^2\rangle_0/dT$ as obtained by various methods are in a good agreement. This demonstrates once more the independence of intermolecular interactions from the configuration of the network chains. For most polymers studied the intrachain energy component may be both positive and negative. According to the isomeric state theory, positive values of energy contributions are found for polymers for which extended conformations are of higher energy [2, 64]. According to this theory one gets for the conformational energy Ec

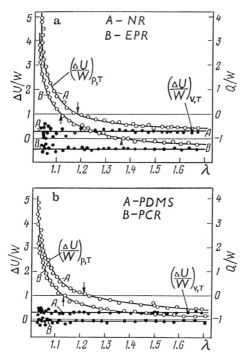

Fig. 6.2a,b. The calorimetrically determined relative entropy and internal energy contributions as a function of extension ratio λ [6, 55]. Room temperature: $\circ$ – $(Q/W)_{P,T}$ and $(\Delta U/W)_{P,T}$; $\bullet$ – $(Q/W)_{V,T}$ and $(\Delta U/W)_{V,T}$. a – NR(A), EPR(B); b – PDMS(A), PCR(B). The hyperbolic solid lines represent $(Q/W)_{P,T}$ and $(\Delta U/W)_{P,T}$ according to Eqs. (6.11) and (6.12) with the following values of parameters α and d $\ln\langle r^2\rangle_0/dT$ (in K^{-1}): NR 6.6×10^{-4} and 8.2×10^{-4}, EPR 7.5×10^{-4} and -14.3×10^{-4}, PDMS 9.0×10^{-4} and 8.6×10^{-4}, PCR 7.2×10^{-4} and 3.4×10^{-4}. The *horizontal solid lines* represent $(Q/W)_{V,T}$ and $(\Delta U/W)_{V,T}$ according to Eqs. (6.6) and (6.7) with the corresponding values of d $\ln\langle r^2\rangle_0/dT$

$$\left(\frac{\Delta U}{W}\right)_{V,T} = T\frac{d\ln\langle r^2\rangle_0}{dT} = -A\frac{E_c}{kT} \tag{6.59}$$

where A is a constant depending on chain symmetry and bond angles. Calculations of the *trans-gauche* energy difference for PE using $(\Delta U/W)_{V,T} = f_u/f = (-0.54)$ led to $E_c = 2\,kJ/mole$ in excellent agreement with the spectroscopic and thermodynamic data. A summary of the values of energy contribution for many polymers and their interpretation according to the rotational isometric state theory was given by Mark [2]. An attempt was made to correlate the value and sign of the energy contribution with some geometric characteristics of chains, in particular with the cross-sectional area of the chain in the crystalline state [65]. An increase of the cross-sectional area seems to cause a change of the sign of the energy contribution from negative to positive, but physical reasons for such a correlation are unclear yet.

Table 6.3. Energy characteristics of instant unloading processes [6]

Polymer	λ	Q(J/g)	Q^*(J/g)	$Q^*/(Q+Q^*)$	$(\Delta U/W)_{V,T}$	d $\ln\langle r^2\rangle_0$/dT $\times 10^3$, K^{-1}
EPR	1.23	−0.035	−0.0092	0.208	−0.39	−1.32
	1.37	−0.115	0.0082	−0.077	−0.43	−1.46
	1.78	−0.583	0.1170	−0.251	−0.40	−1.36
	2.33	−1.870	0.5200	−0.385	−0.42	−1.41
NR	1.45	−0.060	−0.060	0.500	0.27	0.92
	1.97	−0.260	−0.123	0.320	0.22	0.82

Calorimetric determination of the sign and values of the energy contribution can be performed by a quick approach based on the instant unloading of the stretched networks [6]. According to this approach $(\Delta U/W)_{P,T} = Q^* /(Q^* + Q)$, where Q^* is the heat effect resulting from instant unloading of a stretched sample. Table 6.3 tabulates the experimental results for two networks. The data of $(\Delta U/W)_{V,T}$ and d $\ln\langle r^2\rangle_0$/dT for NR and EPR are in a good agreement with data for these networks obtained with other methods.

Interchain Effects. The statistical theory of rubber elasticity predicts that isothermal simple elongation and compression at constant pressure must be accompanied by interchain effects resulting from the volume change on deformation. The correct experimental determination of these effects is difficult because of very small absolute values of the volume changes. These studies are, however, important for understanding the molecular mechanisms of rubber elasticity and checking the validity of the postulates of statistical theory. The interchain effects in polymer networks are reflected in the thermomechanical inversions at low strains, which arise from a competition of intra- and interchain changes. Calorimetric behavior demonstrates this fact very clearly (Fig. 6.3). The points of elastic inversion of heat tabulated in Table 6.4 are in an excellent agreement with the prediction of Eq. (6.18). The value of energy contribution for the only one point of deformation, i.e. the inversion point, coincides with data obtained by a more general method (Fig. 6.2). An equivalent approach can be found in the determination of the elastic inversion of the f–T dependence [27, 56], $\lambda_Q = 2\lambda_f$. Data for NR obtained by both methods agree well. The energy contribution in EPR is negative and according to Eq. (6.21) a thermomechanical inversion of the internal energy should occur which is fully supported by the experimental findings as seen from Fig. 6.3 and Table 6.4.

Kilian [66] also used calorimetric determination of mechanical and thermal energy exchange in isothermal simple elongation for various polymer networks published [6] and demonstrated that it can be described by relations which define thermomechanical properties of van der Waals networks (Fig. 6.3).

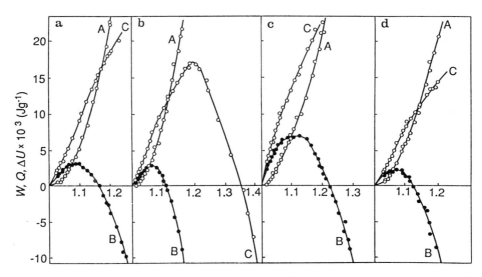

Fig. 6.3a,b,c,d. Mechanical work (A), heat (B) and internal energy change (C) on stretching samples from the unstrained state to λ at room temperature [6, 55]. a – NR; b – EPR; c – PDMS, d – PCR. The *solid lines* have been computed using Eqs. (6.8)–(6.10) with the following values of parameters: NR – C = 0.39 J/g, EPR – C = 0.61 J/g, PDMS – C = 0.61 J/g, PCR – C = 0.33 J/g. (For values of parameters d ln$\langle r^2\rangle_0$/dT and α see Fig. 6.2). The experimental data can also be represented by Eq. (6.52) with the corresponding set of parameters [43]

Table 6.4. Characteristics of thermomechanical inversion of heat [6, 55], internal energy [6, 55] and force [68] and related values of energy contribution

Polymer	Heat inversion		Internal energy inversion		Force inversion	
	λ_Q	$(\Delta U/W)_{V,T}$	λ_U	$(\Delta U/W)_{V,T}$	$\lambda_f \approx 1/2\lambda_Q$	f_u/f
NR	1.165	0.22	–	–	1.075	0.17
EPR	1.105	−0.40	1.35	−0.42	–	–
PDMS	1.235	0.25	–	–	–	–
PCR	1.130	−0.08	2.37	−0.10	–	–
SBR	1.150	0*	–	–	1.070	−0.12**

* 30% styrene; ** 24% styrene

The independence of the relative intrachain energy of the extent of deformation permits to resolve the entropy and energy changes into intra- and interchain components. Typical results of this treatment are shown in Fig. 6.4. It is seen that the prediction of Eq. (6.16) coincides with the experimentally obtained interchain energy and entropy changes only in the region of low strain ($\lambda < 1.3$). At higher extensions, experimental results are higher and the difference between experiment and theory increases with increasing extension. Such deviations are also found for thermoelastic measurements [5].

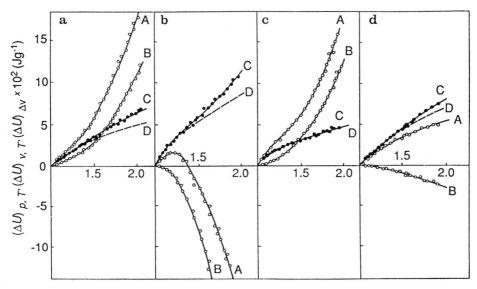

Fig. 6.4a,b,c,d. Intra- and interchain energy changes on stretching samples from the unstrained state to λ at room temperature [6, 55]. a – NR; b – EPR; c – PDMS; d – PCR., A – $(\Delta U)_{P,T}$; B – $(\Delta U)_{V,T}$; C – $(\Delta U)_{\Delta V}$. The *solid curves* C were calculated using Eq. (6.60) with $\gamma' = 0.1$ (EPR), $\gamma' = 0.15$ (NR) and $\gamma' = 0.07$ (PCR). *Lines D* correspond to Eq. (6.16)

An empirical equation which describes the interchain changes $(\Delta U)_{\Delta V}$ and $(\Delta S)_{\Delta V}$ on strain for the four networks has the form [6, 55]

$$(\Delta U)_{\Delta V} = T(\Delta S)_{\Delta V} = C\kappa T \frac{(\lambda - 1)}{\lambda} [1 + \gamma'(\lambda^2 + \lambda - 2)] . \qquad (6.60)$$

Parameter γ' depends on the chemical structure of the polymer chain and is equal to 0.1–0.2. The above equation is equivalent to Eq. (6.42) if $\gamma' = 2\gamma$.

Interchain changes arising from the deformation of polymer networks are a result of strain-induced volume dilatation. Typical results obtained by various methods shown in Fig. 6.5 demonstrate that the statistical theory predicts the volume dilatation and related interchain energy and entropy changes (Fig. 6.4) only at $\lambda < 1.3$. At larger deformations, experimental results show a significant departure from the statistical theory which underestimates the volume changes on strain.

One of the widely accepted approaches to the analysis of the strain-induced volume changes is the use of Eq. (6.46) based on the Mooney–Rivlin equation. Although in some studies [5, 11] it has been claimed that the $\Delta V/V_0$ data agree with Eq. (6.46), Treloar [69] concluded recently that such agreement cannot be a rule. He considered the volume change in considerable detail making use of various functional relations for the stored energy W and came to the following conclusions. First, large discrepancies have been found between experiment and the statistical theory. Second, a considerably closer agreement has been obtained for the three-term function

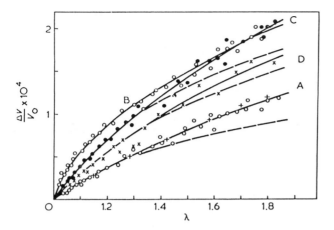

Fig. 6.5. The relative strain-induced volume dilatation of polymer networks as a function of strain [6]. A – NR [o – calorimetric data, + – dilatometric data [70]]; B – PDMS (o); C – EPR (•); D – PCR (×). *Solid curves* correspond to Eq. (6.41) with $\gamma = 0.2$ (EPR), $\gamma = 0.3$ (NR) and $\gamma = 0.15$ (PCR). - - - - lines correspond to Eq. (6.17)

of the Ogden-type [39], although the discrepancies between experiment and theory exceed 20% at large deformation (Fig. 6.6). It is evident that the apparent agreement between experiment and Eq. (6.46) is fortuitous. Hence, all these results clearly indicate that neither the statistical theory nor the phenomenological equations of Mooney–Rivlin or Ogden-type can correctly predict the volume changes accompanying the simple extension of polymer networks. Gee [71] recently reconsidered the dilatation which accompanies the elongation of an elastomer using the most widely acceptable theoretical and experimental stored energy functions and demonstrated that none of these give fully satisfactory predictions of the observed dilatation. He, therefore, concluded that some additional source of volume change must exist and suggested that this to be found in rotational isomerization. On this basis he developed a two-parameter dilatation equation. Data for PB and NR follow this empirical equation rather well.

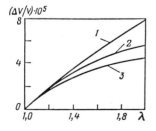

Fig. 6.6. The relative strain-induced dilatation of NR as a function of strain [69]. *1* – the experimental curve according to Price [5]; *2* – the calculated curve according to the theory of Ogden [39]; *3* – the calculated curve according to the statistical theory Eq. (6.17)

In this connection, it is very interesting that the volume and interchain changes obtained by various experimental methods [6, 11, 55] [Eq. (6.60)] agree well with Eq. (6.41) following from the Tobolsky–Shen semiempirical equation of state or the related phenomenological Eq. (6.49). The values of γ determined from the data are rather small (0.1–0.3). As mentioned above, according to the semiempirical approach by Tobolsky and Shen one can formally suggest that the front-factor in Eq. (6.1) is pressure dependent. If it is really so, then the parameter γ for networks can be considered as an experimental coefficient similar to the coefficient of thermal expansion and compressibility [11].

On the other hand, Kilian [43] having analysed the literature data on strain-induced volume dilatation [6, 70] using the van der Waals equation of state (Fig. 6.5) emphasized that only pressure dependence of the interchain parameter, a, is required for a full explanation of the relative volume changes. He arrived at a conclusion that non-crystalline rubbers are anisotropic equilibrium liquids and a higher compressibility of NR was only necessary for fitting the extension data. Hence, on using the van der Waals approach, there is no need to postulate volume dependence of the front-factor as proposed by Tobolsky and Shen.

A final comment should be made about the influence of the incorrect prediction of the interchain changes in the Gaussian statistical theory obtained from the results at constant pressure using the corrections following from the theory. The estimations show [6, 61] that the errors of the intrachain effects following from an inadequate description of the interchain changes by the statistical theory are on the level of experimental accuracy of determination of the energy and entropy component. This explains the insensitivity of the components to all experimental factors which can influence the interchain interaction as well as the condition $C_2 \neq 0$ of the Mooney–Rivlin equation. Nevertheless, Gee [71] emphasized that the correction needed to convert constant pressure thermoelastic data to constant volume is larger than currently calculated.

6.2.2 Thermomechanics at Large Deformations

The thermomechanical behavior of typical polymer networks at small and moderate deformations generally follows the thermomechanics of Gaussian networks. However, at large deformations approaching the maximum extension of the chains, non-Gaussian effects arise from the limited extensibility of the network chains. The existence of the limited extensibility of chains reflects itself as the sharp upturn of the Mooney–Rivlin plot [27, 32] (Fig. 6.7). As pointed out by Mark et al. [72–75], the observed increases in modulus and the appearence of the upturn of the Mooney–Rivlin curves may sometimes be due to internally generated reinforcement of the network structure arising from strain-induced crystallization or from the formation of hard glassy domains in the case of some block copolymers. The principal effect, resulting from both types of reinforcements, is the generation of "physical" crosslinks wich increase the network modulus.

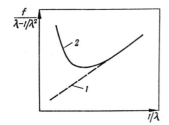

Fig. 6.7. Schematic diagram of stress–strain isotherm illustrating the anomalous increase in the reduce force at high elongation which has been observed for some polymer networks. *1* – according to Eq. (6.44); *2* – experimental behavior

It has been known for a long time that elongation of some networks is accompanied by crystallization [27, 60, 76]. Crystallization results in a decrease in internal energy and volume. This decrease is by one or two orders of magnitude higher than that resulting from the extension of a network without crystallization. Göritz and Müller [60, 76] have systematically studied strain-induced crystallization in NR by deformation calorimetry as a function of temperature and degree of crosslinking. The degree of extension at which crystallization occurs increases strongly with the temperature as seen from Fig. 6.8. Because of this, one can always choose a temperature at which strain-induced crystallization is absent. During crystallization, $(\Delta U/W)_{P,T}$ becomes negative regardless of the sign of the energy contribution. The limited degree of crystallinity which can be reached is strongly dependent on the extent of deformation. For example, at 15°C the degree of crystallinity of NR is about 1% if it is stretched to 300%, and about 15% if it is stretched to 400%. The crystallites formed during elongation are well oriented, with chain backbones parallel to the stretching direction. This conclusion is also supported by investigations of crystallization kinetics of the stretched samples. Similar results were also obtained for other crystallizable networks [77].

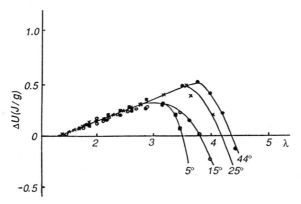

Fig. 6.8. The internal energy changes as a function of strain for NR at various temperatures [60, 76]

In unfilled rubbers, which are not capable of strain-induced crystallization, the upturns on Mooney–Rivlin curves have shown to be absent [72–75]. They disappear also in crystallizable rubbers at elevated temperatures and in the presence of solvent. On the other hand, the upturns do not appear for butadiene, nitrile and polyurethane rubbers if the limited chain extensibility function is introduced in the Mooney–Rivlin expression [78]. Mark [72] has concluded that in the absence of self-reinforcement due to strain-induced crystallization or domains the rupture of the networks occurs long before the limited chain extensibility can be reached.

To support this conclusion, Mark et al. [79–82] have studied elastic properties of model PDMS networks having a bimodal distribution of short and relatively long chains. The short chains are important because of their limited extensibility. The long chains seem to inhibit the growth of rupture nuclei and thereby make the high elongations possible. Stress-strain isotherms in the temperature range 6–150°C exibit the upturns on the Mooney–Rivlin curves which arise from the limited chain extensibility of short chains. The elongations at which the limited chain extensibility becomes descernible are relatively temperature insensitive, which permits the elimination of the strain-induced crystallization. Theoretical calculations have shown that the upturns resulting from non-Gaussian effects begin at approximately 60–70% of the average maximum extensibility of the PDMS chains, which is approximately twice what was generally accepted to be the case [27, 28]. Similar calculations have also shown that the network rupture occurs at 80–90% of maximum chain extension.

The general conclusion about crystallization preceding rupture has been lately confirmed by theoretical considerations using a non-Gaussian theory of rubber elasticity not only for bimodal networks, but also for short-chain unimodal networks [83]. This new theory has been employed for PE and PDMS short-chain networks with chains having 20, 40 and 250 skeletal bonds. It has been demonstrated that similar upturns in modulus at high deformations must also occur in such networks because of the rapidly diminishing number of configurations consistent with the required large values of end-to-end separations of the chains and, hence, a large decrease in the conformational entropy of the network chains. Calculations have also shown that the increase in modulus of PDMS networks should be significantly different from that of the amorphous PE networks having the same number of skeletal bonds and stretched to the same relative length.

Zhang and Mark [84] carried out the standard thermoelastic (force-temperature) measurements for bimodal PDMS networks from 30 to $-50°C$. The constant-length results did not give the expected dependence on λ and $f_u/f = 0.09(\pm 0.02)$, which is significantly less than the values obtained for unimodal long-chain PDMS networks (Table 6.2). The difference is supposed to be due to very limited extensibility of short chains. One can suppose that a network chain near its maximum extensibility can no longer increase its end-to-end separation by conformational changes and deformations of bond angles are required. The energies for these deformations are much greater than those for conformational changes and because of that the energy contribution must increase.

Kilian [85] used the van der Waals approach for treating the thermoelastic results on bimodal networks and concluded that thermoelasticity of bimodal networks could satisfactorily be described adopting the thermomechanical autonomy of the rubbery matrix and the rigid short segments. The decrease of f_u/f was supposed to be related to the dependence of the total thermal expansion coefficient on extension of the rigid short segment component. He also emphasized that calorimetric energy balance measurements are necessary for a direct proof of the proposed hypothesis.

6.2.3 Thermoelasticity of Mesophase Networks

Thermoelastic and photoelastic behavior of liquid crystalline polysiloxane networks containing the mesogenic side molecules has been studied [86]. On cooling, the networks are transformed from the rubbery state first to the nematic liquid crystalline state at T_c and on further cooling into the anisotropic glassy state. At constant force below T_c the length of the sample increases continously, but it decreases below T_c. At constant length of the sample, the stress decreases rapidly below T_c and approaches zero for small elongations of the sample. For unidirectionally elongated samples, the values of CT (C is the stress-optical coefficient) strongly increase some $20\,\mathrm{K}$ above T_c and level off well above T_c. The strong increase of CT is due to a partial orientation of the mesogenic molecules, which is confirmed by X-ray diffraction measurements.

During the last decade, it has been established that some flexible macromolecules without any mesogenic fragments in their chemical structure are able to form the thermodynamically stable thermotropic ordered phases possessing the structure and properties intermediate between the crystalline and amorphous phases [87, 88]. In a most striking form, the tendency to build an ordered state in polymers without mesogens, is displayed by linear elementorganic macromolecules – polysiloxane and polyphosphazenes. Macromolecules of both these classes are very flexible and can form typical molecular networks around room temperatures. Below we will consider the elasticity and thermoelasticity of PDES networks which have been studied lately.

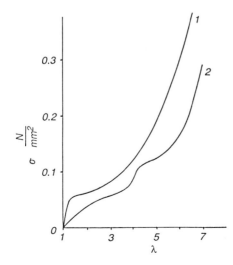

Fig. 6.9. Stress–strain curves of crosslinked PDES film at room temperature [87, 88]. *1* – mesomorphic film; *2* – completely amorphous film. Stretching rate – $0.65\,\mathrm{mm/s}$

The tendency of PDES macromolecules to form the mesophase leads to a peculiar stress–strain behavior of crosslinked samples in the rubbery state. The stress–strain curves of semi-mesomorphic crosslinked PDES films resemble the stress–strain curves of the crystalline polymers with a low degree of crystallinity. Figure 6.9 shows typical curves for slightly crosslinked film at 293 K. The same film which was preliminary amorphosized at 353 K behaves in a completely different manner. At relatively small elongations ($\lambda = 2.3$–3.0 depending on temperature) it elongates as a typical noncrystallizable network. At this initial stage the stress relaxation and hysteresis effects are absent and the curve is insensitive to the rate of deformation. Moreover, this part of the stress–strain curve follows the Mooney–Rivlin equation very well [89]. A rather unusual feature of the deformational behavior consists of a very large increase of the stress with temperature. In fact, the stress at 327 K which is at the final stage of the isotropization is almost twice that at 293 K. Analysis of the values of C_1 and C_2 in Mooney–Rivlin equation showed that it is a consequence of an unusually large decrease of C_2.

A further elongation induces the formation of the mesophase. Its appearence leads to a constant or even decreasing stress. Finally, a sharp increase of the stress occurs on the curve. In general, the shape of the stress–strain curve is determined by the relation of the deformation rate and the rate of the formation of the mesophase. At lower temperatures the formation of the mesophase is a rather fast process and the mesomorphic regions appeared at $\lambda > 2$ seem to be oriented at some angle to the stretching direction. The third part of the curve corresponds to the extension of the oriented mesophase sample. At increased temperatures the stress–strain curve looks like the stress–strain curve of the crystallizable networks. Similar behavior was also established on the PDES sample prepared by means of multifunctional crosslinking agents [90].

The characteristic feature of the thermomechanical behavior of PDES networks is a large decrease of the internal energy ΔU. Typical thermomechanical curves are shown in Figs. 6.10 and 6.11. Even at the initial extensions ($\lambda < 2$) when the formation of the mesophase is excluded according to the stress–strain curves the energy contribution is close to -1.2. It is an unusually high value of the energy contribution in comparison to other networks studied corresponding to a very high negative temperature coefficient of unperturbed chain dimensions $d \ln\langle r^2 \rangle_0 / dT = -4 \times 10^{-3} \deg^{-1}$. Recently this value was reconsidered because it has been established that a well-defined amount of the mesophase appeared at elongations even at temperatures as high as 130°C. This analysis showed that the more realistic value of the energy contribution for PDES macromolecules is $-0.25\,(\pm 0.05)$. It is noteworthy that the sign of the energy contribution in PDES is negative unlike PDMS (Tables 6.1 and 6.2). The negative value of the energy contribution in PDES means that the extended conformations of macromolecules are energetically more advantageous.

At 293 K the absolute value of the $\Delta U/W$ ratio at the initial stage of the stretching curve for the slightly crosslinked network is very high and

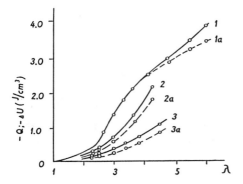

Fig. 6.10. Heat (*1, 2, 3*) and internal energy change (*1a, 2a, 3a*) on stretching PDES films from the unstrained state to λ at different temperatures [48, 50]. 1, 1a and 3, 3a – film with cross-link density 3.2×10^{-6} mole/cm^3; 2, 2a – film with cross-link density 6.5×10^{-6} mole/cm^3. 1, 1a – 293 K; 2, 2a and 3, 3a – 327 K

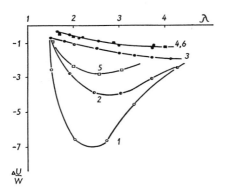

Fig. 6.11. The calorimetrically determined relative energy contributions as a function of elongation [87, 90]. *1, 2, 3, 4* – networks with 15% of a multifunctional crosslinking resin; *5, 6* – networks with 45.5% of a multifunctional resin. Temperatures: 1 – 20°C, 2 – 30°C, 3 – 50°C, 4 – 70°C and 100°C. 5 – 20°C, 6 – 80°C

close to -8. A considerable heat relaxation takes place in this case, although the deformation is completely reversible and the stress relaxation is absent. This thermal relaxation seems to be partly connected with the volume relaxation [89]. In the slightly crosslinked films the content of the mesophase according to the internal energy change can approach 0.7 while in the initial nonstretched samples the content at room temperature is not higher than 0.3–0.4. Hence, stretching promotes the formation of the mesophase and an obvious analogy exists between the mesophase formation and the crystallization of the crystallizable network under stretching.

6.3 Thermomechanical Behavior of Rubberlike Materials

Improvement of the mechanical properties of elastomers is usually reached by their reinforcement with fillers. Traditionally, carbon black, silica, metal oxides, some salts and rigid polymers are used. The elastic modulus, tensile strength, and swelling resistence are substantially increased by such reinforcement. A new approach is based on block copolymerization yielding thermoelastoplastics, i.e. block copolymers with soft (rubbery) and hard (plastic) blocks. The mutual feature of filled rubbers and the thermoelastoplastics is their heterogeneous structure [91]. In spite of their long history, reinforcement mechanisms and elastic properties of elastomers remain the subject of numerous investigations [92–97], but there is only a small number of investigations concerned with thermoelastic and thermomechanical behavior. In this section we will consider the thermomechanical behavior of rubberlike materials with special emphasis on the thermodynamic approach to their reinforcement.

6.3.1 Stress Softening: Thermomechanics and Mechanism

An important feature of filled elastomers is stress softening whereby an elastomer exhibits lower tensile properties at extensions less than those previously applied. As a result of this effect, a hysteresis loop on the stress–strain curve is observed. This effect is irreversible; it is not connected with relaxation processess but the internal structure changes during stress softening. The reinforcement results from the polymer–filler interaction which include both physical and chemical bonds. Thus, deformational properties and strength of the filled rubbers are closely connected with the polymer–particle interactions and the ability of these bonds to become reformed under stress.

Numerous investigations [96, 97] offer different viewpoints concerning the mechanism of stress softening. In one case, this effect is explained by the irreversible rupture of the polymer–filler bonds and some overloaded chain segments which lowers the crosslinking density. A large part of the stored energy ΔW is absorbed during the first extension and spent on the cleavage of bonds. According to another viewpoint, stress softening is caused by chain slippage over the particle surface which leads to a redistribution of the chain length. The extra work is spent during the first extension on the chain friction over the particle surface. Alternatively, stress softening is explained by non-affine irreversible displacement of crosslinks and entanglements during deformation exhibiting a hysteresis loop. It is important to emphasize that all these viewpoints suppose physical processes, mainly slippage, to be operative on the surface of filler particles.

The stress–strain behavior of thermoelastoplastics is as a rule a nonlinear one [98, 99]. It depends strongly on many factors, the most important being the volume fraction of the rigid phase. With a low content of hard blocks (less than 20%), the material behaves as a rubber, while with a high content of hard blocks (30–50%) the stress–strain curve in the first loading often exhibits an initial rapid rise in stress followed by a yielding and cold drawing with neck formation (Fig. 6.12) (see Chap. 7). The initial elastic part of the load-elongation curve is accompanied by absorbtion of heat and the neck formation and its propagation along the sample is accompanied by liberation of heat. The plastic flow and the neck propagation continue to high elongations where a rapid increase in stress occurs. Unlike the cold drawing of plastics in the first unloading of the block copolymers, there is a substantial stress recovery. Thus, after the first loading–unloading cycle, the initially plastic specimen becomes rubberlike and during the second cycle no yielding and necking occurs. Thermal effects accompanying the second cycle are similar to those resulting from the stretching of filled rubbers. This stress softening phenomenon observed in the first loading cycle of thermoelastoplastics was called the strain-induced plastic-to-rubber transition [101, 102].

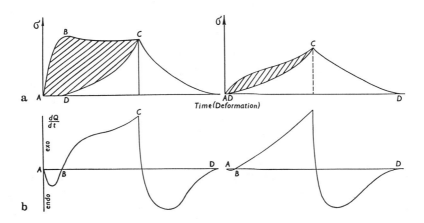

Fig. 6.12. Scheme of the thermomechanical behavior of a well phase-separated thermoelastoplastic [100]. *a* – stress–strain (or time) curves; *b* – heat effect traces versus time. *1* – First loading (*ABC*) and unloading (*CD*) cycle. *2* – Second loading (*AC*) and unloading (*CD*) cycle. The yielding point occurs at *B*. *AD* indicates the residual deformation after the first cycle. *AB* on the dQ/dt-time curve is the *endo*-effect resulting from the initial small-strain deformation *AB*

Quantitative thermomechanical investigations of stress softening were carried out on silicon rubbers [100, 103], SBS thermoelastoplastics [100, 104], segmented polyurethanes [105, 106] and polyester-polyether block copolymers [100, 107]. The results obtained reveal some important difference between stress softening of the filled rubbers and thermoelastoplastics (Fig. 6.13). The mechanical hysteresis for both materials depends on deformation in a similar way. However, the entire hysteresis loop for filled rubbers is converted into heat. The calorimetric data are in agreement with the current concepts explaining stress softening of filled rubbers by molecular mechanisms such as slippage of the chain over the filler particles [95–97] by which the structure of the rubber is transformed without any change of internal energy.

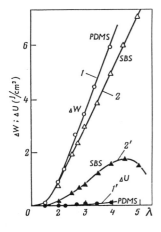

Fig. 6.13. Dependence of the mechanical hysteresis ΔW (*1, 2*) of loading–unloading cycle and the change of internal energy ΔU (*1', 2'*) on deformation [100]. *1, 1'* – filled PDMS rubber [103]; *2, 2'* – star-SBS thermoelastoplastic [104]

On the other hand, stress softening of thermoelastoplastics is accompanied by a considerable internal energy increase. Calorimetric and other results [55, 100, 105–107] seem to suggest a different mechanism of stress softening in filled rubbers and block copolymers. In filled rubber, it includes mainly chain slippage and breakage while breakdown and slippage of filler particles have a minor effect. Stress softening of thermoelastoplastics results in the breakage of a more or less continuos stress-supporting rigid phase and the considerable increase of internal energy is presumably due to this breakdown. An analysis of structural changes and the energy effects shows that absorbtion of energy results from plastic deformation of the glassy or crystalline phase and from the appearence of a new surface after breakage of this phase.

The hysteresis effects during the first loading–unloading cycle of the segmented polyurethanes and polyester-polyether with a high block content (> 50%) are extremely high and can reach 90% of the mechanical work spent on loading [105–108]. The large part of the deformation during such cycles is irreversible. Such behavior is quite similar to cold drawing of solid polymers (see Chap. 7). Hence, in this case, the first cycle is accompanied with stress hardening of the sample instead of stress softening. Such stress hardened elastomeric sample (fibers) resemble from the thermomechanical point of view cold drawn LDPE fibers and films.

Samples of SBS block copolymers prepared by extrusion or compression molding exhibit a long-range structure, consisting of cylindrical PS domains oriented parallel to the extrusion direction [108]. Keller and coworkers [108–110] have termed such a structure "single crystal" and demonstrated that the specimens are exceptionally anisotropic. Small angle X-ray studies of stress softening of the "single crystal" samples demonstrated very obviously that the deformation normal to the cylinders' axes is accompanied by their transformation into a zig-zag structure, while the longitudinal deformation results in the cylinders' breakdown into smaller particles [100, 111, 112]

The most striking feature of the stress softening phenomenon in thermoelastoplastics with hard block contents with less than 40% is its complete reversibility under certain conditions, consisting in a reformation process in the stress-free state which involves healing of cracks and reaching the initial integrity of the hard phase [100, 112, 113]. Figure 6.14 shows the reformation kinetics of the rigid phase in SBS Solprene 411 stress-softened samples. In this experiments the initial sample was stretched to reach total stress softening and the relative hysteresis ($\Delta W/W_{ex}$) was measured where W_{ex} is the mechanical work of extension (see Fig. 6.12). Then the stress softened samples were annealed at different temperatures and the hysteresis loops were obtained again. The reformation kinetics were characterized by ratio $(\Delta W/W_{ex})_2/(\Delta W/W_{ex})_1$. The sample annealed at 100°C (very close to T_g of PS) for some minutes showed nearly total reformation of the original hysteresis loop and there was no reformation process in the sample annealed at −5°C for more than one month. The reformation kinetics in stress-softened samples indicates that this process seems to be controlled by diffusion. Besides hysteresis other properties are also regained during annealing, for example, the initial endoeffect, resulting from the reformation of the stress-supporting hard phase. Finally small angle X-ray data show the reformation of the initial domain structure during annealing.

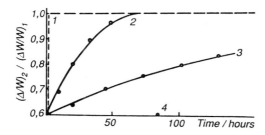

Fig. 6.14. The kinetics of the reformation of the mechanical properties and structure of the rigid phase of Solprene 411 [99, 100]. Temperature of reformation: *1* – 100°C, *2* – 60°C, *3* – 20°C, *4* – 5°C

6.3.2 Energy Contribution

Filled Rubbers.   The reinforcement of filled rubbers is usually determined by the particle size and the surface characteristics of filler particles [92–97]. In some studies an important role of internal energy effects in reinforcement have been emphasized [11]. Hence, thermomechanical measurements provide a very important approach to the study of such reinforcement. In the presence of reinforcing fillers, the elasticity modulus of the elastomers increases in first approximation according to the Guth–Smallwood equation [92–98]

$$E' = E(1 + 2.5\varphi + 14.1\varphi^2) \tag{6.61}$$

where E is the modulus of elasticity of unfilled elastomer and φ is the volume fraction of the filler. Thus, for filled elastomers, Eqs. (6.8)–(6.10) should include parameter φ. Nevertheless, the entropy and energy contributions must not be dependent on the presence of the filler [Eqs. (6.6), (6.7), (6.11) (6.12)]. This, of course, is a consequence of the intrachain character of these contributions in elastomers, because the filler cannot evidently change the energetics of conformational states of chains. It has been suggested that the role of filler consists in increasing the active chains in network [factor ν in Eq. (6.1)] due to network-filler links. In this connection, the determination of the energy contribution in filled elastomers and its comparison with the values obtained for unfilled networks is of great importance.

Figure 6.15 summarizes the results concerning the energy contribution of filled rubbers and thermoelastoplastics. A small amount of some fillers strongly increases the energy contribution which is in full contradiction to the assumed increase in the concentration of active network chains caused by the filler. Curve 2 represents the results for filled PDMS rubber and for PDMS block and graft copolymers. It can be seen that below 20% of the filler or hard phase, the energy contribution is practically independent of the amount of hard phase, but then a considerable increase of $(\Delta U/W)_{V,T}$ is observed. Although in all cases the energy contribution is dependent on the presence of filler, like in unfilled networks, it is independent of deformation.

The dependence of the energy contribution on the filler amount and its reinforcing ability demonstrates that in filled elastomers the energy contribution seems to lose its obvious meaning as a measure of intrachain effects. Galanty and Sperling [116] supposed that the energy contribution in filled elastomers includes the energy of deformation of hydrogen bonds, movement and deformation of filler particle and association or crystallization of chains on the filler surface. To take this into account, it is necessary to introduce some correction. Galanty and Sperling made such an attempt by introducing into the Tobolsky–Shen equation [Eq. (6.39)] a factor which reflects a reinforcing ability of the filler and arrived at the following expression for the energy contribution

$$\frac{f_u}{f} = \left(\frac{\Delta U}{W}\right)_{V,T} = T\frac{d\ln\langle r^2\rangle_0}{dT} + T\frac{d\ln F'}{dT} - \frac{E_f}{RT} \qquad (6.62)$$

where F' is a parameter taking into account the influence of the filler and E_f accounts for the sum of the internal energy contribution by the polymer–filler interaction as well as by the intrachain energy of polymer chain. However, to describe the increase of the elasticity modulus and strength of the filled silicon rubbers, assuming a constant concentration of active chains (parameter ν) is impossible. Therefore, the equations proposed by Galanty and Sperling can hardly be applied to a wide class of filled elastomers. However, regardless of the sign of the intrachain energy changes in an unfilled network, the reinforcement always contributes positively to the energy due to intermolecular interactions (Fig. 6.15). Hence, these results demonstrate indeed an important role of intermolecular energy effects in filled elastomers.

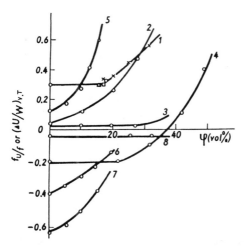

Fig. 6.15. Dependence of the energy contribution on the filler (filled rubbers) or hard phase (thermoelastoplastics) content. *1* – filled silicon rubber [103] Sil-51 (△), Sil-4600 (□); multiblock copolymer of polyarylate-PDMS (○) [117]; graft copolymer of PDMS and AN (×) [117]. *2* – butyl rubber with high abrasion furnace black [114]. *3* – butyl rubber with medium thermal black [114]. *4* – SBR-filled rubber [115]; *5* – aerosil HiSi-filled silicon rubber [117]. *6* – EPR-filled rubber [118, 119]. *7* – plasticized PVC filled with aerosil [120, 121]. *8* – SBS block copolymers [122, 123]

Block and Graft Copolymers. Stress-softened thermoelastoplastics are another class of reinforced rubberlike materials. The stress softening is a consequence of an extensive fragmentation of the initial stress-supporting hard phase during the first extension. Three problems draw the attention to thermomechanics of thermoelastoplastics: the energy contribution to the elasticity of a rubber matrix, the limited chain extensibility at high extensions and the role of hard domains in the hardening at high elongations.

SBS thermoelastoplastics possess the ability to undergo large reversible deformations (up to 2000%), being thus a good model for elucidating the mechanisms of rubber elasticity. Polybutadiens used in these thermoelastoplastics are not capable of crystallizing upon extension, therefore, their thermomechanical behavior can be studied at large deformations. In segmented polyurethanes and polyesters, the soft phase is built of short-chains polyesters which can exibit the limited chain extensibility.

Calorimetric investigations of the thermomechanical behavior of thermoelastoplastics have been carried out in recent years [100, 124–127]. Typical results are shown in Fig. 6.16. Taking into account the role of hard particles in the elasticity of filled networks and block and graft copolymers [98, 123, 128], the energy contribution is expected to be independent of the hard phase content and deformation. For a series of tri- and polyblock copolymers, $(\Delta U/W)_{V,T}$ is independent of the hard block content below 40% (Fig. 6.15, Table 6.5). It is independent of deformation as well. For the PDMS-AN copolymers, the energy contribution increases strongly with the

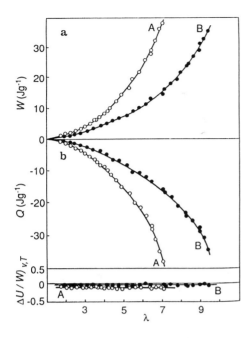

Fig. 6.16. Mechanical work (a), heat (b) and the energy contribution (c) on stretching the thermoelastoplastic elastomers from the unstrained state to λ [6]. A – Solprene 406, B – DST 30

hard block content (Fig. 6.15). In this case, the tendency is the same as in the majority of filled rubbers.

Kilian [43] describes work and heat effects at very high extensions of SBS linear and star thermoelastoplastics published [6] using the van der Waals equation of state with the following set of parameters in Eqs. (6.54)–(6.58): $\lambda_m = 8.5$ (star SBS) and $\lambda_m = 12$ (linear SBS); $a_1 = 0.2$; $a_2 = 0.1$ and $\beta = 2.5 \times 10^{-4}\,\mathrm{K}^{-1}$. He arrived at the conclusion that the non-Gaussian effects at high extensions arise primarily due to the limited chain extensibility.

Very interesting thermomechanical behavior was found for stress-softened segmented polyurethanes and polyesters with a high hard block content. Stress softening of such thermoelastoplastics is accompanied by a considerable residual deformation (approximately 50% of the total deformation) arising from plastic deformation and orientation of hard domains in the stretching direction. These oriented block copolymers exhibit complete elastic recovery and can be used as elastic fibers. Bonart et al. [129–131] have studied the orientational mechanisms and thermomechanics of segmented polyurethanes in considerable detail.

Table 6.5. Energy contribution in block and graft copolymers [100, 117, 124, 134]

Copolymer	Block arrangement	Soft block structure	Hard block structure and content	$(\Delta U/W)_{V,T}$
1. Butadienestyrene				
DST 30	SBS, linear	PB	PS; 0.283	−0.03
Kraton 101	SBS, linear	PB	PS; 0.319	−0.03
Solprene 406	SBS, star	PB	PS; 0.394	−0.13
Solprene 411	SBS, star	PB	PS; 0.340	−0.10
2. Isoprene-styrene				
IST 30	SIS, linear	PI	PS; 0.300	0.25
3. Segmented polyurethanes				
Estane 5707	Multiblock	PTMA	MDI BD; 0.380	0.10
Sanprene E-6	Multiblock	PEA	MDI BD; 0.400	0.20
OBU	Multiblock	PB	MDI BD; 0.175	−0.05
TPBU	Multiblock	PB	MDI TAD; 0.36	−0.30
PEU-18	Multiblock	PTMA/PEA	MDI BD; 0.18	−0.60
PEU-23	Multiblock	PTMA/PEA	MDI BD; 0.23	−0.65
PEU-33	Multiblock	PTMA/PEA	MDI BD; 0.33	−0.25
4. Segmented polysiloxanes				
Siloxane-arylate	Multiblock	PDMS	Ar; 0.18	0.30
Siloxane-arylate	Multiblock	PDES	Ar; 0.18–0.30	−0.25
Siloxane-siloxane	Multiblock	PDES	PASSO; 0.33–0.43	−0.25
TPSU	Multiblock	PDMS	MDI TAD; 0.36	0.30
Siloxane-AN graft copolymer		PDMS	PAN; 0.170	0.30
			0.240	0.40
			0.360	0.60

PTMA – poly(tetramethylene adipate); PEA – Poly(ethylene adipate); MDI – diphenylmethane diisocyanate; BD – butane diol; TAD – triazine-containing diol; PEU – polyester urethane; TPSU – triazine-containing polysiloxyne-urethane; PASSO – ladder polyorganosilsesquioxane; PAN – polyacrylonitril; OBU – oligobutadieneurethane; Ar – polyarylate

Figure 6.17 shows the thermomechanic characteristics of segmented polyurethanes with butadiene soft block [132]. The energy changes in the samples containing less than 50% hard block differ principally from those samples with 50% content or more. In the sample with 42% hard block, the internal energy begins to decrease at a specific elongation and becomes negative. Such behavior is similar to that of typical SBS block copolymers and is controlled by the negative intramolecular energy contribution of PB blocks. In samples with more than 50%, the internal energy increases considerably with elongation, which can arise only as a result of the intermolecular changes in the rubbery component. This means that intermolecular changes dominate the thermomechanical behavior of these samples. This conclusion is quite different from that arrived at concerning the chemically crosslinked networks and tri- and multiblock copolymers with the hard content of less than 50%. Such behavior is very characteristic of not only this particular type of segmented polyurethanes, but has also been observed in block and graft copolymers with PDMS as a rubbery component [117].

Perhaps the most striking example of dramatic changes in thermomechanical behavior with an increasing hard block content has been observed in segmented polyesters. Typical behavior of two multiblock copolymers with 40 and 55 percent of the hard (PBT) block content is shown in Fig. 6.18. Polytetramethyleneoxide, which is the rubbery block in these systems, possesses a rather high negative energy contribution -0.47. Following this value one can estimate the change of the intrachain energy as a function of deformation. The corresponding results together with the calorimetrically determined internal energy changes, as well as the changes according to the statistical theory are shown in Fig. 6.19. It is seen that for the sample with 40% hard block the experimental and estimated data are close only in the region of small and large extensions; in the intermediate region there is a large discrepancy indicating that in this region there is a considerable positive energy contribution, which seems to arise as a result of the intermolecular changes in the rubbery matrix. Nevertheless, the conformational changes dominate.

The situation changes dramatically in the sample with 55% hard block. The energy changes revealed calorimetrically do not follow the theoretical predictions even qualitatively. Indeed, there is thermomechanical internal energy inversion behavior unpredictable by the statistical theory of rubber elasticity: at small strains, the internal energy decreases and then increases considerably at high extensions. Summing up, one may conclude that at present there is sufficient experimental evidence to indicate that the free energy of a strained rubberlike block copolymer with a relatively high hard component content cannot originate only within the chains of of the network. The interchain effects play a considerable role.

At high deformations, non-Gaussian effects cause an upturn in the Mooney–Rivlin plot. For the SBS thermoelastoplastics, a similar upturn is observed at high extensions. The upturn for SBS block copolymers can be explained either by the limited chain extensibility or by the presence of glassy incompressible PS domains [82]. At moderate deformations, these domains are undeformed, but at high deformations they undergo deformation as is evident by electron microscopy. Hence, the hardening of SBS materials resulting from the deformation of domains seems to be a more real reason for the upturns at high extensions than the limited chain extensibility.

One can expect the limited chain extensibility to be effective in segmented block copolymers because of the lower molecular weights of their soft blocks. However, careful thermomechanical investigations have shown that at large deformations, some additional structuring of hard blocks caused by reformation of hydrogen bonds during deformation occurs [100, 124, 133]. This conclusion is in agreement with the result of Bonart et al. [129–131]. Because of this effect, it is difficult to check whether the intrachain energy contribution is independent of the extension ratio and whether the limited chain extensibility of the soft block occurs at large extension. Hence, the problem

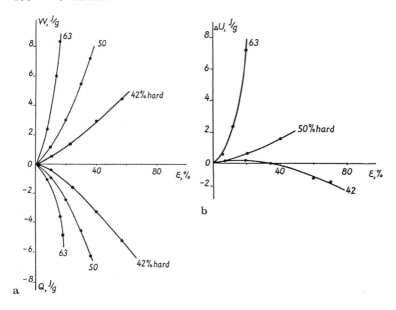

Fig. 6.17. Dependencies of work and heat (a) and internal energy changes (b) on strain for polyurethanes with various hard block contents (indicated near curves) [100, 132]

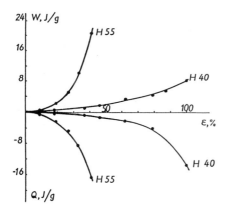

Fig. 6.18. Dependence of work and heat on strain for polyester-type block copolymer "Hytrel" (Du Pont) with 40% (H-40) and 55% (H-55) hard block content [100, 132]

of approaching the limited chain extensibility in block and graft copolymers remains open and requires further thermomechanical studies.

6.3.3 Mesophase Block Copolymers

Incorporation of the mesomorphic components into block copolymers gives a new interesting class of rubberlike materials. A good example of such materials is the multiblock copolymers in which the soft block is mesomorphic PDES and polyarylate or ladder polyorganosiloxane serving as the hard block [134]. The components in these block copolymers are totally incom-

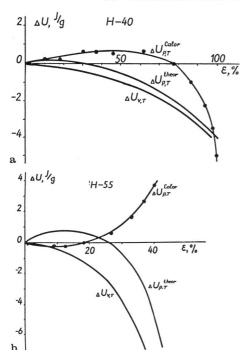

Fig. 6.19. The internal energy changes as a function of strain for polyester-type block copolymers "Hytrel" with 40% (**a**) and 55% (**b**) hard block content [55, 100, 132]. $\Delta U_{V,T}$ and $\Delta U_{P,T}^{theor}$ are computed according to Eqs. (6.5) and (6.10) with the corresponding set of parameters

patible and, therefore, a very good microphase separation exists in them. In multiblock copolymers, molecular weights of their components are as a rule on the oligomer level. In linear PDES the mesophase appears only in polymers with molecular weight above $(30\text{--}40)\times10^3$. Therefore in the initial state soft PDES component in the block copolymers is amorphous. However, the uniaxial stretching of some of the block copolymers is accompanied by appearance of the mesophase, which can be established by means of deformation calorimetry and X-ray analysis. Figure 6.20 demonstrates the dependence of mechanical work, heat evolved during stretching and the internal energy change on elongation. Similar to other block copolymers the values of W, Q and ΔU under elongation depend on the hard block content and temperature. With comparative contents of hard and soft blocks the block copolymers are only capable of relatively small reversible deformations, which are accompanied by an increase in internal energy. Because the energy contribution of PDES is negative it means that at such ratios of the components, the intermolecular changes dominate during deformation. When the content and molecular weight of the PDES block increases, the ability to elongate increases considerably. Now the internal energy decreases upon extension. The form of the curves for samples with the longest PDES block (20×10^3) as well as X-ray data correspond to the appearence of the mesomorphic state under extension at room temperature. At higher temperatures (70°C) the mesophase does not appear. The degree of mesophasity is between 5 and 30% for the block copolymers and about 50% for the additionally crosslinked samples. The effect of the appearence of the mesophase under stretching was

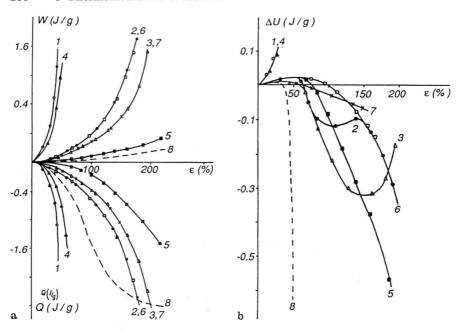

Fig. 6.20. Dependence of work and heat (**a**) and internal energy changes (**b**) on strain for mesomorphic block copolymers [134]. PFSSO – PDES block copolymers with the following molecular weights of the blocks: *1* – 6200–9000; *2, 6* – 8300–15000; *3, 7* – 8300–20400. PAr – PDES block copolymer with the following molecular weights of the blocks: *4* – 4000–9000; *5* – 4600–20400. *8* – chemically crosslinked PDES with 15% of a multifunctional crosslinking agent. 1, 2, 3, 4, 5, 8 – 20°C; 6, 7 – 70°C

completely reversible and on contraction the mesophase disappeared. Such block copolymers seem to be of interest for the development of membrane materials with regulated transport properties.

6.3.4 Random Copolymers

Thermoelasticity of rubberlike random copolymers was studied by Shen et al. [11, 135, 136]. Figure 6.21 shows the dependence of the energy contribution f_u/f on composition for the copolymers of *n*-butyl acrylate and ethylhexyl acrylate. It can be seen that the value of the energy contribution to the rubber elasticity of the copolymers are intermediate between values corresponding to the respective homopolymers. However, this simple correlation does not obtain for other random copolymers such as copolymers of styrene and buta-diene, ethylene and propylene. Shen and coworkers concluded that since the tacticity, *cis-trans* isomerism, and sequence distribution must play an impor-tant role in determining the energy contribution for such a comparison, data on well-characterized copolymers should be used.

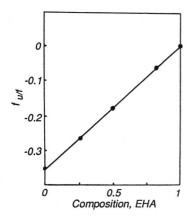

Fig. 6.21. Variation of f_u/f at $40°C$ as a function of composition for the copolymers of n-butyl acrylate and ethylhexyl acrylate [11, 136]

6.3.5 Elastomer Blends

Elastomer blends consisting of two immiscible components are heterogeneous rubberlike materials both components of which are in the rubbery state. Such blends consist usually of either a matrix and a discrete phase or two interpenetrating continuous phases (interpenetrating networks). With homogeneous deformations of such blends the contribution of either component to the thermomechanical behavior of the material is determined by the content of the component and the individual characteristics of its chains.

A comprehensive calorimetric study of the thermomechanical properties of elastomeric blends was carried out on heterogeneous blends NR and EPR [137]. These components are completely immiscible and each of them is characterized by a rather large intrachain energy contribution of opposite sign (see Table 6.2). The energy contribution of the blends at small and moderate deformations, where the strain- induced crystallization of NR is still absent, is an additive function of the composition. The blend containing 60% NR behaves like an ideal rubber since $(\Delta U/W)_{V,T}$ is equal to zero. It means that an increase in the conformational energy of NR is totally compensated by a decrease of the conformational energy of EPR. In spite of the heterogeneous structure of elastomer blends of all compositions, they behave upon deformation like homogeneous networks.

6.3.6 Bioelastomer Materials

It is well known that there exist a number of crosslinked amorphous proteins which in the swollen state exhibit a reversible rubberlike elasticity. Thermomechanical behavior of elastin has attracted special attention because of its widerspread physiological functions in body tissues [138, 139]. Although elastin has some fibrous structure, it is amorphous and swells in water. The thermodynamics of the deformation of elastin was studied using calorimetry and thermoelasticity [139–142]. In a series of microcalorimetric measurements at room temperature in various solvents (water, methanol, ethanol,

n-propanol, formamide), it was found that the heat produced during stretching always exceeded the mechanical work of stretching. It was almost equal to the heat absorbed during contraction. Because the stress–strain curves were also reversible, the swollen system was treated as a thermodynamically closed one working at constant temperature and pressure. The volume change resulting from the deformation was not taken into account. From the calorimetric measurements it was concluded that the elastic force arises from interfacial effects, primarily energetic.

As has been pointed out in some investigations [143, 144], however, the calorimetric results have been misinterpreted because no account was taken of the effect of the heat of dilution accompanying the stress-induced increase in swelling. Recent thermoelastic measurements carried out at constant composition (closed system) rather than at swelling equilibrium (open system) over the temperature range 8–35°C indicated that the elastic force is primarily of entropic origin with the energy contribution $f_u/f = 0.26 \, (\pm 0.09)$ [145]. This value is somewhat larger than that (ca. 0.1) obtained from measurements on open systems [144–147]. The fact that the elasticity of elastin is primarily entropic in origin and that f_u/f significantly exceeds zero and is of the same order of magnitude as in synthetic elastomers, indicates that the rotational state theory of chain conformations can be successfully applied for characterizing thermodynamics of deformation of bioelastomeric materials.

Another area of biology which widely uses the thermomechanical approach is membrane science [148, 149]. In biological membranes which are composite materials made up of both solid- and liquidlike elements the material is anisotropic in three dimensions but can be isotropic in the surface plane. Deformation of a thin membrane material element can be decomposed into: i) a change in thickness, ii) uniform area dilatation or contraction and iii) extension of the material in the surface plane at constant area. The changes in i) and ii) include volume change during deformation. As far as iii) is concerned it may be treated from the point of view that the membrane surface can be considered as a two-dimensional continuum, in some particular cases as a two-dimensional elastomeric surface network. In the early treatment of the elastic free energy changes of deformation of a membrane at constant membrane area based on the classic theory of rubber elasticity it was concluded that the free energy of the red blood cell membrane is totally entropic [150]. However, recent thermoelastic measurements have shown that the red cell membrane is not a simple entropic elastomer and that the dominant change in extension of the membrane is energetic. The energy and entropy contributions to the elasticity of red cell membrane were evaluated and they turned out to be of the order of thermal energy. This illustrates that the elastic extension of red cell membrane is accompanied by small thermodynamic changes in chemical equilibrium, which are much lower than the energies associated with hydrophobic, electrostatic, and metabolic effects [150, 151].

Appendix to Chapter 6

Thermomechanics of the New Models of Rubber Elasticity

During the last 15 years the classical theory of rubber elasticity has been changed significantly. For the phantom noninteracting network whose chains move freely through one another instead of the number of elastically active chains ξ the cycle rank of the network i.e. the number of independent circuits it contains, was introduced (see, e.g. Ref. 15). For a perfect phantom network of uniform functionality $\varphi(> 2)$

$$\xi = \nu(\varphi - 2)/\varphi .\qquad\qquad (A6.1)$$

The behavior of real networks varies between two extremes: the phantom model and the affine model [15]. A number of theoretical models have been formulated recently to describe the behavior of real networks with sterical restrictions resulting from interchain constraints of fluctuations of network junctions. The theoretical treatments can be roughly grouped into four types: the constrained junction fluctuation model [16, 17] in which each network junction is subjected to a domain of constraints; sliplink models [18, 19] in which each network chain threads its way through a number of small rings: tube models [20–22] in which each chain is confined within a tube; and primitive path models [23, 24] in which the chain segments lie along the shortest path between chain ends. According to the models, the free energy of deformation and the stress in the network with constrained chains contains an additive contribution to that describing the phantom network.

In the junction fluctuation model the free energy of deformation for a perfect network is [16, 17]

$$F = \text{vkT}\frac{l_0}{a_0}\left[\frac{\lambda^2 + 2\lambda^{-1}}{3} + r\left(\frac{\lambda^2 + 2\lambda^{-1}}{3}\right)^{1/2}\right]$$

$$\frac{f}{S_0} = \text{vkT}\frac{l_0}{a_0}\left[\frac{2}{3} + \frac{r}{3}\left(\frac{\lambda^2 + 2\lambda^{-1}}{3}\right)^{-1/2}\right](\lambda - \lambda^{-2})$$

$$F = F_{\text{ph}} + F_c = \frac{\text{vkT}}{2}\sum_i(\lambda_i^2 - 1)$$

$$+ \frac{vkT}{2} \sum_i \left\{ \kappa_c \frac{(\lambda_i^2 - 1)(\lambda_i^2/\kappa_c + 1)}{(\lambda_i^2 + \kappa_c)^2} - \ln\left[\kappa_c^2 \frac{(\lambda_i^2 - 1)}{(\lambda_i^2 + \kappa_c)^2} + 1 \right] \right.$$

$$\left. - \ln\left[\kappa_c^2 \frac{(\lambda_i^2 - 1)(\lambda_i^2/\kappa_c + 1)}{(\lambda_i^2 + \kappa_c)^2} + 1 \right] \right\} \tag{A6.2}$$

and the tension in uniaxial strain

$$f = f_{ph} + f_c = f_{ph}(1 + f_c/f_{ph})$$
$$f_c/f_{ph} = [\lambda K_c(\lambda^2) - \lambda^{-2} K_c(\lambda^{-1/2})](\lambda - \lambda^{-2})^{-1} \tag{A6.3}$$

where F_{ph} and f_{ph} correspond to the contribution of the phantom network and F_c and f_c correspond to the constraints. $K_c(\lambda^2)$ and $K_c(\lambda^{-1/2})$ are functions of the parameter κ_c describing the severity of the constraints. If $\kappa_c = 0$, then F_c and f_c are equal to zero and Eq. (A6.2) and (A6.3) go over to those for a phantom network. If $\kappa_c \to \infty$, corresponding to complete suppression of junction fluctuations, Eqs. (A6.3) are transformed into Eqs. (6.1) and (6.2) for the affine network and Eq. (A6.2) to Eq. (6.3). Flory [15–17] suggested that

$$\kappa_c = I \langle r^2 \rangle_0^{3/2} (\mu_c/V^0) \tag{A6.4}$$

where I is a universal interpenetration parameter, and μ_c is the number of junctions in the volume V^0 of the state of reference. Using Flory's definition of the tension–temperature coefficient [15] $d \ln V^0/dT = 3/2 d \ln\langle r^2 \rangle_0/dT$ one can arrive at

$$d\kappa_c/dT = \langle r^2 \rangle_0^{3/2} (\mu_c/V^0)(dI/dT) . \tag{A6.5}$$

We see that, only if the parameter I is temperature independent, the entropic and energetic components of real networks with the sterical restrictions are identical to that of the phantom or affine network.

The free energy of deformation in sliplink model is [19]

$$F = \frac{N_c kT}{2} \sum_i \lambda_i^2 + \frac{N_s kT}{2} \sum_i \left[\frac{(1 + \eta_s)\lambda_i^2}{1 + \eta_s \lambda_i^2} + \log(1 + \eta_s \lambda_i^2) \right] \tag{A6.6}$$

where N_c and N_s are the number of ordinary crosslinks and sliplinks, correspondingly, and η_s is a parameter which measures the ability of a sliplink to slide along the chain passing through it; $\eta_s = 0$ corressponds to the phantom network. The free energy of deformation is attributed solely to entropy change. If the parameter η_s is temperature independent, then this version of the sliplink model corresponds to the ideal entropy-elastic model. If η_s depends on temperature, the entropy contribution is

$$\left(\frac{T\Delta S}{W} \right)_{V,T} = \frac{f_s}{f}$$

$$= - \frac{N_c \sum_i \lambda_i^2 + N_s \sum_i \dfrac{\lambda_i^2}{1 + \eta_s \lambda_i^2} \left[\dfrac{1 + \eta_s}{1 + \eta_s \lambda_i^2} + 0.434 \right] \dfrac{d\eta_s}{dT}}{N_c \sum_i \lambda_i^2 + N_s \sum_i \left[\dfrac{(1 + \eta_s)\lambda_i^2}{1 + \eta_s \lambda_i^2} + \log(1 + \eta_s \lambda_i^2) \right]} \tag{A6.7}$$

We see that in this case the entropic and energetic component of the free energy is dependent on deformation and that the energy contribution is of the intermolecular origin.

The Marrucci random tube model [20] leads to the following expressions for the free energy and the elastic force at simple extension or compression for tubes with a circular cross section

$$F = vkT\frac{l_0}{a_0}\left[\frac{\lambda^2 + 2\lambda^{-1}}{3} + r\left(\frac{\lambda^2 + 2\lambda^{-1}}{3}\right)^{1/2}\right] \tag{A6.8}$$

$$\frac{f}{S_0} = vkT\frac{l_0}{a_0}\left[\frac{2}{3} + \frac{r}{3}\left(\frac{\lambda^2 + 2\lambda^{-1}}{3}\right)^{-1/2}\right](\lambda - \lambda^{-2}) \tag{A6.9}$$

where l_0 and a_0 are the tube length and the tube radius of the initial uncrosslinked polymer, respectively, r is the ratio of tube cross-sectional area before and after curing (r = 1 for a perfect network and r < 1 otherwise) and S_0 is the initial cross section of the sample. One may suggest that both l_0 and a_0 have an identical temperature dependence in isotropic networks and, therefore, the ratio l_0/a_0 seems to be temperature independent. Since r is also a constant, we conclude that the tube parameters of the Marrucci tube model are temperature independent. The Gaylord tube model [21] includes also two parameters having the same meaning as in the Marrucci model. Thus, although the tube models give rise to expressions for the elastic free energy which are able to predict some nonclassic effects, the thermomechanical behavior of the models seems to be similar to the classic entropy-elastic models.

Priss [22] proposed another tube model and obtained the following expression for the elastic force at simple elongation

$$\begin{aligned}f = {} & 2C_1(\lambda - 1/\lambda^2) + 2C_c[-1/\lambda^2 + (3/2)\lambda/(\lambda^3 - 1) \\ & + (\lambda^4 - 4)/2(\lambda^3 - 1)^{3/2}]\end{aligned} \tag{A6.10}$$

where

$$C_1 = (vkT/2)\langle r^2\rangle/\langle r^2\rangle_0 \; ; \quad C_c = \frac{vkTK_0^2}{32\alpha_m^2}\left[\frac{(m+1)\text{sh}2\delta - \text{ch}2\delta - 3}{\text{sh}^4\delta}\right]$$

$$\alpha_m = 3(m+1)/2Nl^2 \; ; \quad \delta = \text{arcch}\,(1 + K_0/2\alpha_m) \; .$$

N is the number of statistical segments in a chain, l is their length and m the number of submolecules. This model includes the front factor only in C_1, but the parameter K_0 is also temperature dependent; it depends on the thermal expansion. Thus, the Priss tube model predicts different temperature dependences of C_1 and C_c. It means that the entropy and energy contribution have to be dependent on λ and include a considerable intermolecular part.

The free energy and the elastic force for simple elongation or compression in the primitive path model is [24]

$$F = \frac{\mu_c kT}{2} \left[J^2 \frac{(1 - \alpha_p)}{(1 - \alpha_p J)} + \log \frac{(1 - \alpha_p J)}{(1 - \alpha_p)} \right] + \mu_c kT[J^2 + \beta_p(J - 1)] \quad (A6.11)$$

$$f = \mu_c kT \left[\frac{2}{3} + \frac{\beta_p}{3J} + \frac{(1 - \alpha_p)}{3(1 - \alpha_p J)} + \frac{\alpha_p J(1 - \alpha_p)}{6(1 - \alpha_p J)^2} - \frac{\alpha_p}{6J(1 - \alpha_p)} \right] (\lambda - \lambda^{-2})$$
$$(A6.12)$$

where $J = [(\lambda^2 + 2\lambda^{-1})/3]$, μ_c is the number of crosslinks, α_p and β_p are parameters independent of deformation but dependent on temperature and density. Only isoenergetic chains are considered. If α_p and β_p are equal to zero Eqs. (A6.11) and (A6.12) transform to Eqs. (6.1) and (6.3) for isoenergetic chains. Using the definition of the entropy and energy components of the free energy, one can demonstrate that both the components are dependent on λ and that the energy contribution again is of intermolecular origin.

Thus, the thermoelasticity of the majority of the new models is considerably more complex than that of the phantom networks. However, these new models contain temperature-dependent parameters which are difficult to relate to molecular characteristics of a real rubber-elastic material. Recent analysis by Gottlieb and Gaylord [25] has demonstrated that only the Gaylord tube model and the Flory constrained junction fluctuation model agree well with the experimental data on the uniaxial stress-strain response. On the other hand, their analysis has shown that all of the existing molecular theories cannot satisfactorily describe swelling behavior with a physically reasonable set of parameters. The thermoelastic behavior of the new models has not yet been analysed.

Finally, recently Abramchuck and Khochlov [26] developed a new molecular theory of elasticity of molecular networks taken into account the orientational ordering of subchains during extension. The thermomechanical properties of the model have not been examined yet.

References

1. Flory PJ (1961) Trans Farad Soc *57*: 829
2. Mark JE (1973) Rubber Chem Techn *46*: 593
3. Mark JE (1982) Rubber Chem Techn *55*: 1123
4. Allen G et al (1971) Trans Farad Soc *67*: 1278
5. Price C (1976) Proc Roy Soc Ser A *351*: 331
6. Godovsky YuK (1981) Polymer *22*: 75
7. Gee G (1946) Trans Farad Soc *42*: 585
8. Guth E, James HM and Mark H (1946) Adv Colloid Sci *2*: 253
9. Shen M (1969) Macromolecules *2*: 358
10. Tobolsky AV and Shen M (1966) J Appl Phys *37*: 1952
11. Treloar LRG (1969) Polymer *10*: 291
12. Shen M and Kroucher M (1975) J Macromol Sci *C12*: 287
13. Bleha T (1981) Polymer *22*: 1314
14. Schwarz J (1981) Polym Bull *5*: 151
15. Flory PJ (1985) Polymer J *17*: 1
16. Flory PJ (1977) J Chem Phys *66*: 5720

17. Flory PJ and Erman B (1982) Macromolecules *15:* 800, 806
18. Marrucci G (1979) Rheol Acta *18:* 193
19. Ball RC, Doi M, Edwards SF and Warner M (1981) Polymer *22:* 1010
20. Marrucci G (1981) Macromolecules *14:* 434
21. Gaylord RJ (1982) Polym Bull *8:* 325; (1983) *9:* 181
22. Priss LS (1981) Pure Appl Chem *53:* 1581
23. Edwards SF (1977) Brit Polym J *9:* 140
24. Greassly WW (1982) Adv Polym Sci *47:* 67
25. Gottlieb M and Gaylord RJ (1983) Polymer *24:* 1644; (1984) Macromolecules *17:* 2024
26. Abramchuk SS and Khochlov AR (1987) DAN SSSR *297:* 385
27. Treloar LRG (1975) The physics of rubber elasticity, (Clarendon 3); Treloar LRG (1979), In: Applied fibre science, p 103, Happey F (ed), Academic London
28. Dusek K and Prins W (1969) Structure and elasticity of non-crystalline polymer networks, In: Adv Polym Sci, vol 6, p 1, Springer Berlin Heidelberg New York
29. Schwarz J (1978) Habilitationsschrift, Technical University Clausthal
30. Schwarz J (1979) IUPAC Internat Symp Macromolecules, Mainz, pp 1370–1373;(1979) Europhysical Conference on Macromolecules, Jablonna, pp 121–122
31. Mark JE and Curro JG (1983) J Chem Phys *79:* 5705; (1984) *80:* 4521
32. Shen M (1970) J Appl Phys *41:* 4351
33. Valanis KC and Landel RF (1967) J Appl Phys *38:* 2997
34. Blatz PJ, Sharda SC and Tschoegl NW (1974) Trans Soc Rheol *18:* 145
35. Sharda SC and Tschoegl NW (1976) Macromolecules *9:* 910
36. Chang WV, Bloch R and Tschoegl NW (1976) Macromolecules *9:* 917
37. Tschoegl NW (1979) Polymer *20:* 1365
38. Tschoegl NW (1982) IUPAC Internat Symp Macromolecules, Amherst, p 862
39. Ogden RW (1972) Proc Roy Soc Ser A *326:* 565
40. Chadwick P (1974) Phil Trans Roy Soc London Ser A *276:* 371
41. Ogden RW (1978) J Mech Phys Solids *26:* 37
42. Gee G (1980) Macromolecules *13:* 705
43. Kilian H-G (1980) Colloid Polym Sci *258:* 489; (1981) *259:* 1084
44. Eisele U, Heise B, Kilian H-G and Pietralla M (1981) Angew Makromol Chem *100:* 67
45. Vilgis Th and Kilian H-G (1983) Polymer *24:* 949
46. Ogden RW (1979) J Phys D Appl Phys *12:* 465
47. Ogden RW (1986) Rubber Chem Technol *59:* 361
48. Chadwick P and Greasy CFM (1984) J Mech Phys Solids *32:* 337
49. Plate NA and Shibaev VP (1980) Comb-Like Polymers and Liquid Crystalls (in Russian), Khimiya Moscow
50. Adv Polym Sci (1984) vol 59, 60, 61, Springer Verlag Berlin Heidelberg New York
51. Jarry JP and Monnerie L (1979) Macromolecules *12:* 316
52. Rusakov VV (1983) In: Mathem Methods for Polymer Investigation, pp 54–55, Puschino
53. Rusakov VV (1984) In: Structural and Mechanical Properties of Composite Materials, pp 37–48, Sci Papers of Ural Sci Center of Akad Nauk SSSR, Sverdlovsk
54. Mark JE (1976) J Polym Sci Macromol Revs *11:* 135
55. Godovsky YuK (1987) Progr Colloid Polym Sci *75:* 70
56. Wolf FP and Allen G (1975) Polymer *16:* 209
57. Gent AN and Kuan TH (1973) J Polym Sci *11:* 1723
58. Allen G, Price C and Yoshimura N (1975) Trans Farad Soc *71:* 548

59. Price C, Evans KA and DeCandia F (1973) Polymer *14:* 339
60. Göritz D and Müller FH (1973) Kolloid Z Z Polym *251:* 679
61. Godovsky YuK (1977) Vysokomol Soedin *A19:* 2359
62. Mark JE and Flory PJ (1964) J Am Chem Soc *86:* 138
63. Price C, Allen G and Yoshimura N (1975) Polymer *16:* 261
64. Birshtein TM and Ptitsyn OB (1964) Conformations of Macromolecules (in Russian), Nauka Moscow
65. Boyer RF and Miller RL (1978) Internat Conference on Rubber, Rubber-78, Kiev, Preprint A3
66. Kilian H-G (1982) Colloid Polym Sci *260:* 895
67. Kilian H-G (1982) IUPAC Internat Symp Macromolecules, Amherst, p 566
68. Shen M and Blatz PJ (1968) J Appl Phys *39:* 4937
69. Treloar LRG (1978) Polymer *19:* 1414
70. Christensen RC and Hoeve CAJ (1970) J Polym Sci A-1 *8:* 1503
71. Gee G (1987) Polymer *28:* 386
72. Mark JE (1979) Polym Eng Sci *19:* 254
73. Mark JE, Kato M and Ko JH (1976) J Polym Sci *C54:* 217
74. Su T-K and Mark JE (1977) Macromolecules *10:* 120
75. Kato M and Mark JE (1976) Rubber Chem Technol *49:* 85
76. Göritz D and Müller FH (1973) Kolloid Z Z Polym *251:* 892
77. Göritz D and Grassler R (1987) Rubber Chem Technol *60:* 217
78. Furukawa J et al (1982) Polym Bull *6:* 381
79. Llorente MA and Mark JE (1979) J Chem Phys *71:* 682
80. Llorente MA and Mark JE (1980) J Polym Sci Polym Phys Ed *18:* 181
81. Andrady AL, Llorente MA and Mark JE (1980) J Chem Phys *72:* 2282; *73:* 1439; *74:* 2289
82. Mark JE (1982) Adv Polym Sci, vol 44, pp 1–26, Springer Berlin Heidelberg New York
83. Mark JE and Curro JG (1984) J Chem Phys *80:* 5262
84. Zhang ZM and Mark JE (1982) J Polym Sci Polym Phys Ed *20:* 473
85. Kilian H-G (1981) Colloid Polym Sci *259:* 1151
86. Kock HJ et al (1983) Amer Chem Soc Polym Prepr *24:* N2, 300
87. Godovsky YuK and Papkov VS (1989) Adv Polym Sci, vol 88, pp 129–180, Springer Berlin Heidelberg New York
88. Godovsky YuK and Papkov VS (1986) Makromol Chem Macromol Symp *4:* 71
89. Papkov VS, Svistunov VS and Godovsky YuK (1989) Vysokomol Soedin *A31:* 1577
90. Godovsky YuK et al (1988) Vysokomol Soedin *A30:* 359
91. Noshay A and McGrath JE (1977) Block Copolymers. Overview and Critical Survey, Academic Press New York
92. Polmanter KE and Lentz CW (1975) Rubber Chem Technol *48:* 795
93. Kraus G (1978) Rubber Chem Technol *51:* 297
94. Warrick EL et al (1979) Rubber Chem Technol *52:* 437
95. Boonstra BB (1979) Polymer *20:* 691
96. Rigbi Z (1980) In: Adv Polym Sci, vol 36, Springer Berlin Heidelberg New York
97. Dannenberg EM (1975) Rubber Chem Technol *48:* 410
98. Holden G, Bishop ET and Legge NR (1969) J. Polym. Sci *C26:* 37
99. Tarasov SG and Godovsky YuK (1980) Vysokomol Soedin *A22:* 1879
100. Godovsky YuK (1984) Makromol Chem Suppl *6:* 117
101. Kawai H and Hashimoto T (1979) In: Contemporary Topics in Polym Science 3, pp 245–266, Plenum Press New York
102. Pedemonto et al (1975) Polymer *16:* 531
103. Papkov VS et al (1975) Mechan polym N3, 387

104. Godovsky YuK and Tarasov SG (1977) Vysokomol Soedin *A19*: 2097
105. Godovsky YuK, Bessonova NP, Mironova NN (1989) Colloid Polym Sci *267*: 414
106. Godovsky YuK, Bessonova NP, Mironova NN and Letunovsky MP (1989) Vysokomol Soedin *A31*: 955
107. Godovsky YuK and Konyukhova EV (1987) Vysokomol Soedin *A29*: 101
108. Folkes HJ and Keller A (1973) In: Physics of Glassy Polymers, Applied Science Publishers London
109. Keller A, Pedemonte E and Willmouth FM (1970) Kolloid Z Z Polym *238*: 385
110. Folkes HJ, Keller A and Odell JA (1977) Amer Chem Soc Polym Prepr *18*: 251
111. Tarasov SG, Tsvankin DYa and Godovsky YuK (1978) Vysokomol Soedin *A20*: 1534
112. Bessonova, NP, Godovsky YuK, Kovriga OV et al (1988) Vysokomol Soedin *A30*: 1690
113. Leblanck JL (1977) J Appl Polym Sci *21*: 2419
114. Zapp RL and Guth E (1951) Ind Eng Chem *43*: 430
115. Oono R, Ikeda, H and Todani I (1971) Angew Makromol Chem *46*: 47
116. Galanti AV and Sperling L (1970) Polym Eng Sci *10*: 177
117. Tarasov SG et al (1984) Vysokomol Soedin *A26*: 1077
118. Pollak V and Romanov A (1980) Collect Czech Chem Commun *45*: 2315
119. Romanov A, Marcincin K and Jehlar P (1982) Acta polym *33*: 218
120. Godovsky YuK, Bessonova VP and Guzeev VV (1983) Mechan polym N4, 605
121. Guzeev VV, Shkalenko ZhI and Malinsky YuM (1981) Vysokomol Soedin *A23*: 161
122. Godovsky YuK, Tarasov SG, Tsvankin DYa (1980) IUPAC Internat Symp Macromolecules, Florence, vol. 3, pp. 235–238
123. Godovsky YuK and Tarasov SG (1980) Vysokomol Soedin *A22*: 1613
124. Godovsky YuK and Bessonova NP (1983) Colloid Polym Sci *261*: 645
125. Lyon RE and Farris RJ (1984) Polym Eng Sci *24*: 908
126. Godovsky YuK, Konykhova EV and Chvalun SN (1989) Vysokomol Soedin *A31*: 560
127. Godovsky YuK, Bessonova NP, Mironova NN and Letunovsky MP (1989) Vysokomol Soedin *A31*: 948
128. Leonard WJ Jr. (1976) J Polym. Sci *C54*: 237
129. Bonart R and Morbitzer L (1969) Kolloid Z Z Polym *232*: 764
130. Bonart R (1979) Polymer *20*: 1389
131. Hoffman K and Bonart R (1983) Makromol Chem *184*: 1529
132. Godovsky YuK et al. (1984) Paper presented at Rubber-84, Moscow, vol A2, Paper A48
133. Godovsky YuK, Bessonova NP and Mironova NN (1983) Vysokomol Soedin *A25*: 296
134. Godovsky YuK, Volegova IA, Rebrov AV et al (1990) Vysokomol Soedin *A32*: 788
135. Cirlin EH, Gebhard HM and Shen M (1971) J Macromol Sci Chem *5*: 981
136. Chen TY, Ricica P and Shen M (1973) J Macromol Sci Chem *A7*: 889
137. Godovsky YuK and Bessonova NP (1977) Vysokomol Soedin *A19*: 2731
138. Partridge SM (1970) In: Balasz EA (ed) Chemistry and Molecular Biology of the Intercellular Matrix, vol 1, p. 593, Academic London
139. Weis-Fogh T and Andersen SO (1970) In: Balasz EA (ed) Chemistry and Molecular Biology of the Intercellular Matrix, vol 1, p. 671, Academic London
140. Weis-Fogh T and Andersen SO (1970) Nature *227*: 718
141. Gosline JM, Yew, FF and Weis-Fogh T (1975) Biopolym *14*: 1811
142. Gosline JM (1987) Rubber Chem Technol *60*: 417

143. Grut W and McCrum NG (1974) Nature *251:* 165
144. Hoeve CAJ and Flory PJ (1974) Biopolym *13:* 677
145. Andrady AL and Mark JE (1980) Biopolym *19:* 849
146. Dorrington KL and McCrum NG (1977) Biopolym *16:* 1201
147. Volpin D and Cifferi A (1970) Nature *225:* 382
148. Evans EA and Hochmuth RM (1978) In: Bronner F and Kleinner A (eds) Current Topics in Membranes and Transport, vol 10, p. 1, Academic New York
149. Evans EA and Skalak R (1980) Mechanics and Thermodynamics of Biomembranes, CRC Press Boca Raton, Florida
150. Evans, EA and Waugh R (1977) Colloid Interface Sci *60:* 286
151. Waugh R and Evans EA (1979) Biophys J *26:* 115

7 Thermodynamic Behavior of Solid Polymers in Plastic Deformation and Cold Drawing

The mechanical work spent on the irreversible deformation of solids is always at least partly dissipated. Therefore, independent of the sign of the heat effect resulting from the elastic deformation, plastic deformation is always accompanied by an exothermic effect. A schematic diagram of thermomechanical behavior of an ordinary solid and a solid polymer is shown in Fig. 7.1. After a small amount of initial cooling, resulting from elastic deformation, evolution of heat accompanying the beginning of plastic deformation occurs. The appearence of plastic deformation is accompanied by heat evolution independent of whether it is localized (necking) or distributed along the sample uniformly. If the plastic deformation is accompanied with a neck formation, which is typical of the cold drawing of the majority of glassy and crystalline polymers well below glass transition or melting point, then the heat generated locally may lead to a considerable local temperature rise. This temperature rise may strongly influence the cold drawing of the sample. At low drawing rates the neck propagates uniformly along the sample. Under certain conditions, however, especially at a high rate of extension for some polymers, instabilities of neck propagation can be observed (self-oscillation phenomenon), which is also closely related with the local thermal effects. Therefore, the first aim of this chapter is a detailed consideration of the local thermal effects resulting from the uniform and self-oscillating regimes of necking.

In the course of plastic deformation and cold drawing of solid polymers some part of the mechanical work is stored as a latent internal energy change (similar to some common metals). The amount of energy stored during cold drawing and plastic deformation of polymers may be considerably higher than corresponding stored energy in metals. Therefore, the second important goal of this chapter is an analysis of the latent energy of deformation and possible physical mechanisms responsible for this process. A valuable additional information can be obtained by analysing the energy state of the cold drawn or plastically deformed samples.

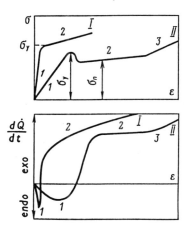

Fig. 7.1. Scheme of the thermo-mechanical behavior of ordinary solids (metal) (*I*) and solid polymers (*II*) at simple extension. *1* – elastic deformation (metal – some parts of percent, polymer – some percent); *2* – plastic deformation of solids and cold drawing of polymers (solids – dozens of percent, solid polymers – hundreds of percent); *3* – plastic deformation of drawn material

7.1 Temperature Effects During Plastic Deformation and Cold Drawing of Glassy and Crystalline Polymers

For qualitative illustration of temperature effects resulting from the cold drawing and plastic deformation of solids a simple model can be used [1]. Let us consider the energy balance of the deformation of a solid cylinder of the length l and diameter d. The temperature of the cylinder is equal to the surrounding temperature T_∞ (Fig. 7.2). Heat transfer occurs through the lateral surface of the cylinder. Along the cylinder heat transfer is carried out due to thermal conductivity. The initial temperature distribution in the sample is uniform and corresponds to temperature $\bar{T}$. The cylinder stretches with stress σ with the deformation rate $d\varepsilon/dt$.

The energy balance of the deformation processes can be expressed in the following form

$$\rho C_p(d\bar{T}/dt) = (dW/dt - dU/dt) - (dQ_\alpha/dt + dQ_\kappa/dt) \qquad (7.1)$$

where ρ is the density. The first term in the right side of the equation corresponds to the dissipated part of the work of deformation, and the second term corresponds to the heat loss through convection and thermal conductivity, respectively. Thus, the change of the average temperature of the sample is determined by the dissipation rate and the intensity of heat transfer. The following limiting cases are obvious. If the rate of the energy dissipation is considerably higher than the rate of heat transfer one can neglect the heat

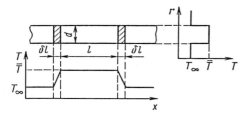

Fig. 7.2. Scheme of the temperature distribution in a model system under cold drawing (for explanation see text)

transfer, which corresponds to the adiabatic condition of deformation. Equation (7.1) in this case may be written in the following form

$$\rho C_p (d\bar{T} = (\sigma - dU/d\varepsilon)d\varepsilon \ . \tag{7.2}$$

In this case temperature $\bar{T}$ depends only on the deformation process.

In the other limiting case when the rate of heat transfer is considerably higher than the rate of the energy dissipation we arrive at the following equation

$$\rho C_p (dT/dt) = -(dQ_\alpha/dt + dQ_\kappa/dt) \ . \tag{7.3}$$

In the steady-state regime

$$\bar{T} = \text{const} = T_\infty \tag{7.4}$$

i.e. deformation is carried out at constant temperature.

The conditions close to adiabatic occur at very high deformation rates of the thick samples especially if the surface of heat transfer is small. The isothermal conditions can be reached at very slow deformation rates especially at a uniform deformation without necking with a high coefficient of heat transfer. In the majority of real cases neither the rate of heat transfer, nor the rate of energy dissipation can be neglected.

From the viewpoint of temperature effects, the deformation with a neck formation is of special interest. In this case heat evolution is localized in a small volume, and the deformation must be accompanied by considerable local heating. The problem of the local temperature rise during necking in solid polymers stimulated a search for experimental methods of registering the temperature rise in the neck formation zone and led to some suggestions concerning the mechanisms of the cold drawing of glassy and crystalline polymers.

There are two limiting cases of the energy dissipation from the point of view of relation of various mechanisms of heat transfer, namely the uniform plastic deformation and cold drawing with necking. At the uniform deformation heat transfer corresponds to convection through the whole surface of the sample. The main role in heat transfer belongs to the term Q_α in Eq. (7.1). In the case of necking during cold drawing the thermal conductivity may be the factor controlling the energy dissipation as well as the dimension of the necking zone [1–3]. The model shown in Fig. 7.1 allows us to estimate the role of the thermal conductivity mode in the whole heat transfer in the case of localization of deformation. According to Eq. (7.1) the relative contribution of thermal conductivity to the hole heat transfer can be expressed in the form

$$Q_\kappa/(Q_\alpha + Q_\kappa) = 1/(1 + 2\alpha^*l\delta l/\kappa d) \tag{7.5}$$

where α^* is the coefficient of convective and radiative heat transfer. For typical values of $\kappa = 0.2$ w/mK and $\alpha^* = 50$ w/mK (heat transfer into a gaseous phase) with $\delta l = d$ for uniform deformation of the sample with $l = 100$ mm we have $Q_\kappa/(Q_\alpha + Q_\kappa) < 2\%$. With the same parameters of heat transfer for the small necking zone of 1 mm the ratio of thermal conductivity in the whole heat transfer is 0.67. It means that during cold drawing in a gaseous phase heat is carrying away from the necking zone mainly by means of thermal conductivity. During cold drawing of polymers in liquids ($\alpha^* \sim 1000$ w/mK) even in the necking zone the main role belongs to the convective heat transfer, while thermal conductivity along the sample is of minor importance. Using this simple model we can consider the temperature effects accompanying cold drawing and plastic deformation of glassy and crystalline polymers.

7.1.1 Uniform Neck Propagation

The problem of the local heating during cold drawing of solid polymers arose in the early 1950s in connection with the hypothesis of consecutive softening of glassy polymers under orientation. Müller and Eckel [4, 5] suggested that the heat generated during yielding and neck formation is enough for heating the material above the glass transition temperature. According to this hypotheses, in the narrow zone of necking, the material is not in the glassy state but in the rubbery state. Therefore, at every moment deformation occurs not in the glassy state but in the rubbery state. The generation of heat is a consecutive process leading to a consecutive softening of the material in the necking zone. The corresponding estimation made for PVC assuming the adiabatic transformation of mechanical work to heat in the necking zone led to the conclusion that at the expance of the heat generated the sample can he selfheated from 20 to 90°C, i.e. above the glass transition of PVC.

Marshall and Thompson [2] used this idea for their explanation of cold drawing of glassy PET. Their calculations permit to illustrate quantitatively the role of the adiabatic heating during drawing. For their calculations they used the deformation curves obtained under conditions close to isothermal (Fig. 7.3). They have suggested that at high drawing rates the drawing is an adiabatic one. The adiabatic deformation curve was constructed under the suggestion that the whole mechanical work spent during cold drawing is converted into heat. The slope of the adiabatic curve (curve 7) is negative in the whole region of deformation. It corresponds to an unstable development of deformation. However, as a matter of fact cold drawing of PET even under such conditions proceeds quite stable. Therefore, Marshall and Thompson have suggested that this stability is a consequence of heat transfer from the necking zone into the neighboring undrawn zone. According to above estimations of the role of thermal conductivity along the sample such a possibility really exists. The estimated temperature changes in the necking zone as a function of draw ratio permitted to evaluate the dimension of the necking zone for PET.

The adiabatic temperature rise is used for explaining the cold drawing of crystalline polymers as well. According to this idea during cold drawing of typical crystalline polymers due to the local heating melting of crystallites occurs with subsequent orientational crystallisation of the melt.

The hypothesis about the local adiabatic heating has stimulated numereous attempts of experimental measurements of the distribution of temperature along the samples which undergo cold drawing [3, 5, 6]. Three methods have been used for direct registering the local temperature rise: thermocouples, colored organic crystals with melting points in the temperature range of interest, and fluorescent phosphors, and IR-radiation method. Typical results obtained with these methods for various polymers are shown in Fig. 7.4. Both the early measurements and modern IR-measurements demonstrate a very strong dependence of the temperature rise upon the deformation rate. There are two regions of the temperature changes. At the drawing rates below 0.1 mm/s the temperature rises do not exceed some degrees independent of the chemical nature and phase state of polymers. Above 0.1 mm/s the temperature rise increases considerably being dependent now upon the chemical nature, phase state and degree of crystallinity of polymers. At the drawing rates normally used in research studies (< 10 mm/s) the temperature rise reaches some dozens of degrees. Nevertheless, it does not exceed 60–70 degrees. Thus, according to the data shown in Fig. 7.4 self-heating of glassy and crystalline polymers during cold drawing can affect cold drawing only at draw rates above 0.1 mm/s.

In technological processes of drawing fibers and films when the drawing rates reach 100–1000 m/min temperature rises can reach a hundred or more degrees. This is shown quite obviously in Fig. 7.5 where dependencies of the temperature rise in PA-6,6 fibers upon drawing rates changing from 56 to 516 m/min are shown. Measurements of the local temperature rises were performed on moving fibers with the help of contact thermocouples. Similar results have been published also by Arakawa [15] for PA-6 fibers. At the constant draw ratio $\lambda = 3.8$ the temperature rise increases from 67°C for drawing rate V = 100 m/min to 110° for the drawing rate V = 1000 m/min.

Taking into account the fact that cold drawing of solid polymers can be performed at very small drawing rates when the local temperature rises are 1–2 orders below those which are needed for softening of glassy polymers and melting of crystalline polymers the hypothesis of subsequent softening or melting can be rejected [12, 16, 17]. There are also other reasons to do this. For example, this approach cannot explain the necking in homogeneously deformed samples when they are cooled during deformation. Besides, this approach is not able to explain the fundamental fact that cold drawing can

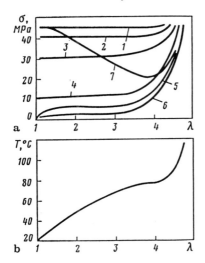

Fig. 7.3. Stress–strain curves for PET at various temperatures (**a**) and dependence of the temperature of the sample upon adiabatic extension (**b**) [2, 7]. Deformation temperature: *1* – 20, *2* – 40, *3* – 60, *4* – 80, *5* – 100, *6* – 140°C. Curve *7* is the calculated curve for adiabatic conditions at the initial deformation temperature 20°C

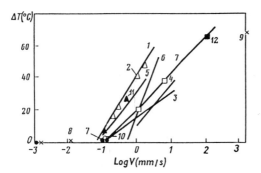

Fig. 7.4. Dependence of the temperature rise during necking or plastic deformation upon the deformation rate. *1* – *i*-PP (IR-radiation) [8]; *2* – *i*-PP (thermocouple, IR-radiation) [9]; *3* – PC (IR-radiation) [8]; *4* – LDPE (thermocouple) [10]; *5* – PET (calculation) [7]; *6* – HDPE (IR-radiation) [8]; *7* – PA-6 (thermocouple – •, thermoindicator – □) [5, 11]; *8* – glass-like rubber SKN-40 (thermocouple) [12]; *9* – PA-6,6 (thermocouple) [1]; *10* – PVC (thermocouple) [5, 11]; *11* – *i*-PP (thermocouple, compression – ▲) [13]; *12* – PA-6 (thermocouple) [15]. All measurements were performed at 20–40°C, measurements for SKN-40 were made at −50°C

be performed more easily (i.e. at lower stress) at lower rates of drawing, i.e. when conditions of drawing are closer to isothermal.

Müller [11, 18] has calculated the temperature rise resulting from cold drawing of PVC at room temperature at drawing rate V = 0.3 mm/s and arrived at the value $\Delta T = 69°C$. This value is about one order of magnitude

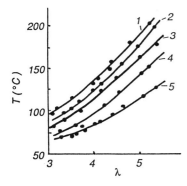

Fig. 7.5. Maximal temperature observed during cold drawing of PA-6,6 fibers as a function of drawing velocity [14]; *1* – 516, *2* – 370, *3* – 211, *4* – 120, *5* – 53.4 cm/s

higher than that which has been monitored during direct measurements using thermocouples. This calculation is based on the idealized assumption according to which all the heat generated during the stable neck propagation at first raises the temperature in the vicinity of the neck zone and only after this it is transferred to other parts of the sample and to the surroundings. Shortcomings of such calculations are quite obvious in the light of the model discussed above.

Nevertheless, Müller [11] raised the question concerning the correctness of the temperature measurements using thermocouples and thermoindicators. Such a question is quite correct assuming that the zone of the local plastic deformation is considerably smaller than the dimension of the thermocouple or phosphorous particles. There are also other experimental uncertainties (see Chapt. 9). The temperature distribution along the polyamide film fixed with a thermoindicator and obtained by calculation of the energy balance differs principally, as one can see from Fig. 7.6. If such a discrepancy really exists it could be a very important factor in our view on cold drawing. Fortunately, recent IR-measurements have been made where the direct contact of a sensor with the polymer surface is absent. As can be seen from Fig. 7.4 the results for PP obtained by various methods in independent studies are in a very good agreement. Therefore, as a result of these temperature measurements one can conclude that all the methods normally used for such measurements give reliable values of the temperature rise averaged on the surfaces with linear dimensions of some fractions of a mm. As far as the local temperature rises in microvolumes with linear dimensions of dozens nm are concerned this problem is still waiting for an experimental study.

In spite of the fact that the local temperature rises cannot play a decisive role, nevertheless, they can effect on the process of drawing as a whole. Ward [7] has analysed the role of the local temperature rises during cold drawing of the glassy PET and established that similar to other polymers the post-yield stress σ_{py} is a linear function of the logarithm of the velocity of neck propagation in the range of velocities 10^{-3} to 10% s^{-1}. On the other hand the yielding stress σ_y has this similar behavior only at low velocities ($<$ 10^{-1}%s^{-1}). At higher velocities the yielding stress decreases with increasing

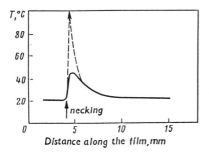

Fig. 7.6. Temperature distribution along PA-6 film during cold drawing with the velocity 1 mm/s [11]. The *solid line* corresponds to experimental observations; the *dashed line* indicates the theoretical distribution, obtained using the energy balance of the deformational process

the drawing rates. This decrease is closely connected with the local heating which promotes the drawing.

In contrast to glassy polymers the application of the local heating hypothesis to crystalline polymers was more stimulating. Although in the early studies [3, 5, 6] have been suggested that due to a local heating the initial crystallites melted and new orientational crystallization took place, however, the corresponding numerical calculation showed [19, 20] that even in the most unfavorable cases the heat generated during cold drawing is not enough not only for complete melting such high melting point polymers as PA ($T_m = 230°C$) and PP ($T_m = 170°C$), but even for melting PE ($T_m = 140°C$). It is very important that the structural and morphological changes resulting from cold drawing are similar in a wide temperature and drawing velocities range. But at low drawing rates again the local heating is so small that drawing conditions are close to isothermal ones. Therefore, the local temperature rise cannot be the only reason of recrystallization of crystalline polymers under drawing. Nevertheless, the idea concerning recrystallization is widely used for explanation of the structural changes resulting from cold drawing [21, 22]. Therefore, the question concerning some other possible physical reasons of the recrystallization process becomes very important.

Another idea concerning recrystallization during drawing is based on the dependence of the melting point of anisotropic crystallites upon their orientation relative acting forces [23, 24]. It has been postulated that if an extensive force is applied along the macromolecules in crystallites then their melting point increases. On the other hand, their extension in all other directions must be accompanied with a decrease of their melting point. It has also been suggested that this melting point decrease may be rather large. Such a mechanism means a phase transition (recrystallization) under stresses. Corresponding estimations showed that in principle such melting point decrease is possible due to a negative pressure which can arise in the surrounding crystallites amorphous regions [25]. In 1980 [27] calculations of a melting point decrease were performed using a model of consecutive complete unfolding of macromolecules in chain-folded crystallites. According to these calculations for PE this mechanism can be accompanied by a considerable decrease of melting point under the action of stresses which are quite comparable with the yield stresses. Juska and Harrison [28] in suggesting a mechanism of plastic deformation of crystalline polymers also concluded that under the action of mechanical stresses, a disordering of crystallites occurs. Then a quick extension of the disordered regions takes place which is accompanied by fibrillization of the material with simultaneous crazing.

A very important aspect of the recrystallization mechanism is the kinetic aspect. For the most widely studied polymers such as PE and PP drawing is normally performed in the temperature range close to the maximum of their crystallization rate [29]. In this case the crystallization rate is extremely fast. Therefore, it is very difficult when investigating such polymers to separate melting and a new crystallization kinetically. For PA-6 cold drawing can be performed easy at room temperature, where the melt must crystallize very slowly because of the closeness of the glass transition temperature. But experimental attempts to find amorphous zones between drawn and undrawn parts in PA-6 and PA-6,6 fibers were unsuccessful [1]. In the author's laboratory this problem has been studied on PBT, with melting point $T_m = 224°C$ and glass transition $T = 40°C$. A sample cut from the undrawn film with a spherulite morphology with degree of crystallinity 40% were drawn in deformation calorimeter at room temperature (thermodynamical parameters of drawing are summarized in Table 7.1, p. 227). All our attempts to get completely amorphous sample after drawing were unsuccessful. X-ray pattern of the drawn sample possesses a very pronounced crystalline texture (Fig. 7.7). DCS study has shown that in the drawn samples crystallization processes are absent during heating and heats of fusion of the undrawn and drawn samples are identical (see Table 7.2, p. 242), which means the identical degree of crystallinity. Thus, one can conclude that during drawing such polymers as PET, PBT, PA at temperatures close to the glass transition temperature the recrystallization mechanism of cold drawing can probabaly be excluded. Peterlin [31] has emphasized recently that even for polymers with relatively low melting points such as PE the mechanical work is too small for complete melting of the crystalline structure, and that the input of the mechanical energy is hardly sufficient for the adiabatic local temperature rise of $40°C$ with nothing left for the melting. The simultaneous action of all temperature effects seems to lead to a considerable increase of the local molecular mobility sufficient for a destruction of the lamellar crystalline structure on small microblocks and the formation of a fibrillar structure from them. Of course, some mostly defect crystallites can melt completely during this process.

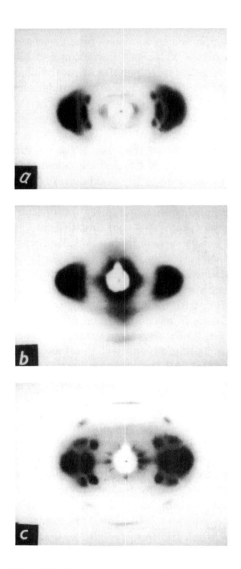

Fig. 7.7. X-ray patterns of drawn polymers. **a** – PBT; **b** – PET drawn at a slow neck propagation; **c** – PET drawn in the self-oscillating regime. The drawing was made at room temperature in each case

7.1.2 Self-Oscillated Neck Propagation

Above we considered the thermal feature of the stable regime of neck propagation. Under certain conditions, however, for some both glassy and crystalline polymers quite another regime of neck propagation can be observed. In this case the neck no longer propagates uniformly with a constant velocity but by a jump-like regime, and the drawing stress fluctuates periodically with time.

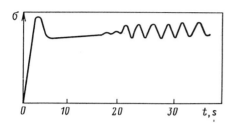

Fig. 7.8. Self-oscillating regime of stretching with the neck formation in glassy PET drawn at a constant rate [33]

The phenomenon of the self-oscillated neck propagation was discovered in PET [32, 33] and PA-6 [3, 32] and was studied in detail for PET by Kargin and Andrianova [33, 34]. In all these studies explanations for the self-oscillated regime of neck propagation were based on the decisive role of the intensive local heating. Therefore, below we consider, first of all, the main features and critical conditions of the appearence of the unstable regime of neck propagation in the light of local heating. The necessary results for such analysis are shown in Fig. 7.9.

Drawing of glassy PET films with a low drawing rate (below 0.5 mm/s) perform with the appearence and stable uniform neck propagation. Local heating during the neck formation is small and the temperature rise does not exceed some degrees (see Fig. 7.4). At a specific relation between the drawing velocity and conditions of heat transfer, characteristic oscillations of drawing force are observed. The local temperature rises, experimentally measured by both thermocouples and thermoindicators, and can reach many dozens degrees. The change of the neck propagation regime is accompanied by a change of the appearance of drawn samples. In contrast to transparent samples resulting from slow drawing with a constant velocity, during unstable regime the neck consists of periodical transparent and nontransparent white parts. The periodicity of the structure is exactly related to stress oscillation, i.e. the transparent areas are related to increasing stress while the nontransparent parts are formed when the stress in an oscillating cycle decreases. With further increase of the velocity of drawing above 10 mm/s the unstable self-oscillated regime disappears and the neck again propagates in a stable uniform manner.

In the self-oscillated regime, the neck formation occurs during the force drop with the velocity 1–2 orders higher than the average velocity of drawing. Because of the very sharp force drop the local heating in this case is especially high. As a result of the local temperature rise the local mechanical compliance increases considerably. Further drawing takes place due to the extension of

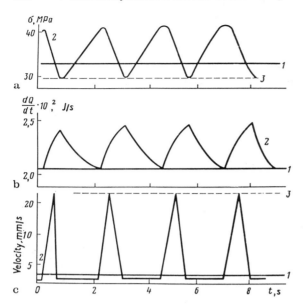

Fig. 7.9. Change in stress (**a**), heat flux (**b**) and real rate of the neck formation (**c**) for various regimes of stretching PET [33, 34]. *1* – stationary regime of stretching with the formation of a transparent uniform neck; *2* – self-oscillation regime of stretching with the formation of a neck with alternating transparent and opaque areas; *3* – non-stationary high-velocity regime characterized by the formation of an opaque neck and by a regime of unsteady heat release

the existing neck until the drawing force again reaches the critical value when the rate of transformation to the oriented state increases sharply and the previous cycle is repeated again. Multiple repetition of such cycles leads to the self-oscillated neck propagation.

The important role of heating in the unstable oscillating regime manifested itself when the conditions of heat transfer were changed. The local temperature rises during the drawing in air are very large due to the minor importance of convective heat transfer. Drawing in a water bath must increase the convective heat transfer which must prevent the appearence of large temperature effects. In fact, during the drawing of PET films in water, the unstable neck propagation did not occur at any drawing velocity. Thus, critical conditions for the appearence of the oscillating regimes depend on, first of all, the drawing rate and heat transfer.

The concept of local heating was used by Barenblatt [35] in his theoretical treatment of the phenomenon of unstable neck propagation. Earlier Barenblatt [36] had developed the theoretical approach to the stable neck propagation during cold drawing of solid polymers based on a combustion theory. Using this early isothermal treatment in the new model the non-isothermal transformation was considered. This theoretical analysis showed that at some relationships between heat transfer and the compliance of the deforming system, a definite range of drawing velocities exists where oscillating neck propagation is actually possible. On the other hand, Matkowsky and Sivashinsky [37] using Barenblatt's model demonstrated that the self-oscillated regime of the neck propagation can occur even in isothermal conditions if the rate of stress diffusion exceeds the rate of diffusion of the oriented material.

Besides the concept of the decisive role of heating on mechanical behavior during drawing of samples at a constant rate, one can suggest another explanation based on the concept of orientational crystallization. Let us, first of all, consider some indirect data. Direct measurements of the temperature rise by means of a thermocouple or thermoindicator fixed the temperature rise during the drawing of PET films in air at about 70 degrees. According to the spectrum obtained with the indicator this high temperature rise corresponds to the white nontransparent areas. According to Fig. 7.4, however, at such rates with stable neck propagation, the local heating must be considerably less and can hardly exceed 20–30 degrees. One can suggest, therefore, that there is one more non-identified source of heating. Such a source may be the orientational crystallization. If it really takes place, then the high local heating is not the reason for the unstable neck propagation, but its consequence. The white areas were indeed semi-crystalline [33]. In the light of this suggestion it is very important that the oscillating regime may occur only in crystallizable polymers (PA, PET, PP). It does not take place in glassy polymers incapable of crystallization, for example, PC [8, 38] even at high drawing velocities (1–10 mm/s) when local heating is very high.

The thermodynamical criterion of orientational crystallization is $\Delta H_{st} <$ 0 (see Table 7.2, p. 242). As can be seen from Table 7.1 (p. 227) during a slow uniform stable neck propagation in PET $\Delta H_{st} > 0$. In a self-oscillated regime, the integral heat to work ratio $Q/W = -1.3$, ΔH_{st} being correspondingly -22 J/g. Thus, the energy balance of drawing in this case unequivocally supports the concept of orientational crystallization. Finally, the direct evidence of the crystalline character of the samples obtained in oscillating regime are X-ray data (Fig. 7.7) The samples obtained by slow stable drawing are characterized by an amorphous texture. The neck formed in an oscillated regime is characterized by a crystalline texture. Heat of fusion of such samples is 48 J/g, which corresponds to a 40% degree of crystallinity. The difference in the stored energy and heat of fusion seems to be a result of a reorganization in the texture during DSC measurements. Thus, the crystalline character of the samples obtained at self-oscillating drawing is proved by direct independent methods.

Pakula and Fischer [38] studied instabilities of the deformation process in cold drawing of PET, PP, PA-6 and PC under various experimental conditions. They reconsidered critically the early results discussed above in the light of their new data. First of all they demonstrated that the samples of PET obtained in the unstable regime at 20 and 40°C with characteristics similar to those used by Kargin and Andrianova [33] were crystalline. At temperatures above 60°C all the samples obtained at any drawing velocities were completely transparent and amorphous. Besides the experiments at constant drawing rates they studied also drawing at constant stress (Fig. 7.10).

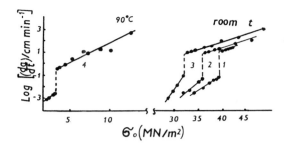

Fig. 7.10. Dependence of the neck propagation rate, measured at constant stress, upon stress [38]. *1* – drawing at room temperature with necking; *2* – drawing through the die at room temperature, apex angle 90°; *3* – as in 2, but apex angle 30°; *4* – drawing through the die at 90°, apex angle 30°

In all cases studied a critical stress exists below and above which different dependencies of the neck propagation rate on the nominal stress are observed. At critical stress, the rate is suddenly increased by a factor about 100 compared with that seen at lower stress. Although the jump in the drawing rate was observed both below and above the glass transition temperature of PET, the critical stress depends on temperature. The samples drawn below critical stress were transparent and amorphous, above – white and crystalline (the X-ray pattern published in [38] is practically identical to that shown in Fig. 7.7). Morphology and mechanical properties of the samples drawn below and above the critical stress also differed considerably. It has been established that the unstable neck propagation takes place even in the case when local heating was small.

Thus, the appearence at some conditions in crystallizable polymers instabilities in cold drawing can be considered as a dynamic transition of disorder–order type closely related to orientational crystallization. Due to the heat release resulting from the local crystallization, a quick drop of stress occurs because of a local orientation. It, in turn, means a quick drop of the local drawing rate and the transition to the low-stress regime. Then stress increases gradually due to an increase in the local rate of orientation. But as soon as the stress reaches some values above a critical point the cycle is repeated again. Having summarized all the results obtained, Pakula and Fischer suggested a simple model based on the existence of a "spinodal transition" from an undeformed state to a deformed one. This approach can explain instabilities of the deformation process in cold drawing without using the concept of heating.

The effect of "jump-like" behavior at plastic deformation is also a very characteristic phenomenon during deformation of metals and alloys at very low (liquid helium) temperatures and at high rates of deformation [41, 42]. One of the hypotheses which explains this "jump-like" deformation is based on the local heating suggestion. At liquid helium temperatures the local temperature rise can be rather high due to a low thermal conductivity and low values of heat capacity. Although a number of criteria have been introduced to describe the thermomechanical instabilities during low-temperature plastic deformation of metals, as a result of numereous experimental investigations [43–45] the conclusion that heating is a consequence of instabilities during plastic deformation rather then their reason was made [41, 46]. This conclusion is quite consistent with that made above, although the instability mechanism in polymers and metals seems to be quite different.

7.2 Thermodynamics of Plastic Deformation and Cold Drawing of Glassy and Crystalline Polymers

The starting point of the thermodynamical analysis of plastic deformation of solids is the energy balance of deformation. Because the first law of thermodynamics is valid for all deformations, whether reversible or irreversible its application to plastic deformation allows one to calculate easily the internal energy changes during deformation resulting from the plastic deformation

$$\Delta U_{st}(\Delta H) = W(1 - Q/W) \tag{7.5}$$

where $W > 0$ for work done on the sample and $-Q < 0$ for heat flow from the sample. An ideally plastic material converts all of the work of plastic deformation into heat ($W = 0$) therefore, the energy state of the sample after deformation is the same as before deformation. Normally, in plastic deformation of real materials only some part of the work of deformation is converted into heat ($W > Q$) while another part is stored by the deformed material (latent or stored energy).

Similar to the thermodynamics of plastic deformation of metals the question of primary importance is the internal energy changes and the energy state of the deformed polymers. Systematic studies of thermodynamics of cold drawing of glassy and crystalline polymers have been carried out in 1960s and 1970s by Müller and coworkers [11, 18] and Godovsky [39, 47] with the help of the deformation calorimetry. Recently new calorimetric data were published [13, 48–50]. The results of all these studies are considered in the next sections.

7.2.1 Plastic Deformation on Uniaxial Extension (Cold Drawing)

During cold drawing of typical crystalline polymers both the work of drawing and the heat dissipated is a linear function of the logarithm of the deformation velocity (Fig. 7.11). For W this dependence is a direct consequence of the linear relation between the stress and the logarithm of the velocity of neck propagation [7]. These experimental dependencies can be represented with the following simple relationships

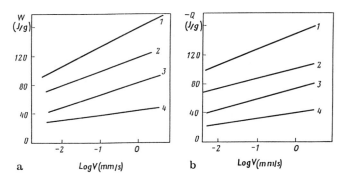

Fig. 7.11. Dependence of **a** work (W) and **b** heat (Q) upon extension rate for *i*-PP (*1*), HDPE (*2*), PA-6 (*3*), LDPE (*4*) drawn in the deformation calorimeter at 294 K [39, 47]

$$W = a + b \log V$$
$$Q = a' + b' \log V \,. \tag{7.7}$$

The relative part of the work of drawing dissipated Q/W is independent of the drawing rate in the range of rates studied (Fig. 7.12). For all the crystalline polymers studied the ratio Q/W < 1 which means that the stored energy $\Delta U_{st} > 0$. According to Eqs. (7.6) and (7.7) this stored energy is also a linear function of the logarithm of the drawing velocity

$$\Delta U_{st} = \text{const} \cdot W = A + B \log V \tag{7.8}$$

The corresponding dependencies are shown in Fig. 7.13a. Because the natural draw ratio is different for various polymers the corresponding comparison is possible only after a normalization (Fig. 7.13b).

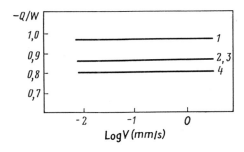

Fig. 7.12. Dependence of heat to work (−Q/W) ratio upon drawing rate for crystalline polymers [39, 47]. *1* − *i*-PP, *2* − HDPE; *3* − PA-6, *4* − LDPE

Thermodynamical characteristics of cold drawing for both the crystalline and glassy polymers are tabulated in Table 7.1. For crystalline polymers the values of ratio −Q/W obtained in various laboratories in the majority of cases are in a good agreement. Müller and coworkers [11, 18] published a considerably lower value of Q/W but only for PA. It is also worth mentioning that according to the data obtained for PA-6 in this work the consid-

Table 7.1. Thermodynamic characteristics of cold drawing and plastic deformation of polymers

Deformed Polymer	Crystal-linity %	Tempe-rature K	Natur. draw ratio	Rate V mm/s	$-Q/W$	ΔH_{st} J/g	$\Delta H_{st}/\lambda$ J/g	Ref.
				Crystalline polymers				
HDPE	61	294	5	1.0	0.85 ∓ 0.05	18.9	3.78	39, 47
	–	296	–	–	0.8–0.9	–	–	8
LDPE	41	294	5	1.0	0.85 ∓ 0.05	7.2	1.8	39, 47
	–	296	5	0.7	0.70	–	–	49
	–	293	–	0.9	0.76–0.87	–	–	11
i-PP	59	294	6	1.0	0.93 ∓ 0.05	11.8	1.97	39, 47
	60–65	300	–	–	0.9	–	–	34
	–	296	–	–	0.9–1.0	–	–	8
PA-6	45	294	3.5	1.0	0.8 ∓ 0.05	17.0	4.85	39, 47
	–	293	–	1.2	0.60–0.62	–	–	11
PPO	50%	294	–	0.11	0.84 ∓ 0.05	–	–	39, 47
PBT	< 40%	293	5.2	0.028	0.95 ∓ 0.05	5.3	1.0	39
PVC	< 10%	293	–	0.34	0.70–0.83	–	–	11
	–	296	–	–	0.60–0.70	–	–	8
				Amorphous polymers				
PC	0	294	2.3	0.015	0.95 ∓ 0.05	–	–	39, 47
	0	293 303	2.3	–	0.65	14.6	6.35	11
	0	296	–	–	0.95 ∓ 0.05	–	–	8
	0	293	1.9 (2.3)	0.083 0.083	0.56 0.56	21 (17.3)	11 (7.5)	49
	0	335	–	0.207	0.48	22		49
PET film	0	294	5.4	0.27	1.12 ∓ 0.05	−14	−3.3	39, 47
fiber		294	5.4	0.03	0.95 ∓ 0.05	6.0	1.1	39, 47
film		294	5.6	0.03	0.95 ∓ 0.05	6.75	1.2	39, 47
fiber		294	5.4	2.0	1.30–1.35	−22.0	−4.1	39, 47
film		300	–					
		319	–					
		330	–	0.17	1.0 ∓ 0.03	0	0	34
		340						
PAr	0	298	1.65	0.083	0.66	19	11.5	49
		298	–	0.207	0.67	17	10.3	49
PSu	0	298	1.8	0.083	0.59	29	16.1	49
		298	–	0.207	0.70	19	10.5	49
PMMA	0	318	1.15	0.207	0.44	11	9.56	49
		335	1.30	0.207	0.40	15	11.5	49

1) In parentheses are shown the values, related to the natural draw ratio equal 2.3, which is necessary for a more correct comparison of data obtained in various studies.

2) Bisphenol A polyarylate Durel DKX-008 Celanese Corporation.

3) Bisphenol A polysulfone Aldrich Chemical Company.

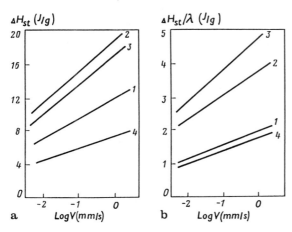

Fig. 7.13. Dependence of the complete amount of the latent energy (**a**) and the latent energy normalized to the natural draw ratio (indicated in parentheses) (**b**) upon the drawing rate at 294 K [39, 47]. _1_ – _i_-PP (6), _2_ – HDPE (5), _3_ – PA-6 (3, 5), _4_ – LDPE (4)

erable decrease of the ratio Q/W from −0.62 to −0.25 was observed when the drawing rate dropped below 1 mm/s. The reason for this decrease is unclear yet. The data tabulated in Table 7.1 show that the highest ability for the energy storage is possessed by PA-6 and LDPE. A comparison of the behavior of LDPE and HDPE shows that the ability to store energy is increased with an increasing degree of crystallinity. It is quite possible, however, that the difference in the chemical structure of macromolecules also plays a role.

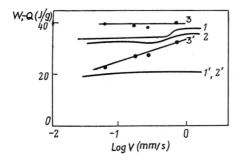

Fig. 7.14. Dependence of work W (_1,2,3_) and heat Q (_1′, 2′, 3′_) on extension rate for PC in the deformation claorimeter at 293 K (1,1′) and 303 K (2,2′). 1,1′ and 2,2′ – from [11], 3,3′ – from [49]

For amorphous polymers the thermodynamics of drawing is more complex. The values of −Q/W obtained for PC in various studies are different. According to Müller [11] and Adams and Farris [49] about 35–45% of the drawing work is stored in samples. Two other works lead to a considerably smaller value of the internal energy change which is only about 5%. The dependence of work and heat upon the drawing rate obtained in [11] and [49] are also different. Figure 7.14 demonstrates a very weak dependence of the drawing work upon the drawing rate which was obtained in both these studies. At the same time in contrast to the work in Ref. [11], Adams and

Farris [49] have obtained a linear dependence of the heat dissipated upon the deformation rate which corresponds to the increase of $-Q/W$ from 0.56 to 0.79 with the drawing rate while in Müller's work it was constant. Adams and Farris have studied the energy balance of cold drawing of amorphous glassy polymers not only at room temperatures but at higher temperatures as well. The stored energy in PC drawn at 335 K was approximately the same as at room temperature but its part in the work of drawing increased. Similar behavior was also found in PMMA.

Very interesting thermal behavior during cold drawing at room temperature was observed for completely amorphous PET films and fibers. At small drawing rates only a few percent of the work of drawing is stored by samples increasing their internal energy. Increasing the drawing velocity leads to the heat dissipated during drawing exceeding the work considerably. At the rate of 2 mm/s drawing was carried out in the striking self-oscillating regime. As demonstrated above, this oscillation is a consequence of a periodic crystallization. The less pronounced, although still quite visible, decrease of the internal energy was observed at the intermediate drawing rates, while the striking oscillated necking was not fixed. The X-ray texture in such samples seems to have an intermediate pattern between the amorphous and crystalline one (see Fig. 7.7).

At drawing temperatures of 75.4 and 77.3°C which are close to the glass transition temperature, Andrianova and coworkers [34] have observed orientational crystallization (Fig. 7.15), which normally occurred at elongations about 160%. The highest value of the heat resulting from the orientational crystallization was only 12 J/g.

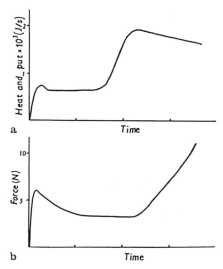

a

b

Fig. 7.15. Change in heat flux (**a**) and force (**b**) on stretching of PET accompanied with orientational crystallization [34]. Temperature of drawing is 75.4°C; rate of deformation is 10 mm/min

It is well established that the cold drawn polymers possess a striking anisotropy of thermomechanical properties (see Chap. 5). Normally the drawn polymers can be redrawn when stress is applied at some angle to the axis of the initial drawing. The deformation calorimetry behavior during re-drawing is generally similar to the drawing of the undrawn polymers. When the angle of redrawing is small, yielding is less pronounced and the redrawing process is less stable. The thermodynamical behavior of glassy and crystalline polymers is similar: when the angle to the axis of the first drawing becomes smaller than 60° both the work spent and heat dissipated increase sharply (Fig. 7.16). Their ratio, however, is independent of the redrawing angle and their values are the following: for angles 90° and 60° $Q/W = -1 \pm 0.05$, for 45° and 30° $-Q/W = -1 \pm 0.08$. Similar ratios are also characteristic for PA-6. Thus, one may conclude that in the limits of accuracy of the calorimetric experiments redrawing crystalline and glassy polymers is not accompanied by any energy changes, because the latent energy of deformation is close to zero.

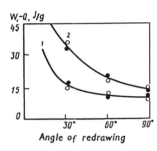

Fig. 7.16. Dependence of the normalized values of work W (○) and heat Q (●) upon the angle of re-drawing of crystalline i-PP (1) and amorphous PET (2) at 294 K [39, 47]

An unusual behavior of PA-6,11 with multiple redrawing at an angle 90° to the axis of the previous drawing was established by Müller [11]. These results are shown in Fig. 7.17. As is seen the ratio Q/W differs considerably from the value -1 and depends upon the number of redrawing cycles in a very peculiar manner. Such behavior means that the latent energy after 5–7 cycles must reach very high values. Müller suggested that this very large amount of stored energy is due to the frozen stresses.

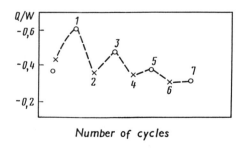

Fig. 7.17. Dependence of heat to work ratio (Q/W) on the number of cycles of re-drawing at the angle of 90° to the previous axis of drawing for PA-6 film [11]

7.2.2 Plastic Deformation on Simple Compression

Data concerning the thermodynamics of plastic deformation of glassy and crystalline polymers on simple uniaxial compression are very limited. Only recently have Oleinik and Nazarenko [48, 50] published systematic results of the deformation calorimetric behavior of typical glassy polymers using the standard Calvet-type calorimeter supplied with a special arrengement for uniaxial compression of small cylindrical samples (see Chap. 9). The small sample sizes and some imperfection of the mechanical loading device do not permit one to study the region of elastic deformations. The region of plastic deformations, however, was studied quite thoroughly. Typical results obtained for atactic PS and PC are shown in Fig. 7.18. As can be seen from these results increasing the value of plastic deformation is at first accompanied by a growth of the stored energy. However, when the values of plastic deformation reach 35–40% the stored energy approaches a constant level. These limiting values of the stored energy ΔU_{st}, as well as the ratio $-Q/W$, are dependent upon the chemical structure of macromolecules as follows: PS — 9.6 J/g and 0.63; PC – 5.4 J/g and 0.78; PMMA – 3.4 J/g and 0.82; and apoxy-amine network polymer — 18.0 J/g and 0.57. It is seen that during plastic deformations on compression, as well as during cold drawing, the stored energy lies between 20 and 40% of the work spent on the deformation. In contrast to the plastic deformation on uniaxial extension, plastic deformation on compression takes place in a rather homogeneous way, therefore, the results obtained can be compared immediately.

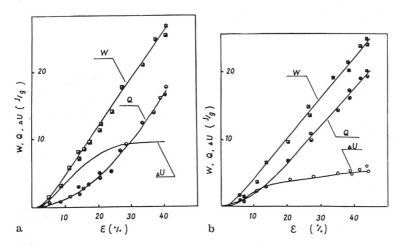

Fig. 7.18. Dependence of heat effect Q, work W and the latent energy ΔU upon the value of plastic deformation resulting from the compressive load–unload cycle in the deformation calorimeter at 30°C [48]. **a** – PS; **b** – PC. The rate of loading and unloading was $10^{-3} s^{-1}$

The only calorimetric study on large plastic deformation under compression was carried out recently with the help of Calvet-type calorimeter made by the researcher [13] (see Chap. 9). Due to anvils placed directly into calorimetric cells, rather high compression stresses can be achieved and the values of plastic deformations were as high as 80%. Typical results for HDPE are shown in Fig. 7.19. In contrast to amorphous glassy polymers the stored energy in HDPE is increased constantly with glassy deformation. The ratio $\Delta U/Q$ is a linear function of deformation being 10–12% at $\varepsilon_p = 20\%$ and about 45% at $\varepsilon_p = 75$–80%. Similar behavior was observed also for isotactic PP.

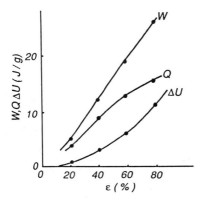

Fig. 7.19. Dependence of work W, heat Q and the latent energy ΔU upon the value of plastic deformation resulting from the compressive load–unload cycle in the deformation of cylindrical samples HDPE with degree of crystallinity 75% at 293 K [13]. The deformation rate was 2 mm/s

Thus, the deformation calorimetric results considered now show that at temperatures well below the melting point plastic deformation of crystalline polymers is normally accompanied by the storage of the internal energy. This latent energy of plastic deformation is a function of deformational conditions and the ratio $\Delta U/W$ may reach 30–40%, which indicates a considerable internal energy change during plastic deformation of crystalline polymers. Although for glassy polymers corresponding results are less definitive they, nevertheless, also show the ability of these polymers to accomodate a large amount of energy during plastic deformation.

The internal energy storage during plastic deformation is not specific polymer behavior. The latent energy is also a characteristic of cold working of metals and alloys and metallic glasses [51, 52]. Therefore, it is very useful to consider briefly the modern viewpoints concerning the main mechanisms of energy storage in plastic deformation of metals. It is generally accepted now that the latent energy of cold work depends on such factors as degree of deformation, temperature, mode of deformation, the purity of metals and some other factors. Under identical conditions the absolute values of the latent energy drops quickly with increasing homologous temperature T/T_m. For example the latent energy of cold work for various metals at room temperature at $\varepsilon = 25\%$ has the following values (in J/mole): steel – 100 ($T/T_m = 0.186$), copper – 39.8 ($T/T_m = 0.217$), lead – 2.5 ($T/T_m = 0.488$) and tin – 0 ($T/T_m = 0.582$) [52]. At high homologous temperatures strengthening of metals after plastic deformation no longer takes place and such metals approach thermodynamically ideal plasticity very closely. The highest values of latent energy are very characteristic of the strengthening deformations at low temperatures (low values of T/T_m) and dynamical (shocked) regimes of deformation.

All the mechanisms of energy storage in the plastic deformation of the strengthening metals suggested up to now are based on the concept of the appearence of new additional accumulations of dislocations and single dislocations stopped at some obstacles and new vacancies. Each of the mechanisms is accompanied by some part of the latent energy, the accumulation of new dislocations being the most powerful source. The emission of the latent energy due to previous cold working if a metal is heated is a multistage process, each of the stages corresponds to the disappearence of the accomulations of dislocations, single dislocations and vacancies. Each of the stages occurs over a specific temperature range and is accompanied by the emission of the corresponding part of the latent energy which can be determined by DSC. If plastic deformation has been performed at high temperatures the absorbtion and emission of the energy take place simultaneously. Therefore, such plastic deformations are not accompanied by strengthening of the metals and the storage of the latent energy. The amorphous metallic glasses are also capable of energy storage in plastic deformation, $\Delta U/W$ being about 4% [53].

This brief consideration indicates that polymers are able to store considerably higher values of energy in plastic deformation then metals. In the next section we will consider the energy state of the cold-drawn and plastically-deformed glassy and crystalline polymers and the suggested mechanisms of the energy storage.

7.3 Thermal Behavior of Cold Drawn and Plastically Deformed Glassy and Crystalline Polymers and the Nature of the Stored Energy

One can suggest a priori that the energy storage mechanisms in polymers are more diverse than in metals. The possible mechanisms seem to include the following:

1. For crystalline polymers, the most powerful mechanism of the energy change seems to be the change in the degree of crystallinity during the deformation. If, for example, during cold drawing of PE the change in the degree of crystallinity is 3%, then the corresponding internal energy

change of the drawn sample will be about 10 J/g. Similar values are characteristic of other crystalline polymers. This source of the internal energy changes may be both positive and negative.

2. The second indispensible source of the internal energy changes both for amorphous and crystalline polymers is the intramolecular conformational transitions. With irreversible deformations, the conformational entropy is decreased, but the conformational energy may both increase and decrease depending upon the sign of the value of $d\ln\langle r^2\rangle_0/dT$. Thus, one can say that the cold drawn or plastically deformed amorphous polymer possesses at least both the latent conformational energy and latent conformational entropy. Both these components are the driving force of the recovery processes on heating the drawn polymers. In contrast to the orientation above the glass transition temperature, the heat dissipated during the irreversible deformation of glassy polymers consists of two parts: the heat resulting from the conformational changes and the heat resulting from the intermolecular friction. The overwhelming part of the drawing work is spent on overcoming the intermolecular friction. This part of the heat effect is, of course, completely irreversible. In contrast, the heat resulting from the conformational entropy changes is the reversible part of the total heat effect. Unfortunately, it is impossible to divide the total orientational heat effect of two components based only on the thermodynamical characteristics of cold drawing. It is quite obvious that $W(T > T_g) = -T\Delta S \ll W(T < T_g)$. Therefore, the reversible part of the heat dissipated can be suggested, as a first approximation, as being equal to the heat effect resulting from the stretching of the polymer in the rubbery state above the glass transition temperature. As can be seen from the data presented in Chap. 6, typical values of work and heat of stretching of unfilled elastomers, stretched three to five times, are of the order of some joules per gram. It supports the suggestion that the reversible part of the heat dissipated during cold drawing is really only a small part of the total thermal effect. Generally, this conclusion seems to be also correct concerning the changes of the conformational entropy and conformational energy of the amorphous regions of crystalline polymers.

3. A very important source of the energy storage may be related to the appearence of internal stresses. Normally such stresses can relax very slowly and their relaxation will be accompanied with the emission of the corresponding energy and with a transition of the deformed polymer to a more stable state.

4. In the deformed solid polymers the internal energy change may be the result of the intermolecular changes, in particularly due to the disrupture of the hydrogen bonds.

5. In plastic deformation of crystalline polymers in crystallites all the elementary molecular processes which are characteristics of plastic deformation of metals, such as the formation and moving of dislocations, gliding of crystallographic surfaces, crystallization and mechanical twinning, are observed [21]. All these mechanisms, however, seem not to

contribute significantly to the latent energy of crystalline polymers similar to their contribution to the latent energy of deformed metals. This seems to be connected, first of all, to the fact that plastic deformation of crystalline polymers is as a rule performed at homologous temperatures $T/T_m > 0.6$, which are closely related to the ratio T_g/T_m, because plastic deformation and cold drawing of crystalline polymers is possible above the glass transition temperature of the amorphous regions of crystalline polymers. As shown above in the plastic deformation of metals at high homologous temperatures the stored energy has very low values. One can suggest that the values of stored energy accumulated due to these mechanisms cannot be important.

6. In recent years in the physics of glassy polymers the solid-like models have been discussed intensively according to which the irreversible deformation is considered as a cooperative process of the appearance and moving some defects similar to those characteristic of crystalline materials [54]. In other words, dislocations in glasses have to be of the Somigliana type and they are responsible for the plasticity of glasses. It is suggested that the energy of these areas of local shearing or gliding is increased relatively to the initial state of the deformed polymers. The appearance of such defects may be responsible for the storage energy.

7. Plastic deformation of glassy and crystalline polymers is accompanied with the appearance of new surfaces (crazing, fibrillization) on the formation of which some part of the work is spent. In spite of the fact that in polymers, high energy consuming surfaces can exist (for example, lateral surfaces of polymers crystallites) simple estimations show that this contribution cannot be important. For example, the appearance of new surfaces due to fibrillization in cold drawing may reach, in extreme cases, some joules per gram.

8. The neck formation during cold drawing is accompanied by the rupture of macromolecules, which can be observed by means of ESR [55]. The number of broken macromolecules is usually small and their concentration is normally below $10^{18}\,g^{-1}$. Therefore, the energy spent on the formation of the new radicals and new end-groups as a result of mechanochemical processes is very small and cannot be taken into account at all. Thermodynamics of the rupturing of the stressed macromolecules will be considered in Chap. 8.

Thus, from the eight mechanisms of energy storage discussed for deformed crystalline polymers only the first four may have some importance. For glassy polymers the main contributions to the stored energy may give the sources 2, 3, 4 and 6. Keeping this in mind in the next sections we will consider the energy state of plastically deformed glassy and crystalline polymers and discuss the particular mechanisms of the energy storage.

7.3.1 Amorphous Polymers

The temperature dependence of heat capacity C_p of plastically deformed glassy polymers measured with the DSC method is characterized by an exothermal drop ("exo-pit", see Figs. 9.19–9.21) [48, 56–58]. Such behavior is a consequence of the release of the excessive energy of the deformed glass on heating. The start of the exo-effect on the DSC-traces of the deformed glassy polymers is observed at temperatures a little above the deformation temperature and the end of the exo-effect is observed near T_g. Similar behavior was observed in typical polymer glasses – PS, PS-copolymers, PMMA, PC, epoxy-amine glasslike networks. The role of the degree of deformation on some characteristics of the C_p exo-anomalies was studied thoroughly for PS deformed plastically under compression [48, 56–58].

Figure 7.20 demonstrates the dependence of the energy stored by samples of PS upon the value of irreversible deformation. The data obtained with various independent methods agree quite well. The agreement between DSC data of various authors is also good. Some quantitative differences among the experimental methods used in [48, 50] are due to the physical aging of samples during DSC-measurements and the measurements of heats of solution. In deformation calorimetry this factor is excluded, because the stored energy is measured directly during deformation.

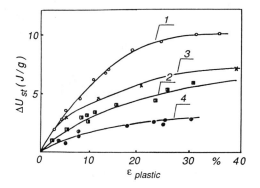

Fig. 7.20. Dependence of the energy stored in plastically deformed at simple compression samples PS on the value of deformation obtained with the help of various experimental methods: *1* – deformation calorimetry [48], *2* – DSC [48, 50], *3* – DSC [57, 58], *4* – calorimetric measurements of heat of dissolution [48]

Müller and coworkers [59, 60] demonstrated the importance of taking into account the physical aging stage in the thermodynamical balance for hot drawn PS and cold drawn PVC. The energy stored during PVC drawing at 313 K according to deformation calorimetry was 14.9 J/g and according to the measurements of heat of dissolution only 3.5 J/g. This difference was due to the physical aging during 12 hours at room temperature. These results emphasize the necessity of knowing the thermal history for a correct consideration of the energy state of deformed polymers by various methods.

Using DSC measurements Oleinik and Nazarenko [48, 50] studied the isothermal enthalpy relaxation of PS samples compressed plastically to 25% (Fig. 7.21). With increasing relaxation time the start of the exo-pit is moved to higher temperatures and the heat released during heating of the relaxed samples decreases. The shape of the DSC traces is also changed: just before the "exo-pit" a small endothermal hump existed increasing in height with increasing annealing time. Similar behavior of deformed glassy polymers was also established in some other investigations [55]. It is very characteristic of polymer glasses only and the reasons for its appearence are not clear yet.

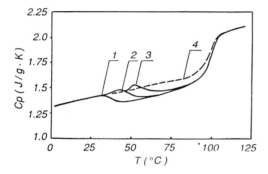

Fig. 7.21. Influence of the time of isothermal relaxation of PS samples after compression to $\varepsilon = 25\%$ on the temperature dependence of dynamical heat capacity [48]. Duration of relaxation (min): *1* – 9; *2* – 210; *3* – 270; *4* – initial undeformed sample. Temperature of deformation and relaxation – 25°C

The complete relaxation of the stored energy in deformed PS samples is possible under elevated temperatures [57, 58]. As can be seen from Fig. 7.22 there are two stages in the total relaxation process. The explanation of the behavior is based on the transformed relaxation spectra of the plastically deformed samples.

The energy stored during plastic deformation of glassy polymers is a function not only of the extent of the deformation, but the temperature of deformation as well. Figure 7.23 shows that the higher the deformation temperature, the higher is the starting temperature of the "exo-pit". With increasing deformation temperature the value of the latent energy of deformation is decreased.

PS is brittle at room temperature, therefore, it is impossible to obtain drawn samples at room temparature. Fortunately, drawn samples may be obtained at elevated temperatures. DSC traces of the samples drawn at 365 K to 9 times their original length showed that in this case evolution of a small amount of latent energy occurred only above T_g [57, 58].

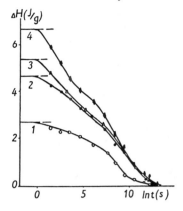

Fig. 7.22. The enthalpy changes during annealing at 70°C as a function of compression for PS [57, 58]. The values of plastic deformation (%): *1 – 5; 2 – 12; 3 – 22; 4 – 40*

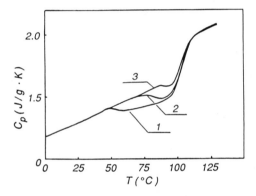

Fig. 7.23. Influence of the compression temperature on the temperature dependence of the dynamical heat capacity of plastically deformed samples od PS [48]. $\varepsilon = 13\%$. Deformation temperature (°C): *1 – 20; 2 – 65; 3 – 80*. Deformation rate $= 10^{-3}\mathrm{s}^{-1}$

Oleinik and Nazarenko [48, 50] assumed that the main mechanism of the energy storage during plastic deformation of polymer glasses is a nucleation and growth of metastable linear dislocation type defects. The decisive role of the shear component in the deformation of such defects has been demonstrated in a comparative DSC study of two samples, one of which was compressed up to 10 kbar and the other was compressed up to 10 kbar and sheared by 200 degrees in Bridjmen anvils [61]. This treatment of the samples with only the hydrostatic pressure resulted in no visible C_p-anomalies while after the treatment under pressure with shearing the corresponding C_p-anomalies were observed.

Bershtein and coworkers [57, 58] assumed that the main mechanism of the energy storage in plastically deformed PS was a decrease of the intermolecular interaction. They arrived at this conclusion as a result of a thorough consideration of DSC and IR data. The energy ΔU_{st} stored in plastically deformed PS was correlated with the change of the coefficient of extinction of the $1602\,\text{cm}^{-1}$ band. They concluded that the changes in ΔU_{st} and in the intensity of this band were similar and that when ΔU_{st} was increased the energy of the intermolecular interaction was decreased. IR studies showed that the contribution of the intramolecular interaction was small. Therefore, the value of ΔU_{st} may be considered as a measure of the intermolecular energy changes.

A more complete quantitative examination was carried out for copolymers of styrene and metacrylic acid [62–64]. In this case the change of the intermolecular van der Waals interaction was very small ($\varepsilon_{1602} = $ const.) and the growth of the enthalpy was completely due to the rupture of the intermolecular hydrogen bonds during plastic deformation. The C_p anomalies observed in DSC traces, in this case, correspond to the reformation of the hydrogen bonds in the course of heating.

Adams and Farris [49] on the basis of deformation calorimetry, thermomechanics and dynamic mechanical properties of plastically deformed films of PC, PSu, PAr and PMMA proposed a simple model to explain their experimental observations. Deformation below T_g is explained in terms of the "disrupture and reformation of secondary molecular interactions. The secondary bonds reform while the macromolecules are perturbed from their equilibrium dimensions, resulting in a quasi-stable energy state for the deformed material". This model was assumed to be applicable to deformation below T_g of typical amorphous polymers.

Kung and Li [65] performed a very elegant investigation concerning the plastic deformation of PS in shear bands. The shear bands of various deformations were obtained by compression of annealed PS prismatic samples. The samples for DSC measurements were cut directly from the shear bands. The DSC traces of such samples were completely different than in the case of normally deformed samples or drawn under elevated temperatures. As is seen from Fig. 7.24 besides the "exo-pit" beginning in the vicinity of $T = 320–330\,\text{K}$, in the case of shear bands an endothermal peak was observed at a temperature about $10\,\text{K}$ higher than T_g. The amplitude of the peak increased with the degree of shearing. After fracture of some samples under shear this peak was diminished. Kung and Li assumed that the material in the shear bands after local shearing is highly oriented and, therefore, seems to have a higher value of T_g. Thus, they attributed the first transition at $373\,\text{K}$ to the glass transition of the normal material and the second one at $383\,\text{K}$ to the very oriented material. It is interesting that the enthalpy of the maximum oriented material ($\gamma = 2.56$) is smaller in comparison with the standard quenched material. This is closely related with the high value of the endothermal peak at $T > T_g$ (Fig. 7.24, trace 2). One can suggest that in this particular case it seems possible to observe seperately inter- and intramolecular stages of the recovery of the deformed PS to the initial state. One can assume that sub-T_g process corresponds to the recovery of the ruptured intermolecular bonds while the endo-peak above T_g seems to be connected

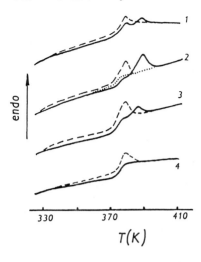

Fig. 7.24. DSC traces of the shear bands of samples of PS [65]. Shear deformation: *1* – 1.61; *2* – 2.56; *3* – shear accompanied by fracture; *4* – rupture of a sample with a shear band. *Dashed lines* – second DSC run after slow cooling; *dotted line* – quenched sample

with the transition of macromolecules from a more extended state to a coiled one, i.e. the number of the higher energy conformations is increased. Thus, in this case two sources of the storage energy during plastic deformation are working: the increase of intermolecular interaction and the change of the conformational state of macromolecules.

The dominating role of the conformational energy changes in plastic deformation has been observed in cold drawing of amorphous PET. As can be seen from Table 7.1, at low drawing rates of amorphous films and fibers, the values of stored energy are 6.0–6.75 J/g. Similar values of the stored energy were also obtained during slow compression of PET: the internal energy change of the samples deformed plastically to 37% was 5.5 J/g [57]. Such samples are characterized by a completely amorphous texture. The internal energy change may be attributed to the changes of intermolecular interactions. In contrast to the behavior at low drawing rates, at higher drawing rates a considerable internal energy decrease was observed. This internal energy decrease may be attributed only to crystallization which was accompanied by the conformational ordering.

Figure 7.25 shows DSC traces of the drawn and undrawn samples. Their behavior is quite different. On heating of the undrawn sample four calorimetric events were observed: glass transition ($T_g = 354$ K), cold crystallization ($Q_c = 28.5$ J/g), recrystallization ($Q_{rc} = 6.3$ J/g) and, finally, melting ($T_m = 528$ K, $Q_m = 39.3$ J/g, the degree of crystallinity is about 25%). The drawn samples behave differently: there is no glass transition region, instead of which a small exo-effect occurs ($Q = 2.5$ J/g), and after that melting is observed ($T_m = 528$ K, $Q_m = 23.5$ J/g, degree of crystallinity is about 15%). Using these data one may conclude that during cold drawing crystallization really took place, which means that some parts of macromolecules during crystallization were converted to the low-energy *trans*-conformations. The lower values of melting points and heat of fusion of the drawn samples seem to indicate that the crystallites formed during cold drawing have a larger con-

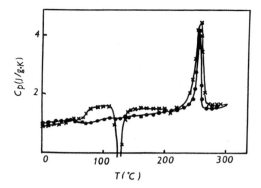

Fig. 7.25. Temperature dependence of C_p for undrawn ($\times$) and drawn ($\circ$) samples of PET [39, 47]

centration of defects. Taking into account the difference in heat of fusion of the drawn sample and the energy released during cold drawing (about 14 J/g) one can suggest that during heating in the drawn sample some reorganization of the crystalline structure takes place. This conclusion is agreed with by the exo-effect in the vicinity of glass transition temperatures and lower values of heat capacity C_p in the temperature range 423–480 K in comparison with the undrawn sample. At even higher drawing rates, the self-oscillating regime of neck propagation is observed which is accompanied by a more intensive crystallization.

Finally, very interesting thermodynamical properties were observed for densified glassy polymers (PS, PVC, PMMA and some other) obtained by cooling of melts under pressures up to 500–600 MPa [67]. In this case, at pressures above 200 MPa, on cooling a linear increase of enthalpy with pressure was observed, the limiting value of ΔH being about 6 J/g. This value is quite comparable with the internal energy change resulting from the plastic deformation. This exceeded internal energy starts to release during heating of samples in DSC in the vicinity of β-relaxation. Such behavior of the densified glasses results in an enhanced concentration of the typical-for-melts conformations of the higher energy frozen at cooling under high pressures. DSC traces of the densified glasses are similar to the DSC traces of plastically deformed glasses. This similarity, however, is only apparent.

7.3.2 Crystalline Polymers

Table 7.1 shows that all crystalline polymers are capable of storing energy on cold drawing. Temperature dependencies of the heat capacity of drawn crystalline polymers in contrast to drawn amorphous polymers show no visible anomalies. Only a lower temperature coefficient of heat capacity was observed for drawn polymers [39, 48]. Therefore, one may assume that the main mechanism responsible for the energy storage during cold drawing seems to be a decrease of the degree of crystallinity. Such a possibility is especially probable for those polymers which in the undrawn state possess a high crystallinity as well as in those cases when the drawing temperature is near to T_g. The results published in the literature demonstrated both the increase and decrease of degree of crystallinity [1, 22].

Assuming that the main contribution to the energy changes during cold drawing is the change in the degree of crystallinity, then on the basis of Table 7.1 one may conclude that, for crystalline polymers studied, a decrease of the degree of crystallinity is observed. The corresponding values for HDPE is 6.5%, for LDPE – 2.5%, for i-PP – 8%, for PA-6 – 10% and for PBT – 4%. These estimations were made using the relation $\Delta U_{st}/\Delta H_m$, where ΔH_m is the heat of fusion of the completely crystalline polymers. These results coincide with above-mentioned suggestion that the changes in the degree of crystallinity are higher for polymers with higher crystallinity (HDPE and LDPE) and when the deformation temperature is close to T_g (PA-6, i-PP, PBT). The decrease of degree of crystallinity of PA-6 by about 7% at a drawing temperature 60°C was also observed in a dilatometrical study [68], which is very close to the calorimetrical value taking into account the difference in the drawing temperature.

A comparison of the energy state of drawn and undrawn crystalline polymers using the heat of dissolution is difficult because their molecular mobility is depressed with crystallinity. Therefore, their melting behavior is of great interest. The thermodynamical characteristics of melting together with the latent energies of deformation are tabulated in Table 7.2. According to the heat of fusion of drawn i-PP its degree of crystallinity is 9% lower than for the undrawn polymer. This value is in very good agreement with the result obtained on the basis of ΔU_{st}. On the other hand, according to the heats of

Table 7.2. Thermodynamic characteristics of melting of drawn and undrawn crystalline polymers [39, 47]

Polymer		T_m K	ΔH_m J/g	Crystallinity %	$\Delta H_m - \Delta H_{m(dr)}$ J/g	ΔH_{st} [1] J/g
HDPE						
	Undrawn	405	177.5	60	−12.6	15.9
	Drawn	411	190.0	65		
LDPE						
	Undrawn	385	118.3	40	−14.7	6.3
	Drawn	385	133.0	45		
i-PP						
	Undrawn	438	85.7	60	13.3	10.1
	Drawn	440	72.4	51		
PBT						
	Undrawn	224	54.5	identical	0	5.3
	Drawn	224	55.0	identical		

1) Correspond to the drawing rate which was used for obtaining the samples for estimates of C_p and heats of fusion.

fusion of HDPE and LDPE, the degree of crystallinity of the drawn samples is higher than for undrawn. Enhanced values of T_m and ΔH_m for drawn PE samples were observed in some investigations [69–72]. Such behavior has been explained by lower values of the enthalpy and entropy of the amorphous regions resulting from a high orientation of chains in these amorphous regions These lower values of the enthalpy and entropy may contribute to the heat of fusion.

A simultaneous consideration of values ΔH_m and ΔU_{st} leads to the conclusion that for PE there seem to be some mechanisms of energy storage during cold drawing which are not connected with the crystallinity decrease. One of the possible mechanisms may be the appearence of internal stresses under drawing. It has been established, for example, that the level of the internal stresses in drawn PE is considerably higher than in undrawn samples with spherulite morphology [73]. These enhanced on 20–30 MPa values of internal stresses are localized in the amorphous regions. The values of the internal stresses have a good correlation with the values of shrinkage during heat treatment of drawn PE samples. In the course of heating these internal stresses can relax and this relaxation may be accompanied by decreasing enthalpy with a simultaneous densification of the amorphous regions.

Although these suggestions need further thorough experimental examination their likelihood is seen from data for PBT (Table 7.2). The values of ΔH_m for undrawn and drawn samples are identical which corresponds to the identical degree of crystallinity. At the same time the stored energy is a rather tangible value, which allow us to attribute it to the internal stresses in amorphous regions. During heating these stresses can relax completely in such a way that at the beginning of melting the enthalpy of drawn and undrawn samples will be the same value.

In contrast to the initial drawing, the re-drawing under various angles goes on without any energy changes (see Fig. 7.16). This seems to indicate that the degree of crystallinity remains constant. According to X-ray studies [74, 75] during the re-drawing of the drawn PE the piles of crystallites due to turning and gliding with only very insignificant destruction are transformed to a new fibrillar structure. However, similar to the first drawing, for the re-drawing there is also the suggestion that the initial crystalline structure is melted completely during necking and new crystallization occurs.

Finally, it is worth mentioning that some polymers such as PE, i-PP, PET and their blends after compression up to 60 kbar and shearing to the angle up to 2000 degrees (approximately 5.5 rotations around its axis) in Bridjmen anvils show unusual thermal behavior [76]. It has been established that after such severe treatment on the DSC traces, besides the usual melting peak, one more endothermal peak 70–100 degrees below the melting peak occurred. The enthalpy change resulting from this peak was 8–20 J/g. Simultaneously the heat capacity jump ΔC_p at T_g disappeared. The second scanning restored the initial picture. It was assumed that this effect resulted from extension of tie-molecules and formation of small crystallites under high pressure and shearing.

7.4 Thermomechanical Behavior of Hard Elastic Fibers and Films

Annealing, near the melting points, of HDPE, i-PP, POM, PA-6 films and fibers is accompanied by the appearence of lamellar morphology with c-texture of lamellar crystallites [77–79]. Such materials demonstrate an unusual mechanical behavior. As seen from Fig. 7.26 the stress–strain curve of the initial sample resembles the typical cold drawing curve of the crystalline polymer. However, the extension is not accompanied by the appearance of any neck. Instead of this, whitening of the sample occurred and it became non-transparent which indicated the appearence of a large number of microcracks. A considerable part of the large deformation happened to be completely reversible. For example, in a typical case for a sample drawn to 100% the reversible part of the deformation was 90%. The second and further deformation cycles are characterized by considerably smaller values of the modulus of elasticity and stresses. After prolonged relaxation in the free state of the deformed samples,

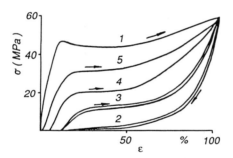

Fig. 7.26. Stress–strain curves of the hard-elastic PP samples at room temperature in air [81, 82]. *1* – first extension; *2* – contraction; *3* – second extension; *4,5* – second extension after relaxation for 24 hours and 1 month, respectively

the initial behavior may be completely recovered. Thus, such behavior resembles not only the cold drawing of crystalline polymers, but the stress softening of thermoelastoplastics as well (Chap. 6).

Thermomechanical behavior of the hard elastic films was studied by means of deformational calorimetry in air [80] and in some liquids [81, 82]. Typical deformational calorimetry behavior which include both the temperature traces and stress–strain curves corresponding to the cycle extension-stop-contraction in air and in propanol are shown in Fig. 7.27. The most important difference in deformational behavior in air and in the liquid is the opposite signs of the heat effect resulting from the contraction of the deformed samples: contraction of the sample in air is accompanied with evolution of heat while in the propanol – with heat absorbtion. As a result of the deformation cycle in air, the sample stores a large amount of internal energy

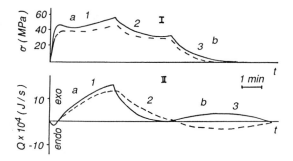

Fig. 7.27. Deformation curves (**I**) and deformation calorimetry traces (**II**) of the hard-elastic i-PP samples in the cycle extension (1) – stop (2) – contraction (3) at room temperature in air ($-$) and in propanol (- - -) [81, 82]

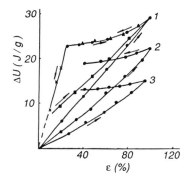

Fig. 7.28. The internal energy changes in hard-elastic PP samples during deformation in air (1), propanol (2) and hexadecane (3) at room temperature [81,82]. $\bullet$ – first extension and contraction; $\blacktriangle$ – second extension without intermediate stopping; $\blacksquare$ – second extension after relaxation in the free state for 1 month; $\times$ – relaxation of the internal energy of the stress-softened sample in the free state

(Fig. 7.28). The part of the mechanical work corresponding to the hysteris loop in such cycles is very large, $\Delta U/\Delta W$ being 57% in air and 48% in hexanol. These values of stored energy are considerably higher than those which have been obtained during the normal cold drawing of PP with necking. The energy stored during deformation in liquids is strongly dependent upon the nature of the liquid. Hexadecan is a plasticator for i-PP while propanol is not. The curve of the internal energy change ΔU during the second and further cycles repeats practically the curve of ΔU change during contraction. Aging in the free state of the i-PP sample after crazing in air is accompanied by a complete recover of its initial structure and mechanical properties. Simultaneously a complete relaxation of the stored energy occurs. Deformation of those samples which only partly recovered during aging is also accompanied by the energy storage but the dependence of ΔU upon the deformation is different in comparison with the initial and stress-softened samples. With increasing temperature the qualitative picture of deformational behavior of hard-elastic PP in air is not changed, although the values of main characteristics are decreased. For example, the increase of internal energy resulting from extension at $20°C$ is $28\,J/g$, at $50°C$ is $14\,J/g$, and at $80°C$ only $8.5\,J/g$.

A consideration of all the thermomechanical results together with X-ray data concerning the structure of crazes and the change of the density of samples during crazing led to the following molecular picture of deformational behavior of the hard-elastic materials [81–84]. At the initial stages of extension the microcracks appear in amorphous regions. The walls of the microcracks are connected with microfibrils built up of the material of the amorphous regions. A further deformation of such a sample is accompanied with an increase of the width of microcracks simultaneously with the drawing of microfibrils from the walls. At this stage, shearing of the crystallite structure takes place resulting in storage of energy. Recovery of the extended sample is accompanied by contraction of microfibrils, "closing" the microcracks and unloading of the crystalline framework. However, even in the completely unloaded sample some microcracks exist and, therefore, the second extension is accompanied at the initial stages with "opening" of the remaining microcracks which is the reason for the stress-softened state. A prolonged relaxation leads to "healing" of all the microcracks, emission of the stored energy and full re-formation of the initial structure.

References

1. Ziabicki A (1976) Fundamentals of fiber formation, Wiley New York
2. Marshall J and Thompson AB (1954) Proc Roy Soc Ser A 221: 541
3. Hookway DC (1958) J Textile Institute 49: 292
4. Müller FH and Jackel K (1952) Kolloid Z 145: 145
5. Jackel K (1954) Kolloid Z 137: 130
6. Müller FH (1954) Kunststoffe 44: 569
7. Ward IM (1971) Mechanical Properties of Solid Polymers, Wiley New York
8. Maher JW, Haward RN and Hay HN (1980) J Polym Sci Polym Phys Ed 18: 2169
9. Schwarz G (1981) Colloid Polym Sci 259: 149
10. Nakamura M and Skinner SM (1955) J Polym Sci 18: 423
11. Müller FH (1969) Thermodynamics of deformation, In: Rheology, vol 5, 417, Academic New York
12. Lazurkin YuS (1954) Doctor dissertation, IFP AN SSSR, Moscow
13. Pokrovsky EM (1985) Candidate Dissertation, Institute of Machinery Moscow
14. Badami DV, Chappel FP, Culpin MF, Madoc Jones D and Tranter TC (1961) Rheol Acta 1: 639
15. Arakawa S (1969) In: Fiber Formation and Development of its Fine Structure, Kagaku Dojin Kyoto
16. Lazurkin YuS (1958) J Polym Sci 30: 595
17. Vincent PI (1960) Polmyer 1: 7
18. Entgelter Ad, Müller FH (1958) Kolloid Z 157: 89
19. Peterlin, A (1971) J Mater Sci 6: 490
20. Meinel G and Peterlin A (1971) J Polym Sci Part A-2 9: 67
21. Wunderlich B (1973) Macromolecular Physics, vol 1, Academic New York

22. Marikhin VA and Mjasnikova LP (1977) Supermolecular Structure of Polymers [in Russian], Leningrad: Khimija
23. Horsley RA and Nancarrow HA (1951) Brit J Appl Phys *2*: 345
24. Kargin VA and Sogolova TI (1953) Zh Ph Ch *27*: 1039; 1208; 1213
25. Baranov VG, Frenkel SYa, Volkov TI and Gasparian KA (1969) Solid State Physics (FTT) *11*: 1220
26. Flory PJ (1956) J Am Chem Soc *78*: 5222
27. Beliaev OF and Zelenev YuV (1980) Vysokomol Soedin *B22*: 471
28. Juska T and Harrison IR (1982) Polym Eng Rev *2*: 13
29. Godovsky YuK (1969) Vysokomol Soedin *A11*: 2129
30. Peterlin A (1979) In: Ultra-High Modulus Polymers, Appl Sci Publ London
31. Peterlin A (1987) Colloid Polym Sci *265*: 357
32. Müller FH and Entgelter Ad (1957) Kolloid Z *150*: 156
33. Andrianova GP, Kargin VA and Kechekjan AS (1970) Vysokomol Soedin *A12*: 2424 (1971) J Polym Sci Part A-2, *9*: 1919
34. Andrianova GP, Popov YuV and Arutyunov BA (1976) Vysokomol Soedin *A18*: 2311 (1978) J Polym Sci Polym Phys Ed *16*: 1139
35. Barenblatt GI (1970) Izv AN SSR, Mech.tverdogo tela, n5, 121
36. Barenblatt GI (1964) Appl Math Mech *28*: 1048
37. Matkowsky BJ and Sivashinsky GI (1979) Quarterly Appl Mathem *37*: 23
38. Pakula T and Fisher EW (1981) J Polym Sci Polym Phys Ed *19*: 1705
39. Godovsky YuK (1976) Thermophysical Methods of Polymers Chracterization [in Russian] Khimija Moscow
40. Andrianova GP, Popov YuV, Artamonova SD and Arutyunov BA (1977) Vysokomol Soedin *A19*: 1230
41. Physical Processes of Plastik Deformation at Low Temperatures [in Russian], Naukova Dumka Kiev 1974
42. Klyavin OV (1975) Doctor dissertation, Ioffe Physical Technical Institute Leningrad
43. Malygin GA (1975) Phiz Met Metalloved *40*: 21
44. Petuchov BV (1977) Solid State Physics (FTT) *19*: 397
45. Estrin Y and Kubin LP (1980) Scripta Metallurgica *14*: 1359
46. Ivanchenko LG and Soldatov VP (1981) Phiz Met Metalloved *52*: 183
47. Godovsky YuK (1972) Doctor dissertation, Karpov Institute of Physical Chemistry Moscow
48. Nazarenko SI (1988) Candidate dissertation, Institute of Chemical Physics USSR Academy of Sciences Moscow; see also Salomatina OB, Nazarenko SI, Rudnev SN and Oleinik EF (1988) Mech of Composite Materials *N6*: 979; (1991) Colloid Polym Sci *269*: 460
49. Adams GW and Farris RJ (1988) J Polym Sci Polym Phys Ed *26*: 433
50. Oleinik EF (1989) Progr Colloid Polym Sci *80*: 140
51. Bever MB, Holt DL and Titchener AL (1973) In: Progress in Material Science, vol 17, pp. 1–190
52. Bolshanina MA and Panin VE (1957) In: Issled Phiz Tverd Tela, pp. 193–233 USSR Academy of Sciences Publish Moscow
53. Chen HS (1976) Appl Phys Lett *29*: 328
54. Escaig B (1984) Polym Eng Sci *24*: 737
55. Kaush HH (1978) Polymer fracture, Springer Berlin Heidelberg New York
56. Prest WM and Roberts J (1981) Ann N Y Acad Sci *371*: 67
57. Bershtein VA and Egorov VM (1990) DSC in Physical Chemistry of Polymers [in Russian], Khimija Leningrad, Ch. 6
58. Bershtein VA and Egorov VM (1984) Solid State Physics (FTT) *26*: 1987
59. Stolting J and Müller FH (1970) Kolloid Z Z Polym *238*: 459; *240*: 790
60. Müller FH (1959) Kolloid Z *165*: 96
61. Oleinik EF (1987) Polymer J *19*: 105

62. Bershtein VA and Pertzev NA (1984) Acta Polym *35:* 575
63. Bershtein VA, Razguliaeva LG, Sinani AV and Stepanov VA (1976) Solid State Physics (FTT) *18:* 3017
64. Anischuk TA, Bershtein VA, Galperin VM et al (1981) Vysokomol Soedin *A23:* 963
65. Kung T and Li J (1986) J Polym Sci Polym Chem Ed *24:* 2433
66. Weitz A and Wunderlich B (1974) J Polym Sci Polym Phys Ed *12:* 2473
67. Prest WM and Roberts J (1984) Contemp Top Polym Sci Proc US-Jap Polym Symp N Y, pp. 855–870
68. Komkov YuA and Shishkin NI (1972) Vysokomol Soedin *B14:* 295
69. Peterlin A and Meinel G (1965) J Appl Phys *36:* 3028
70. Fischer EW and Hinrichsen G (1966) Kolloid Z Z Polym *213:* 28
71. Illers KH (1970) Angew Makromol Chem *12:* 89
72. Peterlin A (1967) J Polym Sci Part C, *N18:* 123
73. Kubat J, Peterman J and Rigdhal M (1975) Colloid Polym Sci *253:* 875
74. Gerasimov VI and Tsvankin DYa (1970) Vysokomol Soedin *A12:* 2599
75. Genin YaV, Gerasimov VI and Tsvankin DYa (1973) Vysokomol Soedin *A15:* 1798
76. Zhorin VA, Godovsky YuK and Enikolopian NS (1982) Vysokomol Soedin *A24:* 953
77. Cannon SL, McKenna GB and Statton WO (1976) J Polym Sci Macromol Revs *11:* 1
78. Park IK and Noetbar HD (1975) Colloid Polym Sci *253:* 825
79. Petermann I and Schultz D (1978) J Mater Sci *13:* 50
80. Goritz D and Müller FH (1974) Colloid Polym Sci *252:* 862
81. Efimov AV, Bulaev VM et al (1986) Vysokomol Soedin *A28:* 1750
82. Bulaev VM (1986) Candidate dissertation, Chemical Department, Lomonosov State University Moscow
83. Efimov AV, Bulaev VM et al (1986) Vysokomol Soedin *A28:* 2341
84. Efimov AV, Bulaev VM et al (1987) Vysokomol Soedin *A29:* 1013

8 Thermal Behavior
of Solid Polymers During Fracture

The fracture of polymers is inevitably accompanied by thermal events. The sources of the thermal events may be both the deformational processes and the rupture of macromolecules. Pure elastic (brittle) fracture of solids is more the exception than the rule [1]. Normally the appearence of the initial sources of fracture – cracks – results in the local plastic deformation. A considerable part of the work, corresponding to the plastic deformation is transformed into heat and this local heat build-up leads to the local rises in temperature during the crack propagation in solids. Unlike the typical low molecular solids in which the plastic deformation at the tip of the cracks is the main mechanism of the local temperature rises under rupture, in polymers one more source of the rise in temperature is possible which is a direct consequence of the chain structure of macromolecules. A part of a macromolecule stressed almost to its load-bearing capacity represents an extremely powerful source of the elastically stored energy. The scission of any single bond in these adequately long segments of the highly stressed macromolecules is inevitably accompanied with dissipation of all the energy stored in all the other bonds. Therefore, this process can also be a very powerful source of the local rise in temperature.

In this chapter we will consider the thermal behavior of solid polymers during fracture with a special emphasis on the heat and temperature effects occurring during fast and slow fracture.

8.1 Thermal and Temperature Effects
Resulting from the Formation and Growth
of Cracks in Solid Polymers

The theoretical strength of the ideal solid is very high and according to the generally accepted estimations is approximately $0.1\,E_{th}$ [2]. The strength of real solids is one–two orders lower and this discrepancy results in the existence of various inhomogeneities, first of all cracks. The explanation of the differences between the theoretical and the real strength of the solids, which display brittle fracture (glasses), is a consequence, according to Griffith, of the widening of a crack. Summing up the resulting decrease dU of elastic energy stored within a stressed sample with stress σ with the surface energy

Γ one can get the free energy in the form [2, 3]

$$F = -Wv + Ws = \frac{\sigma^2}{E}l^2 + 4\Gamma l \qquad (8.1)$$

where 2l is the length of an elliptical crack (the stress σ acts at right angles to the major axis of the crack). The stability criterion has the form

$$\sigma_0 = (2E\Gamma/l)^{1/2} . \qquad (8.2)$$

Above $\sigma > \sigma_0$, $\partial F/\partial l < 0$ and the largest material defect starts propagating irreversibly. For a platelike deformation state a more correct estimation leads to the following expression for the critical stress

$$\sigma_0 = \left[\frac{2E\Gamma}{\pi l(1 - \mu^2)}\right]^{1/2} \qquad (8.3)$$

where μ is the Poisson ratio.

This formula (8.3) has been obtained on the basis of the energy balance of the deformation (the first law of thermodynamics) and, therefore, does not include any suggestions concerning the ideal strength of a solid and about molecular processes during the formation of the free surfaces at the tip of the crack (energy approach). It corresponds to the ideal brittle solids during the fracture of which the only accompanying process is the transformation of the elastic energy into the surface energy of surfaces appearing. Therefore, in the case of ideal brittle fracture the problem of the energy dissipation is actually absent (the problem, however, appears when the quasistatic growth of the cracks in brittle solids is considered using the methods of irreversible thermodynamics [3, 4]).

The class of materials corresponding to brittle fracture postulated in Griffith's theory is the inorganic glasses, in which the fracture surface energy is closed to the surface energy obtained by independent methods. Numerous investigations of fracture of other classes of materials such as metals, ceramics, solid polymers showed [1, 2, 5, 6] that the relation $\sigma_0 \sim 1/\sqrt{l}$ is really qualitatively fulfilled in these materials but the values of the surface energy Γ, obtained on the basis of experimental findings differed significantly (some orders) from the ones calculated theoretically. The further modifications of this simple approach suggested by Irvin and Orovan have led to the conclusion that the experimentally obtained values of Γ relate not only to the brittle appearence of the new surface, but some other processes, first of all plastic deformation. It is well known that some metals which are capable of plastic deformation in normal mechanical tests behave in brittle manner during crack formation. During their fracture the plastic deformation is localized in a very thin layer near the surface of a crack. The values of Γ estimated using Eq. (8.3) are as a rule two–three orders higher than the theoretical values which can be estimated by taking into account only the molecular forces. Therefore, instead of the value Γ in Eq. (8.3) it is necessary to use some effective value Γ_{eff}, which includes also the work of plastic Γ_{pl} deformation. As a result Eq. (8.3) may be written in the form

$$\sigma_0 = \frac{2E(\Gamma_s + \Gamma_{pl})}{\pi l(1 - \mu^2)} = \frac{2E\Gamma_{eff}}{\pi l(1 - \mu^2)} . \qquad (8.4)$$

In contrast to the Γ_s the value of Γ_{pl} cannot be estimated theoretically at present. Because during the growth of cracks usually $\Gamma_{pl} \gg \Gamma_s$ the latter can be neglected and, therefore, the Γ_{eff} is actually controlled by the work of plastic deformation. Hence, Eq. (8.4) in contrast to Eq. (8.3) describes the materials, which are characterized by a quasi-brittle, or brittle-plastic fracture.

Whether the fracture of solids is brittle or quasi-brittle depends on the stress state at the tip of the crack [2, 3]. The stress state and the character of the fracture are closely related to the molecular forces at the tip of the crack which, in turn, are determined by the competition between the shear strength and the rupture strength. The integral energy approach to the change of elastic energy resulting from the introduction of a crack into it can be used in the local form in order to analyse the energy state at the tip of the crack. Irwin (see, for example, [2]) has introduced the critical stress intensity factor K_c which is a complete characteristic of the local deformation and fracture at the tip of cracks. For the plane deformation K_c is

$$K_c = \sqrt{\frac{E\Gamma_{eff}}{\pi(1 - \mu^2)}} . \qquad (8.5)$$

The dimension of the crack and the macroscopic stress are absent in this relation which underlines the decisive role of the tip of the cracks and K_c in this local approach. K_c determines the conditions of the crack's growth: after K_c reaches the value determined by Eq. (8.5) a quick crack propagation occurs. Thus, in contrast to Eq. (8.3), Eq. (8.5) has a local character.

Unlike the ideal brittle fracture of solids, the quasi-brittle fracture should be accompanied with significant thermal effects, resulting from the local plastic deformation during the crack appearence and propagation. In the case of the fast-moving cracks the heat generated in plastic deformation may lead to a considerable local temperature rise. The corresponding estimations will be given below.

Similar to other solids fracture of solid polymers may also be either brittle or quasi-brittle [5–7]. The values of surface energy found experimentally for solid polymers are also considerably (two–four orders) higher than the theoretical values. It has been suggested that the main contribution to the effective surface energy Γ_{eff} gives plastic deformation in the surface layers of microcracks, which may be identified by means of morphological studies of the fracture surfaces. Besides this contribution arising from plastic deformation some contribution to the fracture energy can also give chain ruptures of the stressed macromolecules at the tip of growing cracks as well as mechanical losses, resulting from the visco-elastic behavior of solid polymers [7]. Theory of crack growth in the visco-elastic materials is described in [8, 9]. Hence, the effective surface energy on fracture may be represented by a sum of various contributions. Kausch [5] concluded that the most important contributions to the material resistance will include specific surface energy, plastic deformation, elastic retraction of stressed molecules and chemical reactions including chain scission. The relation between these contributions may be obtained only by the experimental method.

8.1.1 Estimation of the Temperature Rise at the Tip of Propagating Cracks

The work resulting from the plastic deformation taking place at the tip of a running crack is dissipated as heat most of which flows into the material. The temperature rise due to this heat flow is strongly dependent upon the ratio of the plastic work transformed into heat, the thickness of the heated zone, and crack velocity. There have been considerable efforts directed at the theoretical estimation of the local temperature rise at the tip of propagating cracks [10–14]. However, the values of the local temperature rise which have been arrived at are extremely sensitive to some details of the models used for the calculations. A rough estimate of the temperature rise as a function of the velocity of crack propagation can be made using the dependence of the temperature rise resulting from the plastic deformation during cold drawing (Fig. 7.4). Typical values of the crack velocities normally used for the theoretical estimations are some hundreds m/s. Under such velocities the heat dissipation may be considered as an adiabatic process and even a rough extrapolation of any of the dependencies shown in Fig. 7.4 leads to the conclusion that the temperature rise under such conditions must reach some hundreds of degrees. This conclusion is in a full agreement with the theoretical calculations [10–14]. Levi and Rise [11] have arrived at the value 1200 K for a crack velocity of 250 m/s for PMMA. Weichert and Schönert [13] arrived at a similar value (900 K) for a crack speed of 200 m/s. On the other hand, Kambour and Barker [10] as a result of their calculations concluded that the maximum temperature increase at the tip (210 K) is reached in PMMA at velocities above 200 m/s. Finally, Williams [12] predicted a significant temperature increase for velocities above 0.2 m/s for PMMA the maximum rise being 115 K. Hence, the differences in the theoretical estimations of the temperature rise at the tip of the moving cracks are as much as 10 times. This consideration demonstrates quite obviously that the values obtained vary greatly depending upon the details of the model employed. Following Weichert and Schönert [14] let us consider the main principles used in such estimations.

The problem of a finite heat source moving at high speed through a specimen while the heat flows into the material and is partially radiated away is rather complex and no close form solution is available. In order to make the calculations tractable some simplifications are employed. The assumptions used are: (1) the heat is produced at the tip of a crack moving with the speed c which can be approximated by a rectangular layer of a thickness 2δ and length d (Fig. 8.1); (2) the physical properties of the material, in particular thermal conductivity and thermal diffusivity are suggested as being constant;

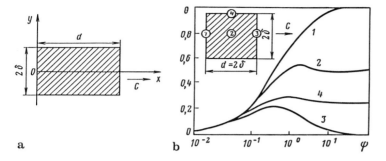

Fig. 8.1. Schematic diagram for calculating the temperature distribution at the crack tip [14]. (**a**) The zone, where the generation of heat occurs. (**b**) Dependence of the normalized temperature function θ upon the velocity parameter of the crack growth ψ for various positions of the heat source

(3) the classical equation of thermal conductivity (Fourier law) is valid. In the coordinate system shown in Fig. 8.1a which is moving together with the heat source, the temperature distribution is steady-state. This distribution can be obtained from the following expression

$$T(x, y) = T_0 + \frac{Q}{2\delta C_p \rho} \cdot \theta \left(\frac{x}{\delta}, \frac{y}{\delta}, \frac{d}{\delta}, \psi \right) , \quad \psi = \frac{c\delta}{2a} \qquad (8.6)$$

where θ is a temperature function normalized according to the dimensions of the heat source and velocity parameter ψ. The solution of this equation in terms of θ may be represented as a function of the velocity parameter in the form of a generalized diagram (Fig. 8.1b) taking into account the position of the heat source into the heat generation zone. According to the diagram there are three separate zones. At $\psi < 0.1$ there is the area of the isothermal states. Into this zone $\theta \ll 1$ and does not depend upon the coordinate of the source. The temperature gradients into this zone are very small and the isotherms can be represented by circles. At $0.1 < \psi < 10$ a transient zone occurs. In this zone with increasing ψ, the values of θ at first increase in all areas of the zone, but in the centre of the zone (point 2) and at the tail of the zone (point 1) this increase is considerably higher than at the front surface (point 3).

Approximately at $\psi = 0.3$ the temperature at the front surface reaches its maximum and after that begins to drop, while the temperature at points 2 and 1 continues to increase. Finally, at $\psi > 10$ the adiabatic regime occurs, when at the front surface parameter θ equal zero, and in the center of the zone it is 0.5 of the value at the tail. Hence, the maximum value of θ for all zones occurs at the tail of any zone.

The heat generated at the tip of a moving crack, is the source of temperature increase in the vicinity of the crack. The maximum temperature rise at the distance $y \gg 2\delta$ for a rather fast moving crack can be estimated from the equation

$$T_{max} - T_0 = (2\pi e)^{-1/2} \left(\frac{1}{C_p\delta}\right)\left(\frac{Q}{y}\right) \tag{8.7}$$

which follows from the differential equation of heat conduction with the instantaneous finite heat source [15]. Thiq equation shows that the maximum temperature rise is inversely proportional to the distance y between the heat source and the point of measurement. The time after which the maximum temperature rise is reached can be determined from the relation

$$\tau = y^2/2a . \tag{8.8}$$

These equations can be used for determining the heat output per unit surface area Q for the materials which are ruptured in the quasi-brittle regime (metals, organic glasses). For such materials the surface fracture energy is $10^3 - 10^5$ J/m^2. For inorganic glasses which best of all other materials correspond to the brittle fracture mechanism, this value is of the order of 10 J/m^2.

As has been mentioned above in the estimation of the temperature rise at the tip of moving crack the heat output Q and the surface fracture energy is similar for both the brittle and quasi-brittle behavior. Although the main contribution to the value of Γ_{eff} gives the work of plastic deformation, it is known that for polymers it is only partly transformed into heat (Chap. 7). The ratio of the transformation is between 0.6 and 0.9. Therefore, a more correct estimation of the temperature rise was obtained in [14], where this ratio was taken into account.

At very high velocities of crack propagation parameter ψ turns out to be large and the conditions at the tip of the fast-moving cracks may be no doubt adiabatic (Fig. 8.1b). In this case, Eq. (8.6) transforms to a simple relation

$$T_{max} - T_0 = \frac{Q}{2\delta C_p\rho} . \tag{8.9}$$

The only parameter in this equation is the thickness of the heated zone 2δ. As is seen from the diagram shown in Fig. 8.2 for fast-moving cracks, that is under condition $\psi \gg 1$, if the thickness of the zone is small the local temperature rise really can reach many hundreds of degrees. It is necessary to add, that in these calculations, a very intensive source of the heat generated, namely the heat resulting from the rupture of elastically overstressed chains, is once more not taken into account. Below we will demonstrate that including this factor into consideration may significantly increase the local temperature rise during the fast propagation of cracks.

8.1.2 The Local Thermal Processes During the Crack Growth in Solid Polymers

It is now well established experimentally that in many polymeric materials three stages of crack propagation behavior can be distinguished: stable crack behavior, slow crack growth and rapid crack propagation [5]. There are three zones of temperature change corresponding to the crack behavior

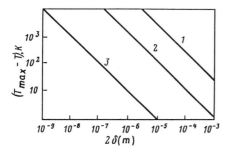

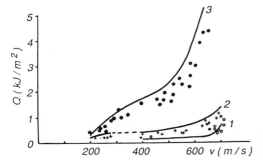

Fig. 8.2. Dependence of the maximal temperature at the tip of moving crack upon the dimension of the zone where heat is generated at $\psi \gg 1$ [14]. *1* – steel; *2* – PMMA; *3* – inorganic glasses

(Fig. 8.1b). The largest local heating during the crack growth must naturally occur in the adiabatic zone of crack growth. The growth rate of the adiabatically propagating cracks can be estimated from the expression for the temperature function θ [Eq. (8.6)]. For typical values of the thermal diffusivity a $= 10^{-3} \text{cm}^2/\text{s}$ and the thickness of the plastic zone $\delta = 10\,\mu\text{m}$ the lower value of the adiabatic crack propagation ($\psi > 10$) is c $= 200\,\text{m/s}$. Hence, the maximum local temperature rise can be expected at the crack propagation velocities close to the sound velocity in solid polymers.

Döll [16] has studied the heat output associated with the plastic work at the crack tip as a function of crack velocity for PMMA of various molecular weights. He suggested that only 60 percent of the plastic work is converted into heat. Using this energy relation and the measured curves of crack velocity as function of dynamic strain energy release rate the heat output was predicted and compared with the measured heat output values. Figure 8.3 shows these predicted heat curves as functions of crack velocity for PMMA of various molecular weights. The measured heat output values Q are also shown. It is seen that there is a very good agreement between the predicted and measured heat output curves both in order of magnitude and in their general shape: the slope of the curves increases with increasing crack velocity.

Fig. 8.3. Heat output as a function of crack velocity [16]

Table 8.1 summarizes the values of the local temperature rise measured experimentally with the help of various methods for some solid polymers. These results, first of all, show a very good agreement with the theoretical estimations demonstrated above. They prove the fact that at high velocities of crack propagation in solid polymers the temperature rise in the vicinity of tip of the crack is many hundreds of degrees higher than the temperature of the polymer as a whole. These local temperature rises are so high that they exceed not only the transition temperature from the solid into the liquid- or rubber-like state but are sufficient to cause thermal degradation of poly- mers. In spite of a very short duration of the local temperature rise (to the order of hundreds of μsec) such thermal microexplosions must be accompa- nied with characteristic morphological changes of fracture surfaces and with the appearence of some products of thermal degradation. Morphological and mass-spectrometric studies of the rupture of PMMA and PS [17, 18] are quite consistent with this conclusion.

Table 8.1. The temperature rise accompanying the fracture of polymers

Polymer	Velocity of crack, m/s	Method	Temperature rise, ΔT, K	Zone dimension $\delta, \mu m$	Ref.
PMMA	200–600	IR	500	0.5	18
	140	T.couple	230	0.8	19
PS	380–550	IR	400	0.5	18
PET					
1. Undrawn	40	IR	≥ 250	20*	20
2. Undrawn	0.9–1.5	IR	180 ± 15	–	22, 23
3. Drawn ($\lambda = 4$)	0.9–1.5	IR	170 ± 15	–	22, 23
4. Undrawn radiated	0.9–1.5	IR	120 ± 20	–	22, 23
Inorganic glass	170	Light radiation	3000	0.002	14

* The actual dimension of the zone is not known, the value indicated seems to be considerably increased and therefore the value of the temperature indicated is apparently considerably lower than the real one.

The local temperature rise is responsible for the local heat evolution Q, generated during the propagation of cracks. This value is equal or propor- tional to the surface fracture energy Γ_{eff}. The magnitude of Q per a unit surface of rupture may be calculated using Eq. (8.9). Typical curves Q and ΔT as functions of the width of the hot zone δ are shown in Fig. 8.4. Q varies from 8×10^3 to 2×10^4 J/m^2 for the values of δ ranging from 20 to 100 μm for nitrocellulose. The corresponding estimations for typical glassy polymers such as PMMA and PS give values $Q = 10^3$–10^4 J/m^2 which are of the same order of magnitude [16, 19, 20]. Similar results are obtained on the basis of analysis of the dependence of the stress upon the crack dimension [6] which is an independent support of high values of the surface fracture energy of solid polymers.

Tomashevskii et al. [20–23] have considered the values of Q obtained by measuring the local temperature rise by IR radiation from the position of two contributions, namely the transformation of plastic work into heat and rupture of overstressed fragments of macromolecules in a small region ahead of the moving crack. Generally, the crack growth process consists of

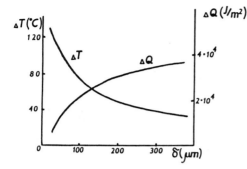

Fig. 8.4. Local temperature rise ΔT and heat evolution ΔQ as functions of the width of the hot zone δ for nitrocellulose [26]

two stages: local stressing of a small zone ahead of the crack up to the level of high stress close to the theoretical strength and rupture of overstressed bonds in this zone. Both stages must be accompanied by heat evolution. On stressing, this is due to internal friction between fragments being displaced during this plastic deformation of macromolecules. At the second stage it results in the formation of vibrationally and rotationally excited states of interatomic bonds in the ruptured macromolecular fragments. Kambour and Barker [10] suggested that about 25% of the actual heat is produced during the first stage, that is during craze formation, and 75% of it is produced upon the release of the stored elastic energy, that is, after the craze has broken. Their particular estimation for PMMA has shown, that of the total temperature rise of about $140°C$ approximately $30°C$ corresponds to the first stage and about $110°C$ corresponds to the second stage. Hence, the expected thermal effects resulting from the overstressed bond scission have to be of the order of $(5–10) \times 10^4 \, J/m^2$.

To estimate experimentally the real ratio of the contributions Tomashevskii et al. [20–23] studied the local temperature rise in PET films during a rather slow rupture with the velocities of crack propagation in the vicinity of some m/s, that is, well below the adiabatic crack propagation ($> 200 \, m/s$). Besides the value of the maximum temperature rise in the rupture zone (Table 8.1) the authors determined the width and temperature profile of the hot zone and having used these data they estimated the ratio of the two contributions to the value of surface fracture energy. As mentioned above, in the adiabatic zone of crack propagation, polymers are ruptured rather "brittly". The width of the plastic zones is rather thin in this case. If the crack velocity is considerably slower than the sound velocity in a polymer and its rupture occurs in the transient rate-temperature zone (Fig. 8.1b), the resistance to the crack formation may increase [5, 14] as a result of an increase of the width of the plastic zone and fracture of a polymer may become a more plastic one. The corressponding estimations showed [14] that the maximum resistance to the crack formation for a typical glassy polymer (PMMA) is in the vicinity of $1 \, m/s$. The temperature profile of the fractured zone in PET was studied using such rates.

A typical temperature profile of the fracture zone is shown in Fig. 8.5. It is seen that the highest temperature rise is monitored at the crack edge and then a considerable decrease (about twice the initial temperature rise) occurs. Such a temperature profile is in quite good accord with the suggestion that a considerable contribution is made by the fractured overstressed macromolecular fragments to the heat dissipation and to the temperature rise. This statement was also proved by some indirect facts, based on a comparison of the temperature profiles of fractured zones of the undrawn and drawn samples and non-radiated and radiated samples [21–23]. A quantitative analysis of the temperature profile showed the ratio of the thermal effects resulting from the plastic deformation and scissions of stressed macromolecules is 0.45 to 0.55, correspondingly. This estimation is in very good agreement with calorimetric results (see below). Finally, it is interesting to note, that the local temperature rises resulting from the plastic deformation correlate very well with the temperature effects accompanying cold drawing of PET (Fig. 7.4). Tomashevskii et al. [21] concluded on the basis of such estimations that the length of the overstressed section must be approximately 130 Å, that is, consists of about 100 bonds, and the minimum width of the ruptured zone should be 1–3 μm.

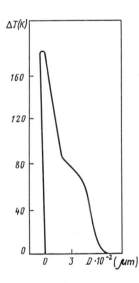

Fig. 8.5. The temperature profile of the hot zone during fracture of PET [22]. D corresponds to the distance from the edge of the crack

Hence, the most important conclusion which can be arrived at as a result of inertialess detection of IR radiation accompanying fracture of solid polymers consists of the statement that either slow or fast loading is accompanied by a considerable heat evolution. Both the transformation of plastic deformation into heat and the elastic energy dissipation process arising from the rupture of overstressed macromolecular fragments in a polymer contribute significantly to the exothermal effects. First of all, heat evolution due to molecular rupture can locally accelerate the degradation of macromolecules in stressed polymers and stimulate the formation of submicroscopic discontinuities and growth of micro- and macrocracks. Then, these results demonstrate quite obviously that the main rupture process occurs not at the temperatures at which the experiments are conducted but at considerably higher temperatures. This conclusion agrees with the suggestion concerning the microexplosion in polymer fracture. Finally, the effect of dissipation of elastic energy after macromolecular rupture sheds a new light on the abnormally high values of surface fracture energy which is usually ascribed only to the plastic deformation of a polymer material in the fracture zone.

8.2 Energetics of Chain Rupture in Stressed Polymers

Even in early considerations, it was recognized that a macromolecule or its adequately large part stressed to its load-bearing capacity is an extremely powerful source of elastically stored energy [24]. Loading of any macromolecules by an external force results in the storage of elastic energy in every one of its chemical bonds. The stored energy is higher the closer their tensile stress to the limit of bond strength. When non of the stressed bonds are broken, the macromolecule is in equilibrium and is capable of producing external work. If, however, only one stressed bond is broken, the equilibrium is disturbed immediately and all the energy stored in the both parts of the broken macromolecule must be dissipated as heat. The mechanical work spent on the dissociation of one molecular bond and formation of two radicals is only a small part of the whole elastically stored energy. If the broken parts are localized in a very small volume which seems to be a rule in crystalline polymers, the dissipation of the stored energy may look like a microexplosion. Starting from these qualitative suggestions and following Tomashevskii [25] let us consider the energetics of elementary scissions of stressed macromolecules.

8.2.1 Energetics of Elementary Scissions of Stressed Macromolecules

In order to estimate the energy characteristics of the fracture process at the molecular level, let us consider the behavior of overstressed macromolecular segments. The elastic energy of a tensilely stressed chemical bond, the interatomic interaction of which is supposed to follow the Morse potential, can be expressed as follows [25, 26]

$$w_{el} = \frac{D}{4} \left(1 - \sqrt{1 - \frac{f}{f_{th}}} \right)^2 \tag{8.10}$$

where D is the bond dissociation energy, $f_{th} = kD/2 = 300\,\text{GPa}$ is the limit of the bond strength, k is the coefficient in the Morse expression. The elastic energy stored in a chain of n bonds $W_{el} = n\,W_{el}$. Without external stress, the energy required for dissociation is taken from heat, therefore, the molecular degradation process is an endothermal process and the internal energy of the broken chain increases by the value D. Although the degradation of stressed macromolecule becomes easier, nevertheless, as long as the tensile stress remains lower than the limit of bond strength an additional external energy source is needed for rupture in this case as well. It means that the bonded and broken states of a macromolecule are separated by an energy barrier (Fig. 8.6). The difference between the total stored elastic energy of a segment and the dissociation energy of one bond may be considered as the thermal effect of the degradation process:

$$Q = D - W_{el} = D - \frac{nD}{4} \left(1 - \sqrt{1 - \frac{f}{f_{th}}} \right)^2 . \qquad (8.11)$$

Figure 8.7 shows the typical character of the thermal effects in terms of D for segments of various length n = 100 and n = 10. When f = 0, the thermal effect of chain scission is positive (Q > 0) and equal to D. The energetics of scission of the stressed chains depends upon the level of stresses: although with small stresses it is an endothermal process, at a high level of stresses it becomes an exothermal one (Q < 0). At such stresses, the degradation of macromolecules is an energetically favorable process because it is accompanied by a decrease in internal energy. For segments long enough their degradation is accompanied by exothermal effects even if the stress is not very high. According to Eq. (8.11) and Fig. 8.7 at a specific tensile stress W_{el} = D and the thermal effect vanishes. This case corresponds to the Griffith energy criterion of fracture as applied to the elementary chain scission. However, it can be seen that the Griffith criterion is fulfilled at stresses which are considerably lower than the molecular strength limit [25, 26].

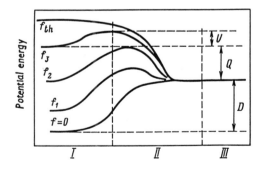

Fig. 8.6. Energy transitions upon rupture of an overstressed macromolecule $f_1 < f_2 < f_3 < f_{th}$ [25]

The thermal effects resulting from the chain scission can be related to the kinetic characteristics of fracture. The corresponding estimations showed that at room temperature and normal times of loading (290 K, t = 100 s), the effective rupture of macromolecules will occur if the activation energy for stressed macromolecules is reduced to 75 kJ/mole. This decrease in the energy barrier takes place in the case of a 100-unit chain built of C-C bonds with the dissociation energy D of 330 kJ/mole at a tensile stress of $0.6\,f_{th}$. Under these conditions the thermal effect, resulting from the chain scission

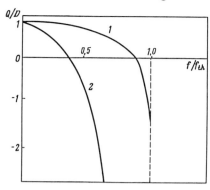

Fig. 8.7. Dependence of the thermal effects upon force during rupture of chains with various numbers of units [25, 26]. *1* – n = 10; *2* – n = 100

will reach $10^3 \, \mathrm{kJ/mole}$, which is about three times higher than the bond dissociation energy. With fast loading or low temperatures, when $f \rightarrow f_{\mathrm{th}}$, the thermal effects will exceed the dissociation energy by a factor of 30. Similar estimations are also given by Kausch [5].

Thus, under high stresses the macromolecular degradation assumes the character of thermal microexplosions [5, 17, 26], the consequences of which are quite obvious. It may accelerate bond dissociation in the vicinity of the primary place of rupture and succeding chemical reactions. Therefore, besides chain free-radical reactions, chain energy processes can develop. The ruptured segments of macromolecules go over to vibrationally and rotationally excited states and the dissipation of excessive energy may cause large local overheating.

The heat effect is not the only energetic result of the scission of macromolecules. As a result of this scission the internal energy of the polymer should be increased [26, 27]. The rupture process is accompanied by the formation of free radicals and new groups with double bonds which appear during chemical reactions initiated by free radicals [5, 17]. The formation of each radical increases the energy by $D/2$. If N_{rad} radicals are stabilized in a polymer, the increase in the energy will be $DN_{\mathrm{rad}}/2$. The other chemical products are groups with double -C=C bonds, which are formed during the distribution reaction of radicals and in the decay of internal macroradicals. The formation of such groups implies the substitution of π-components of double bonds for single σ-bonds. The difference between the component dissociation energies $D_{\mathrm{C=C}}$ and $D_{\mathrm{C-C}}$ increases the energy of the system by approximately 80–100 kJ/mole. If the number N_{gr} of new groups with double bonds appears in a polymer, the energy will increase by $(D_{\mathrm{C=C}} - D_{\mathrm{C-C}})N_{\mathrm{gr}}$. Typical values of free radical concentration as measured by EPR in polymers, for instance in PA-6, before breakdown reaches 10^{17}–$10^{18} \, 1/\mathrm{cm}^3$ and the total number of newly formed double bonds – $10^{19} \, 1/\mathrm{cm}^3$. Hence, the energy responsible for the appearence of radicals is of the order of some tenth parts of $\mathrm{J/cm}^3$, and "belonging" to double bonds is of the order of 1–$2 \, \mathrm{J/cm}^3$. Therefore, the mechano-chemical destructive processes accompanying polymer fracture do change significantly the energy state of a stressed polymer. It should be mentioned finally, that one more source of increase in the internal energy is irreversible plastic deformation which normally accompanies the rupture of macromolecules. As shown in Chap. 7, typical maximum values of the stored energy during plastic deformation of crystalline and glassy polymers reach 0.4–0.5 of the mechanical work spent. Therefore, this source may be of significant importance. Below we will estimate the corresponding values according to data of deformation calorimetry obtained for drawn PA-6

Now a very important question arises: under what conditions may the considered process of dissipation of elastic stored energy take place in polymers. Analysis of the thermomechanical behavior of drawn polymers (Chap. 5) shows that the rupture of adequately long overstressed segments is most probable for highly drawn crystalline polymers. In such polymers the tie-molecules connecting adjacent crystallites may be overstressed considerably during elastic deformation. The length of the overstressed tie-molecules, the rupture of which is accompanied by dissipation of the stored elastic energy, depends upon the intermolecular interaction, morphology and rigidity of macromolecules. Because the intermolecular forces in polymers are considerably weaker than the intramolecular ones the length of the overstressed segments must be rather long, which is a necessary condition for the exothermic fracture.

The micromechanics of the deformational behavior and scission of the tie macromolecules in highly drawn crystalline polymers was considered for the first time by Chevychelov [28, 29] and a further extension of the treatment was made by Kausch [5]. The theory was developed for the Hosseman-Bonart microfibrils (Fig. 5.13). In the theory, the following problems were considered: 1) what are the maximum stresses which can transfer a completely extended tie chain before the parts of the macromolecule embedded in two different crystalline layers can begin to pull away from the crystallites; 2) at what distance in the periodic intermolecular potential of the crystallites is the stress in the tie molecule affected; 3) in what manner are these stresses dependent upon the intermolecular interaction in crystallites. Without going into too much detail about the calculations, let us consider only some results which are necessary for the following estimates. For typical values of the dimensions of the crystallites and amorphous regions in PE and PA-6 the following estimates have been obtained. The stress limits which can be applied to the extended tie molecule in the amorphous region through the ideal crystallites are 7.5 GPa for PE and 22.4 GPa for PA-6. The higher values of the static strength of the PA-6 crystallites are due to the presence of strong hydrogen bonding. Because of this difference the stresses in PA-6 crystallites disappear at a distance of about 3.5 nm, while for PE at a distance of about 5 nm. The value of the limit stress for PE is about 30% of the molecular strength, while for PA-6 it is close to the molecular strength. The typical contour length of the macromolecule embedded in two adjacent crystallites including the amorphous region consists of 200 C-C bonds. The length of the tie part of the molecule consists of some dozens of C-C bonds. Suggesting that the stresses disappear close to the centres of the crystallites one can arrive at a length of the overstressed segments consisting of about a hundred C-C bonds.

According to the above calculations (Fig. 8.7) the rupture of such a segment is an exothermal process. It is especially applicable to PA-6 crystallites which are more rigid.

Hence, these estimations show that the rupture of drawn crystalline polymers will be accompanied by extremely powerful local exothermal processes resulting from the scission of overstressed tie molecules. Although it is a more difficult task to estimate the length of the overstressed segments in the undrawn crystalline and amorphous polymers, nevertheless, there are grounds for suggesting that they also consist of about 100 C-C bonds [17, 21].

Chavychelov's theory showed that at the high axial stresses and large relative deformations of the amorphous regions the main mechanism of elasticity of crystalline chains in the drawn crystalline polymers consists of pulling the tie molecules against the intercrystalline potential and out of the crystalline lamella. This pulling out should be accompanied by local exoeffects. Unloading of the loaded specimen should also be accompanied by heat evolution if the pulled out segments return. In Chap. 5 a similar mechanism of elasticity of the drawn crystalline polymers was considered at the supermolecular level, when at a considerably lower level of stresses it was suggested as reversible displacement of the microfibrils.

8.2.2 Experimental Data on the Energetic Effects Resulting from the Degradation of Macromolecules in Stressed Polymers

It is now generally accepted in the literature that the rupture of chains occuring during the tensile deformation of drawn polymers governs macroscopical failure processes including irreversible changes of tensile properties, the formation of various defects such as microvoids and cracks [5, 17]. It is a well-established experimental fact that major differences in subsequent loading-unloading cycles are observed between the first and the second loading whereas only minor differences occur between the second loading curve and those obtained for subsequent deformation cycles. The conclusions based on the data are that the drastic changes in the sample structure and properties occur namely during the first loading. From the energetic point of view this means that in the energy balance of the first loading the contribution resulting from the chain scission should exist while during the second and subsequent loadings the energy balance should be governed only by the elastic response of strained segments. It is necessary to keep in mind, however, that such a structure of the energy balance should be characteristic not only of the pure elastic stretching and scission of macromolecules. If the chain scission during the first loading is accompanied by the appearance of the irreversible (plastic) deformation it also may give the contribution to the energy balance similar to that responsible for the chain scissions. Therefore, this approach needs very careful control of the appearance of the irreversible deformation during the first loading and the corresponding separate estimation of the contributions resulting from the plastic deformation and from the chain scissions. The only experimental study of such energetical effects was carried out by calorimetric measurements of highly drawn PA-6 fibers [27]. These PA-6 fibers, drawn to 5.5 times their original length at 210°C, were repeatedly stretched in the Calvet-type calorimeter at room temperature. The specimen was stretched to some definite deformation, then the load was removed, after which it was reloaded to the same deformation in the same manner. In both cases the work spent on the extension-contraction cycle (mechanical hysteresis) and the total thermal effect during the cycle was measured. Figure 8.8 shows the corresponding energy characteristics. It is seen that the energy characteristics for the first and the second cycles are equal only when the stress is below about 400 MPa. On further deformation the situation changes drastically: both the mechanical work and the thermal effect upon the first loading, when the molecular rupture occurred, turned out to be a few times higher than those upon the second and subsequent loadings when the molecular ruptures were not involved in the energy effects. It is very important to emphasize that the difference occurred only at the stage of loading because the unloading stages for all the cycles were energetically equivalent. The energy characteristics above 400 MPa resulting from the irreversible events during the first loading are shown in Fig. 8.9. Two very important features are seen. First of all both mechanical work and heat evolution grow with increasing tensile stress exponentially. This is seen when the data are replotted in semilogarithmic coordinates. As an example, Fig. 8.9 shows the plot $\log \Delta Q = \varphi(\sigma)$ from which it follows that $\Delta Q = \Delta Q_0 \exp \beta\sigma$, where σ is the applied tensile stress. Furthermore, the mechanical work always exceeds the heat evolution, so that some energy ΔU remains in the polymer. The growth of ΔU is also exponential upon the stress. It should be noted that the concenration of free radicals also exponentially increases under extension of these PA-6 fibers above 400 MPa during the first loading [17]. Therefore, the stress dependence of the energy parameters measured correlates qualitatively with the corresponding dependence of the number of bonds ruptured in the polymer during the first loading.

To reach a final conclusion it is necessary to consider the role of plastic deformation which occurred above 400 MPa during the first loading. A very thorough analysis of the length of the samples deformed showed that after prolonged annealing at room temperature the length of the deformed samples was totally reversible below a stress of 350 MPa. Above this stress, the residual (plastic) deformation occurred and its dependence on the upper tensile deformation during the first loading is shown in Fig. 8.10. The dependence is qualitatively similar to that found for the energy characteristics (Fig. 8.9) Hence, it is quite obvious that the energy parameters of the irreversible changes accompanied the first loading of the samples include contributions both from the plastic deformation and from the bond ruptures.

The contribution of the plastic deformation may be easily estimated assuming that the total plastic deformation occurs at the maximum stress reached during the first loading. The corresponding estimations showed that for the highest stresses in the first loading the work of plastic deformation

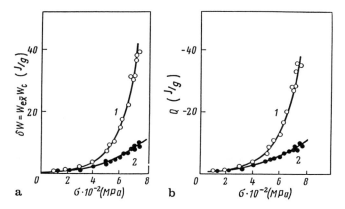

Fig. 8.8. Mechanical hysteresis $\delta W = We - Wc$ (**a**) and heat Q dissipated (**b**) as functions of stress for highly drawn PA-6 fibers [27]. *1* – first cycle; *2* – second and subsequent cycles. Characteristics of the fibers: diameter – 0.27 mm, tensile strength – about 0.8 GPa, elongation at rupture – 16–17%. Deformation rate – 0.116 mm/s

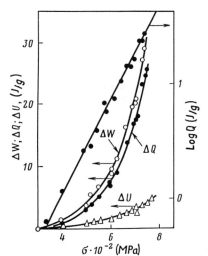

Fig. 8.9. Energy effects due to molecular ruptures upon loading of the PA-6 fibers [27]

$W_{pl} = 13$ J/g, which is about 45% of the total mechanical hysteresis. Taking into account the data concerning the cold drawing of PA-6 (Table 7.1) one may assume that 80% of the work of plastic deformation is dissipated and 20% is spent on the increasing of the internal energy of the sample. Hence, the energy parameters, resulting from chain scissions in the preruptured state are as follows: $\Delta W_{rup} = 16$ J/g, $\Delta Q_{rup} = 14.5$ J/g and $\Delta U_{rup} = 1.5$ J/g.

Let us use these energy characteristics for estimating some fracture parameters of PA-6. First of all, let us estimate the number of broken chains. Assuming that ΔU_{rup} results completely from formation of the end groups with the double -C=C bonds and suggesting that every scission requires

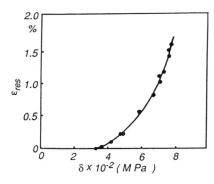

Fig. 8.10. Dependence of the residual strain ε_r on the upper strain ε_{up} for PA-6 [27]. Recovery time – 24 hours, T = 295 K, relative humidity – about 60%

1.7×10^{-19} J (100 kJ/mole) then the concentration of chain rupture at the upper deformation limit will be $N_{rup} = \Delta U_{rup}/1.7 \times 10^{-19} = 1.0 \times 10^{19}$ g^{-1}. The dependence of N_{rup} upon stress is similar to that of ΔU_{rup} on stress and in the preruptured state the concentration of ruptures is reached about a 1×10^{19} g^{-1} which is in good agreement with the concentration of ruptures determined on the same samples from the decrease in the molecular weight [27].

It is also of great interest to estimate the length of the segment the rupture of which in one bond is accompanied by the dissipation of the total energy stored in this segment. Using the value of ΔW_{rup} spent on the elastic extension of such segments and the number of ruptures it is possible to estimate the average value of the mechanical energy, stored in every segment $\Delta W'_{rup} = \Delta W_{rup}/N_{rup}$. The results of such estimations are shown in Fig. 8.11. Having analysed this dependence one can arrive at a very interesting conclusion: the amount of mechanical energy stored in every ruptured segment is practically independent of the stress applied to the sample. Hence, the ruptures of macromolecules in the stressed samples occur at some definite local stress on the segments independent of the external stresses. As can be seen from Fig. 8.11 $\Delta W'_{rup} = 15 \times 10^{-19}$ J. It means that the energy stored in the ruptured segments is higher than the dissociation energy of C-C bonds (about 4×10^{-19} J) and especially the energy of transition to the stable end groups (1.7×10^{-19} J). Hence, the rupture of such overstressed segments is really exothermic process in quite accord with the estimation given above (p. 260). It is also seen from the fact that the ratio $\Delta Q_{rup}/\Delta U_{rup}$, as well as $\Delta W_{rup}/\Delta U_{rup}$ is about 10. IR-investigations showed [17], that the local stresses on such segments reach 15–20 GPa. These values correspond to the ratio $f/f_{th} = 0.5$. Then according to formula (8.10) the value of elastic energy of overstressed bonds $\Delta W = 1 \times 10^{-20}$ J. Accordingly, n = $\Delta W'_{rup}/W = 100$. The estimated length of the overstressed segment corresponding to this n is about 13–15 nm. This value is in very good agreement with the structural data: the long period for the sample studied is 12 nm [27].

Thus, from all these estimations one can conclude that there is a chain character to the mechanical destruction process, when free radicals resulting from the rupture of the first molecule lead to the fast rupture of tens to hundreds of adjacent stressed molecules [17]. Due to this "group" dissipation

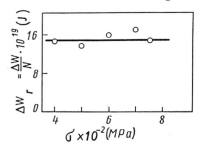

Fig. 8.11. Mechanical energy stored in one rupturing fragment of macromolecule of PA-6 [27]

process, a very high local heat evolution can occur. This heat evolution is localized in the weak amorphous interlayers and therefore the local overheating resulting from heat evolution can be classified as peculiar thermal microexplosions. They will accelerate the formation of submicroscopic cracks, which can be identified from small angle X-ray results [17, 30]. According to such X-ray analysis, the number of submicrocracks in similar samples of PA-6 in the prerupturing state is about $5 \times 10^{16}\,\text{cm}^{-3}$, their diameter being close to 15 nm [30]. Therefore, it is now possible to estimate the local temperature rise resulting from the "microexplosions". Assuming the microadiabatic character of the microexplosions one can arrive at the following result

$$\Delta T = -(\Delta Q_{\text{rup}})/(N_{\text{cr}} V_{\text{cr}} \cdot C p) = 750\,\text{K} . \tag{8.12}$$

where N_{cr} and V_{cr} is the number of cracks and their volume, respectively. Hence, this estimation shows that the local overheating reaches a very high level. Nevertheless, these estimates are in good agreement with IR-measurements of the local temperature rise accompanying the propagation of fast-moving cracks in the solid polymers considered in the previous section.

Frank and Wendorff [31, 32] have studied tensile deformation behavior and chain scission for PA-6 and PA-6,6 fibers using similar cyclic stress–strain experiments. The strain dependence of the elastic, inelastic and plastic components of deformation energy was determined. Heat effects resulting from the deformation were not measured. In general, the agreement between energy characteristics for PA-6 fibers obtained in these studies and in [27] is good. The elastic and anelastic deformation energy are found to increase continuously with increasing strain. The energy component which has been attributed to irreversible deformation, was found to increase strongly with increasing strain predominantly above a critical value of about 11% for PA-6 fibers. The plastic energy component was correlated with free radical formation. Therefore, it has been concluded that the irreversible deformation energy be predominantly controlled by the occurrence of chain scission events. However, the comparison of results of plastic deformation energies obtained experimentally and from calculations based on the concentration of free radicals showed that the macroscopically determined deformation energy is larger by at least one order of magnitude than values calculated from the ESR data. A rather rapid recombination of free radicals was suggested as being one of the many possible reasons for this discrepancy.

8.3 Self-Heating During Cyclic Deformation and Thermal Fatigue Failure

During the cyclic deformation of a perfectly elastic material there is no temperature change throughout a fatigue test, since no energy is dissipated in the sample tested. Therefore, in this case the material remains at a uniform temperature equal to that of the environment. Most real materials, however, and especially polymers, show mechanical hysteresis and a rise in temperature may occur as a result of energy dissipation. The resulting temperature at any point in the sample at any time is determined by the competition between heat generated and heat loss. This temperature rise may be especially significant for polymer materials because of the low values of their thermal diffusivity and conductivity. At high frequencies of cyclic deformation the energy generated becomes considerably larger than the heat which can be lost by the sample due to conduction or radiation. As a result of this steady self-heating, thermal fatigue failure can occur.

The problem of self-heating during the cyclic deformations and thermal fatigue failure can be considered on the ground of simple suggestions [5, 30, 33, 34]. An amount of energy ΔW which is dissipated per cycle in a sinusoidal cyclic loading of a viscoelastic solid polymer is

$$\Delta W = \pi \sigma_0 \varepsilon_0 \sin \delta \qquad (8.13)$$

where σ_0 and ε_0 is stress and strain amplitude, correspondingly, and δ is the phase angle between the stress and strain. The outflow of heat over the surface of the sample is determined by the relation

$$Q = \alpha S(T - T_0) \qquad (8.14)$$

where T is the mean temperature of the sample surface, T_0 the temperature of the environment, S is the surface area and α the coefficient of convection at the sample surface. Therefore, the complete energy balance is

$$C_p \rho (dT/d\tau) = k \omega \pi \sigma_0 \varepsilon_0 \sin \delta - \alpha S(T - T_0) \qquad (8.15)$$

ω is the frequency of the cyclic deformation, $dT/d\tau$ the rate of increase of temperature in the sample, and k the coefficient, characterizing the ratio of the hysteresis losses transformed into heat. The level of the hysteresis losses and correspondingly the power of the internal source of the temperature rise is dependent upon temperature through the temperature dependence of the mechanical properties of polymers (loss tangent, loss modulus and storage modulus).

According to Eq. (8.15) two regimes of self-heating during the cyclic deformation may exist. At low freqencies and low stress levels, the temperature increase of the sample generally approaches a finite value. After reaching this steady-state regime, mechanical energy input and thermal energy loss are in equilibrium. From the physical point of view it means that the heat generated during cyclic deformation is completely transferred into the environment, although the temperature of the sample is $T > T_0$. At high frequencies, or high stress amplitude the heat loss is not high enough to transfer all the heat generated to the environment and the temperature of the sample increases progressively during deformation. Therefore, the second non-steady-state regime corresponds to the thermal fatigue failure of polymers. In this regime $\Delta T = f(T)$, i.e. $T = T_0 + f(\tau)$. Failure of the sample in this case occurs through thermal softening due to the approach of the temperature of the sample to the glass transition temperature or melting point. Hence, the dependence of self-heating on time (number of cycles) corresponding to these two regimes shows that there are two types of fatigue fracture of solid polymers. In the steady-state regime the time to fracture depends upon the action of stress, and in a non-steady-state regime it depends upon the time which is needed to reach the critical value of the temperature rise, which, in turn, is closely related to the glass transition temperature or the beginning of the decrease of crystallinity. Theoretical analysis of the hysteresis heating of rigid polymers taking into account its localization at the tip of propagating crack is given by Barenblatt at al. [35].

Figure 8.12 shows typical dependencies of temperature as a function of the number of cycles corresponding to these two different regimes at 15°C for POM [33]. At constant stress amplitude of 32 MPa a stable temperature of about 30°C is reached after some 10 cycles. At this temperature mechanical energy input and thermal loss are in equilibrium. At higher stress amplitude the mechanical hysteresis heating leads to a progressive increase of the sample temperature and an excess temperature of about 50°C the failure occurs. Similar results have been found for various polymers and various deformation modes [36–38].

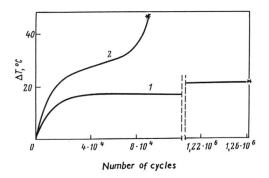

Fig. 8.12. Self-heating during the cyclic deformation of POM [33]. *1* – steady-state regime (mechanical failure), $\sigma_0 = 32$ MPa; *2* – non-steady state regime (thermal failure), $\sigma_0 = 37.5$ MPa, frequency – 50 Hz, $T_0 = 15$°C

References

1. Liebowitz H (ed) (1972) Fracture, an advanced treatise, vol 1, Academic New York
2. Kelly FRS (1973 2) Strong Solids, Clarendon Oxford
3. Kachanov LM (1974) The Grounds of Fracture Mechanics, Nauka Moscow
4. Rice JR (1978) J Mech Phys Solids *26:* 61
5. Kausch HH (1986) Polymer fracture, 2nd ed., Springer Berlin Heidelberg New York
6. Berry JP (1972) In: Liebowitz (ed) Fracture, an advanced treatise, vol 7, Academic New York London
7. Bartenev GM (1966) Mechanika Polym *N5:* 700
8. Williams ML (1965) Int J Fracture *1:* 292
9. Schapery RA (1975) Int J Fracture *11:* 141
10. Kambour RP and Barker RE (1966) J Poly Sci, Part A-2, *4:* 359
11. Levy, N and Rice JR (1969) In: Argon AS (ed) Physics of Strength and Plasticity, Mit Press Cambridge, p. 277
12. Williams JD (1972) Int J Fracture *8:* 393
13. Weicher R and Schonert K (1978) J Mech Phys Solids *22:* 127
14. Weicher R and Schonert K (1978) J Mech Phys Solids *26:* 151
15. Carslow HS and Jaeger JC (1959 2) Conduction of Heat in Solids, University Press Oxford
16. Döll W (1976) Int J Fracture *12:* 595
17. Regel VR, Slutsker AI and Tomashevskii EE (1974) Kinetic Nature of the Strength of Solids [in Russian], Nauka Moscow
18. Fuller KNG, Fox PG and Field JE (1975) Proc R Soc London *A341:* 537
19. Döll W (1972) Kolloid Z Z Polym *250:* 1066
20. Egorov EA, Zhizhenkov VV, Savostin AYa and Tomashevskii EE (1975) Solid State Phys (FTT) *17:* 111
21. Tomashevskii EE et al (1975) Int J Fracture *11:* 803
22. Egorov EA, Zhizhenkov VV, Bezladnov SN, Sokolov IA and Tomashevskii EE (1980) Vysokomol Soedin *A22:* 582
23. Egorov EA, Zhizhenkov VV, Bezladnov SN, Sokolov IA and Tomashevskii EE (1980) Acta Polym *31:* 541
24. Patrikeev GA (1958) Dokl Akad Nauk SSSR *120:* 339
25. Tomashevskii EE (1970) Solid State Phys (FTT) *12:* 3202
26. Tomashevskii EE, Egorov EA, and Savostin AYa (1975) Int J Fracture *11:* 817
27. Godovsky YuK, Papkov VS, Slutsker AI, Tomashevskii EE and Slonimskii GL (1971) Solid State Phys (FTT) *13:* 2289
28. Chevychelov AD (1966) Vysokomol Soedin *8:* 49
29. Chevychelov AD (1966) Mechanika Polym *N5:* 664; (1967) *N1:* 8
30. Tamuzh VP and Kuksenko VS (1978) Fracture Micromechanics of Polymer Materials, Zinatne Riga
31. Frank O and Wendorff JH (1981) Colloid Polym Sci *259:* 1047
32. Frank O and Wendorff JH (1988) Colloid Polym Sci *266:* 216
33. Ratner SB and Korobov VI (1965) Mechanika Polym *N3:* 93
34. Ratner SB, Korobov VI and Agamalyan SG (1966) Phys Chem Mekhanika Mater *5:* 88
35. Barenblatt GI, Entov VM and Salgannik RL (1966) Eng J Mechan of Solids *N6:* 47
36. Tauchert TR and Afzal SM (1967) J Appl Phys *38:* 4568
37. Oberbach K (1973) Kunststoffe *63:* 35
38. Zilvar V (1971) J Macromol Sci-Phys *B5/2:* 273

9 Experimental Methods
and Instrumentation

Three types of measurements are usually used for the study of the thermal behavior of polymers under deformation: the temperature changes resulting from the deformation of the sample, the temperature dependence of the stress or force and direct calorimetric measurements of heat effects in various deformation modes.

9.1 Measurements of the Temperature Changes During Deformation

The temperature changes during the deformation of various materials can be simply measured by thermocouples. The method was first employed by Joule in the 19th century to measure the temperature changes accompanying the rapid, essentially adiabatic deformation of rubbers and metallic wires, and since then it has been widely used for the same purpose. Using conventional thermocouples, temperature changes of magnitude 0.01 K and even smaller can be easily measured. Two types of thermocouple attachment are used: on the surface of the sample and in the body of the sample. Both methods have some shortcomings. The attachment on the surface needs a very good thermal contact between the thermocouple and the sample. The attachment in the body of the sample can change the stress field in the vicinity of the thermocouple and, consequently, the local temperature can change.

If the elastic material is loaded or unloaded very quickly (instantaneously) then the elastic deformation is adiabatic and the temperature changes observed can be used for various calculations based on the Thompson formula (Chap. 5), as well as for the estimation of the heat effects resulting from the elastic deformation. Reaching the ideal adiabatic conditions during deformation is not a simple task. There are some experimental factors (such as, for example, heat transfer through wires, heating of the thermocouple and some others) which can be taken into account in the course of such measurements.

More widely, this type of measurement is used for measuring thermal behavior resulting from plastic deformation of solid polymers, first of all, to obtain the temperature distribution in the zone of necking during cold drawing of glassy and crystalline polymers, as well as for observing the temperature rise at the tip of a crack propagating through solid polymers. The use of thermocouples for the registration of the temperature distribution is complicated due to the striking local character of plastic deformation. Therefore, in this case, besides thermocouples, colored organic crystals including liquid crystals with melting points in the temperature range of interest, fluorescent phosphorous and especially IR-radiation methods are also used. In the following sections typical instruments and experimental results are considered.

9.1.1 Elastic Deformations (Isoentropic Measurements)

The classical curve of the temperature changes resulting from the elastic extension of a piece of natural rubber was obtained for the first time by Joule [1, 2] (Fig. 9.1) and was then repeated in many other measurements [3, 4]. The measurements of the temperature changes in all these investigations were performed by miniature thermocouples fixed on the surface of the samples. Such a method of attachment of the thermocouples was also used in some studies of thermoelasticity of solid polymers [5]. In a typical arrangement, a very thin thermocouple wires are placed between two pieces of materials to reach a good thermal contact and decrease heat transfer through the wires. Often thermocouples are embedded in samples [6]. The problem of optimal attachment of thermocouples to samples, as well as the time of deformation which is necessary to reach the adiabatic conditions are considered in [7].

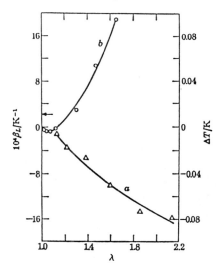

Fig. 9.1. Thermophysical results of Joule [2] for cross-linked NR. The linear thermal expansion coefficient at constant pressure and force, β_L, as a function of extension ratio, λ (a); the temperature change, ΔT, for an adiabatic extension from the unstretched state to extension ratio, λ. The arrow indicates the value of β for the unstretched state

As seen from Fig. 9.1 the initial cooling of the natural rubber turns to heating on further extension. Using these data the entropy changes resulting from the elastic stretching can also be estimated.

The quick elastic stretching of solid polymers such as PMMA, PE, i-PP is accompanied by cooling and contraction – with heating [5, 7]. In the majority of cases the temperature changes are a linear function of deformation in full accord with the Thompson equation (5.13)

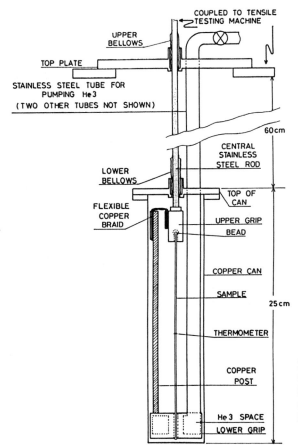

COUPLED TO TENSILE
TESTING MACHINE

UPPER
BELLOWS

TOP PLATE

STAINLESS STEEL TUBE FOR
PUMPING He3
(TWO OTHER TUBES NOT SHOWN)

60 cm

CENTRAL
STAINLESS
STEEL ROD

LOWER
BELLOWS

TOP OF
CAN

FLEXIBLE
COPPER
BRAID

UPPER GRIP

BEAD

COPPER CAN

SAMPLE

25 cm

THERMOMETER

COPPER
POST

He 3 SPACE

LOWER GRIP

Fig. 9.2. The low-temperature insert, showing the way
in which the sample is strained
and the method of obtaining
low temperature [8]

A very elegant setup for measuring thermoelastic effects in glasses at very
low temperatures was developed at the Cavendish Laboratory by Wright and
Phillips [8] A schematic drawing of the setup is shown in Fig. 9.2. The long
rod samples 1 mm in diameter and 20 cm in length are slotted into two copper grips, which are thermally linked by the copper post. The lower grip is
placed in a helium 3 chamber which enables temperatures below 1.5 K to
be attained. The temperature can be regulated in the range 0.5–17 K. To
produce controlled deformation of the samples a tensile testing machine (Instron 1253) was used. The deformation was measured by means of a linear
variable differential transducer and the force was measured with a load cell.
The temperature changes resulting from elastic deformation of the samples
were measured with the thin thermometer consisted of carbon-impregnated
paper bonded to the middle of the sample. The carbon thermometer was
calibrated after each measurement using a standard thermometer. The temperature resolution was 0.2 mK at 10 K and 0.03 mK at 1.3 K. The setup was
used for measurements of thermoelastic temperature changes resulting from
the deformation of PMMA rods in simple stretching and sinusoidal strains in

the range of 0.01–10 Hz in the temperature interval 0.3–10 K. Values of the Gruneisen function were also determined and analysed.

Butjagin and coworkers [9] developed an experimental setup for measuring the temperature changes accompanying the deformation of polymers by means of IR radiation, which is characterized by a rather high integral sensitivity and extremely low thermal inertia. A schematic diagram of the setup is shown in Fig. 9.3. It consists of two main parts: a loading device and IR bolometer. The IR bolometer is sensitive in the 0.3–25μm range. Two thermoensitive resistance wires are used: working and compensating. Both resistance wires are introduced into a Wheatstone bridge, which can work either with alternating or with continuous current. In the first case an oscillograph is used which leads to a very small time constant ($\sim$ 20 ms) of the setup. In the second case a recorder is used. Time constant of this scheme is 1–2 s.

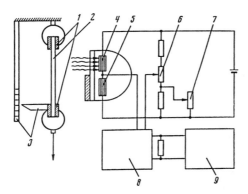

Fig. 9.3. Schematic diagram of the experimental set-up for IR-radiation experiments [9]. *1, 2* – sample with clamps; *3* – transducer for strain measurement; *4, 5* – bolometer with two thermosensitive resistances; *6, 7, 8* – electronics; *9* – recorder

The IR-bolometer can be calibrated by means of a polymer film into which a thermocouple and a small heater are embedded. Using this film the dependence of the intensity of thermal radiation on the temperature during steady state radiation can be obtained. This dependence is a linear one for the range of temperature change of some degrees. The sensitivity of the setup with various bolometers was $(1–3) \times 10^{-3}$ deg/mm of the recorder scale. For calculating the thermal effects the following expression can be used

$$dT/d\tau = (1/\tau_g)(T - T_0) ; \quad \tau_g = (C_p\rho/2\gamma)\delta \qquad (9.1)$$

where T and T_0 is the temperature of the sample and of the environment, respectively; τ_g is the characteristic time of reaching equilibrium; γ is the constant of heat transfer; δ is the thickness of the sample. Other values have their usual meaning. τ_g can be determined by means of calibration

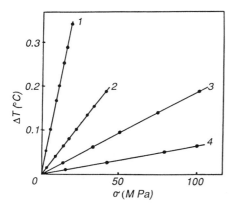

Fig. 9.4. Temperature changes as a function of elastic stress during simple extension of polymer films [9]. *1* – PA-6 (stretching perpendicular to the drawing axis); *2* – epoxy polymer ED-5; *3* – PET; *4* – densely crosslinked glassy network BF-4

experiments with films of various thickness. The value of $C_p\rho/\gamma$ is fairly constant for various polymers. Typical dependencies of temperature changes upon stress for various polymers are shown in Fig. 9.4. These results agree with the data obtained by thermocouple measurements.

However, these reversible thermomechanical effects can be observed experimentally not only during sudden uniaxial stretching of solid and rubber-like polymers but during rapid compressive uniaxial or hydrostatic pressure. A series of such measurements was performed by Müller and coworkers [10–12] in the 1950s on plastics but their results were more qualitative then quantitative. Haward and coworkers [6, 7] and later Rodriguez and Filisko [13, 14] developed experimental setups and performed a series of measurements of reversible thermomechanical effects in solid polymers. Figure 9.5 shows diagram of the experimental equipment for measuring the temperature during rapid compressive deformation. A specially-modified Instron capillary rheometer with a barrel diameter of 9.525 mm was employed in the experiments. A fixed load was applied to the plunger with a plunger speed of 40 mm/min. The total force applied to the sample was measured by the compressive load cell. The adiabatic temperature changes during rapid compressive deformation were measured with a thermocouple, the measuring junction of which was embedded at the center of the cylindrical polymer sample. The temperature difference ΔT between sample and the reference junction which was placed on the outside surface of the barrel was monitored as a function of time with this arrangement. Experiments can be performed at various temperatures. For thermoelastic measurements samples with a constant aspect ratio L/D of 2.31 were used.

An increase in temperature was observed in PMMA and HDPE samples upon the rapid compressive load application. Rapid removal of the stresses was accompanied with a decrease in temperature (see Fig. 5.2).

In contrast to the uniaxial rapid extension or compression when only moderate strenses can be applied to avoid plastic deformation or rupture of the samples under investigation the rapid application of hydrostatic pressures at different temperatures permits measuring the adiabatic temperature changes as a function of pressure including high pressures. A schematic di-

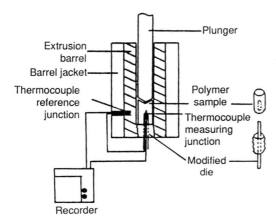

Fig. 9.5. Schematic diagram of experimatal equipment for measuring the temperature changes during rapid compressive deformation of polymer [14]

agram of the experimental equipment for such measurements is shown in Fig. 9.6. Measurements were performed on the cylindrical samples 6.5 mm in diameter and about 90 mm in length contain a thin hole along the axis for the measuring junction of a thermocouple. The sample is placed into a high pressure vessel. The reference junction of the thermocouple is fixed on the outside of the vessel but being in good thermal contact with its surface. Such a scheme enables one to monitor only the temperature changes resulting from deformation of the sample. The sensitivity of the temperature measurement was about 0.1°C, the temperature of the vessel was maintained to an accuracy +0.5°C. The high pressure vessel is filled with mercury which serves as the working liquid. Pressure in the system is applied by means a special water system. The pressure was measured with two Bourdon gauges, each with a maximum of 345 MPa and a sensitivity of 3.45 MPa/division. The pressure application was practically instantaneous, therefore the resulting temperature changes were close to adiabatic. The increase in temperature was found to be the same as the decrease in temperature for the same applied pressure (Fig. 9.7). This reversibility [ΔT (heating) = ΔT (cooling)] was observed throughout all experiments. The main experiments were performed on PMMA (Chap. 5).

9.1.2 Temperature Rise in the Necking of Polymers

When a glassy or crystalline polymer specimen yields under tension by the process of cold drawing heat is generated in the necking zone. The first attempts at measuring the local heating during neck formation directly were made by Müller and coworkers [10, 15, 16] using thermocouples. The typical temperature distribution during necking is shown in Fig. 7.6. These researchers have performed many measurements of the temperature changes accompanying the neck formation using various methods of attachment of the thermocouples to samples and concluded that such measurements can be considered only as qualitative. Nevertheless, these methods were widely used after their investigations [17].

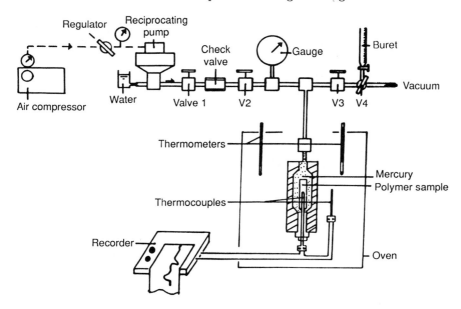

Fig. 9.6. Schematic diagram of the experimental equipment for measuring adiabatic temperature changes during application of hydrostatic pressures [13]

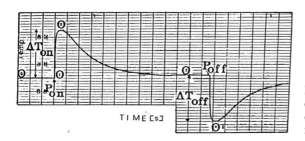

Fig. 9.7. Typical recording of thermoelastic effcts corresponding to a delayed release of pressure [13]

In search of a more exact method of monitoring the temperature distribution along the sample during cold drawing, colored organic crystals with melting points in the temperature interval of interest and fluorescent phosphorous were used [12,18]. The color of some organic crystals and phosphorous is dependent upon temperature, therefore, this property can be used for determining the temperature distribution in the necking zone. However, some of these methods lack sensitivity [19, 20] and can interfere with the deformation process. Nevertheless, Müller [12] concluded that these methods can be used for the correct estimation of the temperature rise during necking.

All the above-mentioned methods of temperature measurements during necking are based on the direct contact of temperature sensors with samples. New possibilities opened in this field with the application of infrared sensors [9, 21–23]. Maher, Haward and Hay [21] studied the temperature distribution in the necking zone during cold drawing of PC, PVC and *i*-PP using the AGA Thermovision System 680, consisting of two separate units; the infrared camera and the thermal picture display unit, monitoring the temperature picture of the object studied. A typical diagram of the temperature profile observed during necking in PC is shown in Fig. 9.8. Qalitatively this temperature distribution is similar to that monitored by means of thermocouples (see Fig. 7.6).

9.1.3 The Temperature Rise in the Zone of Rupture of Polymers

During the fracture of polymers a part of the stored energy of deformation may be converted into heat which may lead to a local temperature rise. In a very striking form this behavior is observed at the tip of fast-moving cracks in glassy and crystalline polymers. For experimental observation of the temperature rise in the vicinity of moving cracks the same methods as those for the investigation of the temperature profile in the necking zone are used. Application of thermocouple technique permits us to estimate only the total heat evolved during cracking [24–26]. The total heat Q evolved can be estimated using the following experimental scheme. The thermocouple made of very thin wires (2–25μm diameter) is embedded in the sample 0.5 mm from the expected crack path. The reference junction of the thermocouple is also placed on the surface of the sample but rather far from the path to compensate for any thermoelastic effects. For simultaneous monitoring of the temperature profile as a function of time after fracture and the velocity of the moving crack the two-beam oscillograph can be used. The total heat Q evolved can be estimated using the temperature rise ΔT produced at a given distance y from the crack plane at a known time t after fracture according to the equation

$$Q = \rho c (4\pi at)^{1/2} \Delta T \, \exp(y^2/4at) \qquad (9.2)$$

where Q is the heat evolved on both crack faces per unit area of crack. Two disadvantages of the method exist. Firstly, from each specimen only one value of Q can be obtained. Secondly, the reliable positioning of the junctions is not a simple task, therefore, some part of measurements is unreliable.

The color sensitivity to changes of temperature makes liquid crystals useful for monitoring the surface temperature of the polymer rupturing during deformation [26, 27]. Fuller and coworkers [26] used a mixture of cholesteryl oleyl carbonate and cholesteryl nonanoate for this purpose by spreading it across the specimen where the crack was to run as a layer about $100\,\mu$m thick. The color distribution produced after fracture on either side of the crack path was recorded and the temperature distribution corresponding to the color distribution was estimated. The values of heat evolved during the crack propagation can be evaluated using the same relation (9.2). A typical dependence of heat evolved on fracture surfaces upon crack velocity for PMMA obtained with both methods is shown in Fig. 9.9.

Neither the thermocouple technique nor the liquid crystal technique possesses the necessary spatial resolution and time response fast enough to obtain an estimate of the temperature within the deformed region itself. Therefore, the technique which is most widely used now for this purpose is the use

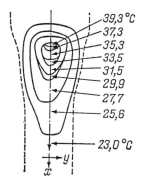

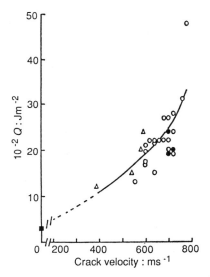

Fig. 9.8. Diagram of temperature profile of a necking PC specimen [21]

Fig. 9.9. Q, heat evolved on both fractures per unit area, plotted as a function of crack velocity for PMMA [26]. ○ – liquid crystal results, error in Q ≈ 20%; ● – liquid crystal results, error in Q ≈ 35%; △ – thermocouple results: error in Q ≈ 15%; ■ – value of Γ for initiation of crack growth. Error in crack velocity ≈ 7%

of an IR detector, which offers the advantages of high sensitivity, fast response and ability to perform remote measurements. A number of arrangements for infrared detector experiments have been developed [26, 28–30]. A schematic diagram of the unidirectional IR-microscope based on the quick registration of the exceed equilibrium infrared radiation is shown in Fig. 9.10A. The magnified image produced by the mirror objective is fixed in the IR-detector 2, which has an indium antimonide element ($150 \times 150\,\mu$m) cooled to 77 K. Its response is proportional to the power of the radiation flux. Corresponding electrical signals from the detector were recorded with a dual-beam oscilloscope. With this IR-microscope one can scan a linear area of 1.2 mm with a frequency of 200 pictures/s. The response time of the detector is about $5\,\mu$s, the area response is $20\,\mu$m. It is impossible to increase this area response to any extent because this value is closely related to the $5.6\,\mu m$ wave to which the detector has a maximum sensitivity. Hence, this arrangement can be used for observing the radiation signals from rather large heated zones and cannot be used to study brittle fracture when the heated zones may be less than $1\,\mu m$. During plastic non-brittle fracture which is typical for the majority of polymers this setup can be used for registering both the maximum temperature and the temperature profile of the hot zone and its evolution. The fracture of the sample is performed by dynamical loading. Typical velocities of the dynamical loading were $1-1.5\,m/s$. The force applied to the sample is recorded by the second beam of the oscilloscope.

The study of the temperature profile of the hot zone may be performed by two methods. The first method consists of linear scanning a long $1.2\,mm$ perpendicular to the fracture zone. Before fracture the IR microscope is focused on the centre of the intended fracture zone (point P). The second method consists of "autoscanning". In this case the microscope is focused on point Q on the sample surface which is placed $0.5-1.0\,mm$ from the suggested fracture zone. After fracture the hot end of the sample is moving quickly through the area visible to the microscope and its radiation profile is monitored with the oscilloscope. Using the oscilloscope traces, one can estimate the transversal cross-section of the hot zone.

A schematic diagram of IR radiation impulses accompanying the fracture of solid polymers is shown in Fig. 9.10B. This impulse can be divided into three typical parts: (a) the front edge with a duration of t_1 owing to the formation of hot rupture surfaces when the crack extends by $10\,mm$ with the velocity close to that of sound; (b) the drop of the pulse by one half due to the removal of one part of the fractured specimen beyond the diaphragm; (c) a prolonged tail of the pulse t_2 resulting from the heat transfer from the surface of the rupture and cooling. The amplitude of the impulse can be determined using the sensitivity of the IR detector and the area under the impulse is proportional to the total radiation energy. Typical parameters of the impulses obtained upon breakdown of various polymers are listed in Table 9.1.

Table 9.1. Characteristic of IR pulses from the hot surfaces of cracks [29]

Polymer	Pulse amplitude mv	Radiation power mw	t_1 microsec	t_2 sec
PET	45	7	250	0.8
Nitrocellulose	12	1.9	25	0.3
Acetyl cellulose	5	0.8	–	–
PA-6	2	0.3	–	–
PMMA	2	0.3	–	–

The intensity of the IR impulses monitored during fracture of polymers permits us in principle to determine a local temperature rise and the total heat evolved during crack propagation. For this purpose it is necessary

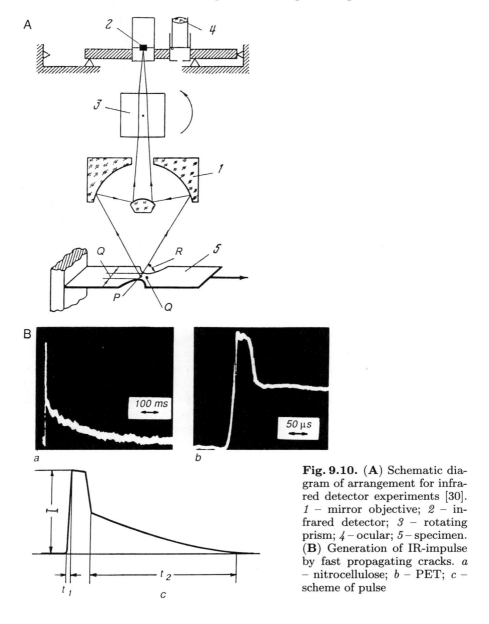

Fig. 9.10. (**A**) Schematic diagram of arrangement for infrared detector experiments [30]. *1* – mirror objective; *2* – infrared detector; *3* – rotating prism; *4* – ocular; *5* – specimen. (**B**) Generation of IR-impulse by fast propagating cracks. *a* – nitrocellulose; *b* – PET; *c* – scheme of pulse

to know the radiation ability of the polymer under consideration which can be computed or determined experimentally. The initial temperature rise in the fracture zone is a function of I/δ, where δ is the thickness of the hot zone on the crack edge. Although determination of the dimension of this heated region at the edge of the crack is not a simple task, nevertheless, IR detector technique enables us to estimate the temperature rise quite explicitly. Experiments on PET showed that a temperature rise corresponding to the impulses parameters of about 215 K occured with this polymer. Because the temperature rise is simply related to the heat evolved per unit area of crack, knowing the temperature rise one can easily estimate the total heat effect $\Delta Q = \rho C_p \delta \Delta T$. According to such estimates, the absolute values of Q are rather high being 10^3–10^4 J/m^2 for PET. They indicate a considerable evolution of heat during polymer fracture.

9.2 Temperature Dependence of Stresses (Isometric Measurements)

The adiabatic thermomechanical measurements on NR carried out by Joule showed that the temperature decreased for small strains, but increased for strains above 13–14% extension (Fig. 9.1), which was referred to as the adiabatic inversion point for NR. Although these experimental results agreed rather well with the theoretical predictions, they were disregarded in quantitative studies. They rather played the role of visual evidence of the entropic nature of rubber elasticity. For precise determination of the entropic and energetic components of the elastic restoring force in rubbers, another approach was found. According to this approach, the total force f can be resolved into entropic and energetic components defined by [31]

$$f = f_u + f_s = (\partial U / \partial L)_{V,T} + T(\partial f / \partial T)_{V,L} \qquad (9.3)$$

Experimental determination of the components of the elastic force thus requires measurements of the changes in force with temperature at constant volume and length. The constant volume requires the application of hydrostatic pressure during measurements of the force–temperature coefficient for an external compensation of very small volume changes resulting from the extension of rubbers. Although this experiment is extremely difficult to perform, the experimental technique for such investigations was developed [32, 33].

Instead of measuring the force–temperature dependence at constant volume and length, one can measure this dependence at constant pressure and length, but in this case it is necessary to introduce the corresponding corrections. The corrections include such thermomechanical coefficients as the isobaric volumetric expansion coefficient, the thermal pressure coefficient or the pressure coefficient of elastic force at constant length [32–34].

The measurements of the temperature dependence of the restoring force are usually carried out in the temperature range 350 ± 100 K. Determination of f_u requires a rather large extrapolation of experimental results and it restricts the accuracy of this method. This type of thermoelastic measurement requires equilibrium conditions. Most widely, this method is used for simple elongation and seldom for compression [35] and torsion [36, 37].

A typical experimental arrangement used for thermoelastic studies of polymer networks on elongation is shown in Fig. 9.11 [38]. The network sample, fixed between two clamps, is placed into a thermostatically controlled

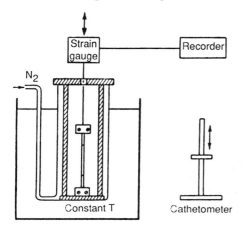

Fig. 9.11. Schematic diagram of typical apparatus used to measure the stress of an elongated polymer network as a function of strain and temperature [38]

chamber filled with nitrogen. The upper clamp is connected to a very sensitive strain gauge which can record the values of the force f as a function of elongation and time. The values of the sample length, required to characterize its deformation can be determined by means of a cathetometer or traveling microscope. Often, instead of the force f the value of stress, $f^* = f/A$ is used, which is simply the force per unit area of the undistorted sample at some convenient temperature. Finally, the value of the reduced force $[f^*] = f^*/(\lambda - \lambda^{-2})$, which is the stress devided by the strain function used in the molecular theories of rubber elasticity is also used.

A classical series of stress–temperature curves for typical networks obtained at constant pressure and constant elongation is shown in Fig. 9.12 [39]. At small elongations the value of the restoring force decreases with temperature, then an inversion occurs and after this inversion, the restoring force begins to increase with temperature. The determination of energy contribution f_u on the basis of such measurements requires the correction term $\alpha T/(\lambda^3 - 1)$ which is strongly dependent upon elongation. At high elongations this term turns out to be small and can, in some cases, be neglected. But at very small elongations it is extremely important to determine this term correctly while estimating the energy contribution. A very important feature of the thermoelastic measurements on torsion is that the correction term is independent of deformation [see Eq. (6.36)].

Although traditionally the thermodynamic treatment of the deformation of elastomers has been centered on the force, the alternative condition of keeping the force (or tension) constant and recording the sample length as a function of temperature at constant pressure is even simpler [33, 40]. This type of thermoelastic measurement is based on the relation

$$\beta_L = -(\partial \ln L/\partial \ln f)_{T,P} \cdot (\partial \ln f/\partial T)_{L,P} \qquad (9.4)$$

Since in the elastic region $\partial \ln L/\partial \ln f)_{T,P}$ is always positive, the force–temperature coefficient at constant length and pressure must be of opposite sign to β_L. For rubbers, at some extensions, the isometric inversion

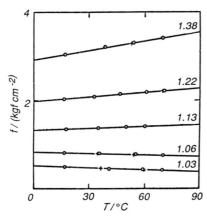

Fig. 9.12. Thermoelastic data for NR network [31, 33]. The stress is expressed for relative to the undeformed cross section and each curve is characterized by the value of the elongation

$\partial \ln f / \partial T)_{\mathrm{L,P}} = 0$ must occur since β_{L} of the isotropic sample is always positive. For solids, such measurements correspond to the determination of the coefficient β_{L} of an elastically stretched sample which, however, does not differ from the usual coefficient of thermal expansion.

9.3 Deformation Calorimetry (Isothermal Measurements)

9.3.1 Gas Calorimeters

The method of deformation calorimetry was first developed and applied to polymers by Müller in 1957 [41, 42], who also reported the results of numereous thermomechanical studies of rubberlike and solid polymers and drew attention to this thermomechanical approach. In his early investigations, Müller used a gas calorimeter based on the principle of a gas thermometer which measures the change in pressure of a gas in which the sample is deformed. The pressure changes in the gas were caused by temperature changes in the sample during deformation. Heat transport also occurs between the sample and the chamber wall which is maintained at constant temperature. Heat flux resulting from the deformation of the sample changed the pressure in the working chamber which activated an electric feedback mechanism attached to the differencial manometer produce the same heating effect in the comparison chamber as is produced by the sample during the deformation. The pressure difference in the working and comparison chambers are maintained at zero. Thus, the principle of this instrument is based on the measurements of the compensation effects rather then on direct measurements of heat effects resulting from the deformation. This instrument can be used for the simultaneous recording of the stress–strain dependence at simple elongation and thermal effects accompanying the elongation or contraction of the sample. The gas calorimeter possesses some shortcomings. First of all, it is difficult to measure endothermal effects (cooling). There must be a very rigorous proportionality between the gas pressure and heat flux during the whole measurement. Such a proportionality actually exists only at very small time lags between the gas pressure and temperature changes. Both measuring and comparison chambers must be under identical temperature conditions.

In spite of these difficulties and the rather complex construction of the instrument and some other sources of uncertainty, some researchers have used similar instruments in their studies of thermodynamics of the deformation of polymers [43–46]. The most perfect version of such an instrument was developed by Lyon and Farris [46]. Schematic drawing of the deformation calorimeter is shown in Fig. 9.13. The instrument consists of two hermetic cylindrical chambers with 1.9 cm diameter and 12 cm length in one of which the sample is fixed. The sample can be stretched by means of a thin Invar pull wire which passes through an air-tight mercury seal. The reference cell is identical to the measuring cell including the pull wire, so that there is no change in volume due to motion of the wire during deformation. Changes in sample volume due to the Poisson effect turned out to be very small compared with the total volume of the sample cell and therefore do not contribute to the pressure change. Any differential pressure which occurs is a result of heating or cooling of the gas surrounding the sample. The differential pressure between the two chambers is monitored during deformation. To avoid convective streams into the measuring chamber there are polyester film partitions placed horizontally every 4 cm. These partitions made pressure changes independent of the axial positioning of the sample or heater.

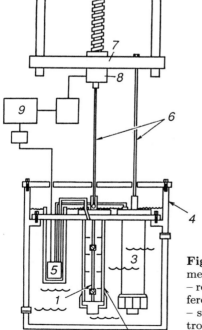

Fig. 9.13. The Müller's gas deformation calorimeter [46]. *1* – sample; *2* – measuring chamber; *3* – reference chamber; *4* – thermostate; *5* – the differential pressure transducer; *6* – pulling wires; *7* – stretching unit, including load cell 8; *9* – electronic recorder

The calorimeter also contains a load cell and displacement transducer. The load cell contains a dynamometer up to 5 kg. The mechanical system can perform the deformation with the rates in the range from 0.027 to 27 cm/min. The analog outputs from the load cell, pressure transducer, and displacement transducer are connected to an analog-to-digital converter. The work of deformation is calculated from the force–displacement data and the heat of deformation is calculated from the pressure–time data.

Lyon and Farris [46] performed a more complete analysis of this gas calorimeter than the analysis given by Müller. They demonstrated that the differential pressure is connected with the heat flux through the following linear integral equation

$$\Delta P(T) = \int_0^t K(t - \xi)\partial Q/\partial\xi d\xi \tag{9.5}$$

where ξ is dependent on time. This equation can be transformed to a new one

$$Q(t) = C \int_0^t \Delta P(\xi)d\xi + C\tau\Delta P(t) \tag{9.6}$$

which is identical to the Tian–Calvet equation (see below) and in which temperature is replaced by the gas pressure. In this equation C is a constant independent of the position of the source of heat and its properties for the construction of the calorimeter described above, K and τ are constants of the calorimeter. Thus, the relationship between gas pressure and heating history is described by a convolution integral, the solution of which yields equations for calculating the rate of change of heat $dQ(t)/dt$ and the total heat $Q(t)$ at any time during the deformation process.

Figure 9.14 demonstrates typical calorimetric curves for various modes. The time constant of the calorimetric cells is 7–8 s, the limited thermal flux 84 μw, the minimal amount of heat $\sim$ 0.42 mJ. The accuracy of the heat measurements in interval 10–500 mJ is $\pm3\%$.

Another type of deformation calorimeter which is based on differential thermometry and uses a gas flow as a heat transfer medium has been developed by Duvdevani et al. [47]. A test sample is fastened in the test chamber and surrounded by the flowing gas, introduced at a constant controlled temperature and flow rate (Fig. 9.15). An electric heater, placed close to the sample, causes a slight temperature increase in the flow gas. Heat evolved or absorbed by the sample during deformation tends to change this predetermined temperature difference. In order to keep this difference constant, the electrical power input to the heater is adjusted, the change in power being proportional to the thermal effect resulting from the deformation. A stretching device permits one to adopt various stretching modes including deformation at a constant rate and sinoidal deformation. The sensitivity of the instrument is strongly dependent on the gas flow rate and its type. A higher flow rate and turbulence improve the time constant but also increase the noise level. Consequently, the flow rate must be optimized according to specific requirements. Under typical measuring conditions, this type of

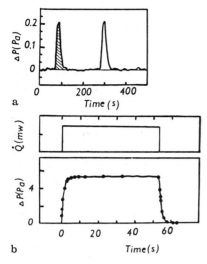

a

b

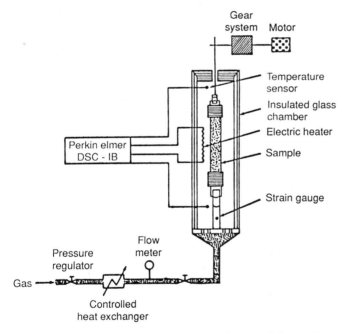

Fig. 9.14. Typical recorded plots of differencial pressure–time for two heat pulses of 3 cal (**a**) and differencial pressure–time for a rectangular heat pulse (**b**) [46]. Points correspond to calculation with $\tau = 7.1\,\mathrm{s}$

Fig. 9.15. Schematic diagram of a gas deformation calorimeter [47]

gas-flow deformation calorimeter is capable of recording heat fluxes smaller than 20 mJ/s with a sensitivity limited by the noise level of approximately 0.6 mJ/s. This type of deformation calorimeter is rather difficult to operate and it has not been widely applied in thermomechanical studies.

9.3.2 Heat-Conducted Deformation Calorimeters (Tian–Calvet Type)

Heat effects resulting from the deformation of solid and rubber-like materials are as a rule very small. For determining small thermal effects the most suitable calorimetric principle is the Tian–Calvet principle [48]. For thermomechanical studies of polymers this principle was used for the first time by Godovsky [49, 50], who developed a fully automatic deformation microcalorimeter for the simultaneous recording of the thermomechanical behavior of rubbers and glassy and crystalline polymers in the form of films, fibers and strips with uniaxial deformation. A schematic diagram of the deformation calorimeter is demonstrated in Fig 9.16a. The instrument consists of two parts: microcalorimeter and mechanical loading system with a dynamometric assembly. The calorimeter consists of two cells, the working dimensions of the calorimetric cells are as follows: diameter – 10 mm, length – 125 mm. The temperature difference between measuring and reference cells is measured by means of two thermobatteries, each of them including 810 copper-constant thermojunctions made electrochemically. The thermobatteries cover the whole surface of the cells. A sensitive amplifier and electronic recorder are used for automatically detecting heat fluxes. The microcalorimeter is placed in a thermostat with a working temperature range of 20–80° (Fig. 9.16b).

The dynamometric assembly which includes an automatic bridge-type dynamometer with a sensitivity of 0.5 g/mm allows one to record force and deformation which can be used for determining the work of deformation. The loading device permits one to use 8 linear rates of deformation in the range from 0.0146 to 1.865 mm/s.

According to the theory of the method [48] the relation between the power of a relatively slow thermal process taking place in the calorimeter measuring cell, and the temperature difference between the measuring and comparison cells is as follows (Tian equation)

$$dQ/dT = K\left[\Delta T(t) + \tau d\Delta T(t)/dt\right] \qquad (9.7)$$

where K and τ are the constant of the heat flux and time constant, respectively, which can be determined experimentally. The total heat resulting from the heat process for time t can be determined using the integral equation

$$Q(t) = \int_0^t (dQ/dT)dt = K\int_0^t (\Delta Tt) + \tau(d\Delta T(t)/dt)dt = KS(t) \qquad (9.7)$$

where S(t) is the area under the experimental curve temperature–time. The calorimeter can be calibrated by Joule heat by the procedure suggested by Calvet [48]. The maximum sensitivity of the first instrument at room temperature was 3×10^{-6} J/s. Later the sensitivity was improved, by using a more sensitive amplifier, to 2×10^{-7} J/s. The time constant of the empty calorimetric cells was about 30 s. This value is about an order of magnitude

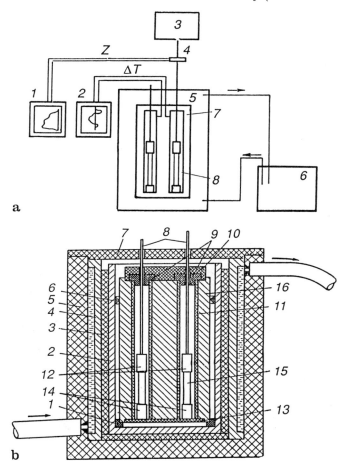

Fig. 9.16a,b. (a) Block scheme of Calvet-type deformation calorimeter [49]: *1, 2* − recorders; *3* − loading device; *4* − force transducer; *5, 6* − thermostating system; *7* − microcalorimeter with measuring and reference cells; *8* − sample. (b) Detailed construction of the calorimeter: *1* − teflon ring; *2* − copper cylinder; *3* − plastic isolation; *4, 5* − thermostate; *6* − teflon ring; *7* − isolating cover; *8* − pulling invar wires; *9, 10* − teflon covers; *11* − ceramic rings with thermobatteries; *12* − upper sample holders; *13, 16* − calorimeter's base and walls; *14* − lower sample holders; *15* − sample

lower than in standard Calvet-type calorimeters. The total time constant of the cell with the sample and sample holders placed into the cell is somewhat higher than for the empty cell. Its value can be determined experimentally by embedding tiny electrical heaters into the samples of various dimensions. The erorr for both heat and work measurement is estimated to be ±3 percent.

The calorimeter is capable of working not only in the normal mode described but in the ballistic mode and permits one to detect thermal effects of less than 1 s duration. Many thermal measurements in deforming polymers can be carried out in the ballistic mode. According to the theory of the method [48], if the thermal process lasts for a time t < $0.2\tau_{eff}$ (τ_{eff} is the effective time constant of the cell with a specimen), then $\Delta T/Q = \text{const}$ to an accuracy of 1% and is time-independent (ΔT is the maximum of the ballistic curve peak, Q is amount of heat). The experimental ballistic curve corresponds to a very sharp peak (see Fig. 9.14a). The calorimeter allows one to detect thermal effects of deformation of less than 1 second duration. The minimum heat pulse that can be detected in a deformed polymer is about 4×10^{-4} J.

In the case when the measurements are carried out in the ballistic mode, it is important to estimate the characteristic time in which the equilibrium temperature distribution across a specimen becomes established. It can be estimated from ratio δ^2/a, where δ is the specimen thickness and a is the thermal diffusivity (for solid polymers and elastomers a $\sim 10^{-7}$ m/s). For $\delta = 0.01$ cm this time is about 0.1 s, and for $\delta = 0.03$ cm it is about 1 sec.

The use of ballistic conditions in measurements permits one to analyse, even though roughly, two consecutive processes one of which is a ballistic process, for example, rapid tension of elastic material to a constant elongation followed by stress relaxation $Q = Q_{el} + Q_{rel}$. In this case the integral thermal effect can be determined as the area under the thermal effect curve and the thermal effect of elastic tension Q_{el} according to the peak maximum. Using this approach one can estimate the heat effect resulting from the stress relaxation. Experiments show that the delay in the peak maximum after the supply of ballistic thermal power has stopped is approx. 5 s. Therefore, in the first approximation, the contribution of the relaxation heat into the thermal effect of elastic deformation can be ignored.

The calorimetric experiments with the deformation calorimeter can be performed in various deformation modes, first of all in elongation-contraction mode in a non-stop regime or with the intermediate stop.

The above-described method of deformation calorimetry has found rather wide application. Modifications of the original design were constructed [51–56] and applied for investigating the thermomechanical behavior of polymers and polymer composites. At the same time, the commercial Calvet-type calorimeters were used in thermomechanical experiments on rubbers in the uniaxial mode [57–59] and torsion [60, 61] and in the compression mode of glassy polymers [62]. In our laboratory together with the above-mentioned hand-made instrument the deformation calorimetric system with the standard Calvet-type calorimeter DAK-I-I made in the USSR is also widely used. A very important feature of this Calvet-type calorimeter is the possibility of operating in the so-called "compensation regime", in which the calorimeter time lag (time constant) is decreased considerably. This feature is especially important for measuring the thermal effects resulting from mechanical deformation of polymers. Such a regime in commercial Calvet-type calorimeters supplied with "Setaram" (France) is not possible.

Schematic drawing of typical thermal processes resulting from the deformation of polymers are shown in Fig. 9.17.

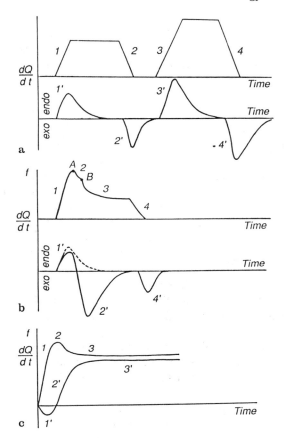

Fig. 9.17. Typical examples of thermo-mechanical behavior of solid polymers at various deformations. (**a**) Elastic deformation: *1, 1'* and *3, 3'* – extension, *2, 2'* and *4, 4'* – contraction. (**b**) Elasto-plastic deformation: *1, 1'* – elastic extension; *2, 2'* – plastic deformation (necking), *3, 3'* – stress relaxation. *4, 4'* – contraction. (**c**) Cold-drawing: *1, 1'* – elastic extension, *2, 2'* – necking, *3, 3'* – cold-drawing at a constant rate

9.4 Calorimetric Methods for Investigating the Energy State of Deformed Polymers

The energy state of deformed polymers can differ significantly from the undrawn state. In particular, the internal energy of plastically deformed glassy and crystalline polymers is normally different to undrawn polymers. To compare the energy states of drawn and undrawn polymers some calorimetric and thermomechanical methods are used.

9.4.1 Dissolution Calorimetry

For the most general case, namely a partly crystalline polymer at temperatures below both the melting point and glass transition of amorphous regions the heat of dissolution of such polymers can include three contributions

$$\Delta H_s = \Delta H_m + \Delta H_f + \Delta H_{ea} \qquad (9.9)$$

where ΔH_s is the heat of solution; ΔH_m is the heat of mixing of the liquid-like amorphous polymer with the solvent used for dissolving; ΔH_f is the heat of fusion; ΔH_{ea} is the excessive heat of the glassy amorphous regions of crystalline polymers. Only the first contribution ΔH_m is determined by the interaction of macromolecules with the molecules of the solvent. Two other contributions are independent of the solvent and are determined by the energy state of polymers. Therefore, measurements in various solvents with different concentrations have to give the identical values of $\Delta H_f + \Delta H_{ea}$. ΔH_m has always a positive value in the range 20–200 J/g. ΔH_m for nonpolar polymers is as a rule a small positive value of the order of 10 J/g. ΔH_{ea} is the value most difficult to determine and the following is known. Dissolution of glassy polymers is always accompanied by heat liberation, the values of which can reach 60–80 J/g [64]. The exothermic heat of solution is a linear function of the glass transition temperature. The largest values of the exothermal effects are observed for glassy polymers with very high glass points such as polyarylat ($T_g = 600\,K$) [64]. Such large values are a consequence of their nonequilibrium state. These suggestions are also valid for the amorphous regions of crystalline polymers. For glassy amorphous polymers when $\Delta H_f = 0$, only $\Delta H_m + \Delta H_{ea}$ in the right hand of Eq. (9.9) remains.

The contribution ΔH_{ea} is very sensitive to the physical structure of polymers. Therefore, between the parameters determining this structure (thermal history, the method of preparation of samples, molecular weight and others) and the heat of solution a direct relationship exists. This relationship gives us the possibility for determining the latent energy of drawing of deformed samples on the basis of heats of solution. Cold drawing of crystalline polymers may be accompanied by the change of the degree of crystallinity and, therefore, the value of ΔH_f may also change proportionally.

Experimental application of the method was demonstrated by Müller [12, 65] for drawn PVC (Fig. 9.18) and Nazarenko [62] for drawn PS. For determining heats of solution any suitable calorimeter can be used [64, 66].

9.4.2 DSC-Measurements

Determination of the stored energy in plastically deformed polymers can be performed by means of DSC. The idea of using this method was first applied to plastically deformed metals and alloys by Clarebrough et al. [67] and since then is widely used for this purposes not only in metal physics but also in polymer physics. Müller [11] seems to be the first person who drew attention to the anomalous behavior of heat capacity C_p in the region of

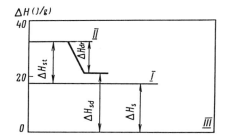

Fig. 9.18. Energy diagram illustrating the method of determination of the stored energy of orientation of a PVC sample according to solution calorimetry [12, 65]. *I* – initial solid sample state; *II* – drawn sample state; *III* – solution state. ΔH_{st} – energy stored during cold drawing; ΔH_{sd} heat of solution of drawn sample; ΔH_s – heat of solution of undrawn sample; ΔH_{dr} – heat of recovery process before solution experiment. Heats of solution were measured at 313 K in chlorobenzen

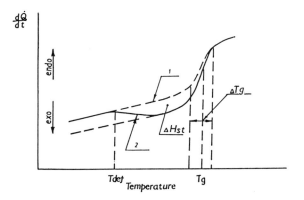

Fig. 9.19. Typical DSC trace of a plastically deformed sample of glassy polymers [62, 68]. *1* – undeformed sample; *2* – deformed sample (see text)

the glass transition temperature in drawn glassy PS compared to undrawn sample. However, these results were only qualitative. Recently DSC method is widely used for determining the latent energy of deformation of drawn polymers due to the comparison C_p–T dependencies of drawn and undrawn samples [68, 69].

Figure 9.19 illustrates the "exo-pit" observed for plastically deformed glassy polymers. As a result of a comparison of the curves for drawn and undrawn polymers, one can conclude that at approximately the temperature of drawing, DSC traces show exothermic C_p-anomalies (exo-pit) the area under the anomalies being closely related to the the latent energy of deformation. Typical experimental results for amorphous and slightly crystalline polymers are shown in Figs. 9.20 and 9.21. The samples studied were deformed with various deformation modes: simple uniaxial stretching and compression, tor-

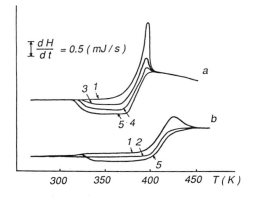

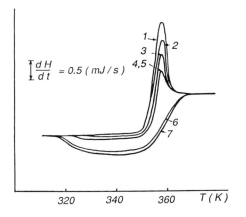

Fig. 9.20. DSC traces of PS (**a**) and copolymer of styrene with 26% of methacrylic acid (**b**) [68]. *1* – initial state; *2* – quenched state; *3, 4, 5* – after plastic deformation under compression to 2, 20 and 45%, respectively

Fig. 9.21. DSC traces of PVC samples [68]. *1* – initial undeformed sample; *2, 3, 4* – plastically deformed samples at uniaxial drawing to 2, 19, and 40%, respectively; *5* – plastically deformed sample to 40% at torsion; *6* – plastically deformed to 40% at uniaxial compression; *7* – rolled sample with plastic deformation equal to 40%

sion, cold rolling. The effects observed after these deformation modes are qualitatively the same for all samples: deformation results in a decrease of the amplitude of the peak near the C_p jump at T_g which is a very characteristic of samples annealed near T_g and the appearance of the corresponding C_p anomalies near T_g. By comparing DSC traces of the samples with different energy states one can estimate rather accurately ($\pm 5\%$) the difference in the enthalpy of these samples. As the baseline for such a comparison it is very suitable to choose the DSC trace for samples quenched directly in PSC instrument, the trace being obtained in every measurement during the second scan. This DSC trace is independent of the thermal history of the samples.

Plastic deformation and cold drawing of crystalline polymers can also be accompanied by some energy changes. The difference in the C_p–T curves for these polymers may be observed, first of all, in the temperature interval of fusion.

References

1. Joule JP (1857) Proc Roy Soc *8:* 564
2. Joule JP (1858) Phil Mag *15:* 538
3. James HM and Guth E (1943) J Chem Phys *11:* 538
4. Dart SL, Anthony RL and Guth E (1942) Ind Eng Chem *34:* 1340
5. Anisimov SP, Volodin VP, Orlovsky IYu and Fedorov YuN (1978) Solid State Physics (FTT) *20:* 77
6. Trainor A and Haward RN (1974) J Mater Sci *9:* 1243
7. Gilmor IM, Trainor A and Haward RN (1978) J Polym Sci Polym Phys Ed *16:* 1277
8. Wright OB and Phillips WA (1984) Phil Mag B *50:* 63
9. Butyagin PYu, Garanin VV and Kuznezov AP (1974) Vysokomol Soedin *A16:* 333
10. Binder G and Müller FH (1961) Kolloid Z *177:* 129
11. Müller FH (1959) Kolloid Z *165:* 96
12. Müller FH (1969) Thermodynamics of deformation. Calorimetric investigation of deformation processes. In: Rheology, Academic New York, vol 5, p 417
13. Rodriguez EL and Filisko FE (1982) J Appl Phys *53:* 6536
14. Rodriguez EL and Filisko FE (1986) Poly Eng Sci *26:* 1060
15. Müller FH and Jackel K (1952) Kolloid Z *129:* 145
16. Jackel K (1954) Kolloid Z *137:* 130
17. Andrianova GP, Kechekyan AS and Kargin VA (1971) J Polym Sci A-2 *9:* 1919
18. Brauer P and Müller FH (1954) Kolloid Z *135:* 65
19. Müller FH (1954) Kunststoffe *44:* 569
20. Müller FH (1956) Kautschuk u Gummi *9:* 197
21. Maher JW, Haward RN and Hay JN (1980) J Poly Sci Polym Phys Ed *18:* 2169
22. Schwarz G (1981) Colloid Polym Sci *259:* 149
23. Springer H, Schenk W and Hinrichsen G (1983) Colloid Polym Sci *261:* 9 (1983)
24. Döll W (1972) Kolloid Z Z Polym *250:* 1066
25. Döll W (1973) Engng Fracture Mech *5:* 259
26. Fuller KNG, Fox PG and Field JE (1975) Proc Roy Soc Lond A *341:* 537
27. Kobayashi A, Munemura M, Ohtani N and Suemasu H (1982) J Appl Polym Sci *27:* 3763
28. Egorov EA, Zhizhenkov VV, Savostin AYa and Tomashevskii EE (1975) Solid State Physics (FTT) *17:* 111
29. Tomashevskii EE, Egorov EA and Savostin AYa (1975) Int J Fracture *11:* 803
30. Egorov EA, Zhizhenkov VV, Bezladnov SN, Sokolov IA and Tomashevskii EE (1980) Vysokomol Soedin *A22:* 582 Polymer Mat
31. Treloar LRG (1975 3) The Physics of Rubber Elasticity, Clarendon Oxford
32. Allen G, Kirkham MJ, Padget J and Price C (1971) Trans Farad Soc *67:* 1278
33. Price C (1976) Proc Roy Soc Ser A *351:* 331
34. Sharda SC and Tschoegl NM (1976) Macromolecules *9:* 910
35. Wolf FP and Allen G (1975) Polymer *16:* 209
36. Boyce PH and Treloar LRG (1970) Polymer *11:* 21
37. Gent AN and Kuan TH (1973) J Polym Sci *11:* 1273
38. Mark JE (1976) J Plym Sci Macromol Revs *11:* 135
39. Anthony RL, Caston RH and Guth E (1942) J Phys Chem *46:* 826
40. Shen M (1969) Macromolecules *2:* 358
41. Engelter A and Müller FH (1958) Kolloid Z *157:* 89
42. Engelter A and Müller FH (1958) Rheol Acta *1:* 39
43. Morbitzer L, Hentze G and Bonart R (1967) Kolloid Z Z Polym *216/217:* 137
44. Foster HO and Benner RE (1965) Proceedings of the Fourth International Congress on Rheology, Interscience New York, Part 2, p. 121

45. Göritz D (1986) J Polym Sci Polym Phys Ed *24:* 1839
46. Lyon RE and Farris RJ (1986) Rev Sci Instrum *57:* 1640
47. Duvdevani IJ, Biesenberger JA and Gogos CG (1969) Polym Eng Sci *9:* 250
48. Calvet E and Prat H (1956) Microcalorimetry, Masson Paris
49. Godovsky YuK, Slonimsky GL and Alekseev VF (1969) Vysokomol Soedin *A11:* 1181 (1969)
50. Godovsky YuK (1976) Thermophysical Methods of Polymers Characterization [in Russian] Khimiya Moscow
51. Molchanov YuM and Molchanova GA (1970) Mechan Polym *N5:* 579
52. Sergeev YuA, Feinberg RZ and Michailov NV (1972) Vysokomol Soedin *A14:* 250
53. Andrianova GP, Arutyunov BA and Popov YuV (1978) J Polym Sci Polym Phys Ed *16:* 1139 (1978)
54. Miller D and Höhne GWH (1980) Thermochim Acta *40:* 137
55. Pokrovsky EM (1985) Thesis, Institute of Metallurgy Engineering, Moscow
56. Mironzov, LI, Privalko VP, Antonov AI et al (1983) Composite Polymer Materials *18:* 34
57. Price C, Evans KA and De Candia F (1973) Polymer *14:* 339
58. Göritz D and Müller FH (1974) Kolloid Z Z Polym *252:* 862
59. De Candia F, Romano G, Russo R and Vittoria V (1982) J Polym Sci Polym Phys Ed *20:* 1525
60. Allen G, Price C and Yoshimura N (1975) Trans Farad Soc *71:* 548
61. Price C, Allen G and Yoshimura N (1975) Polymer *16:* 261
62. Nazarenko SI (1988) Thesis, Institute of Chemical Physics USSR Academy of Sciences, Moscow; see also Oleinik EF (1987) Polymer J *19:* 105
63. Maron SN and Filisko FE (1972) J Macromol Sci Part B *6:* 413
64. Volynskaya AV, Godovsky YuK and Papkov VS (1979) Vysokomol Soedin *A21:* 1059
65. Stolting J and Müller FH (1970) Kolloid Z Z Polym *238:* 459; (1970) *240:* 790
66. Hemminger W and Höhne G (1984) Calorimetry (Fundamentals and Practice), Verlag Chemie Weinheim Deerfield Beach, Florida, Basel
67. Claerbrough LM, Hargreaves ME, Michell D and West GW (1952) Proc Roy Soc *A215:* 507
68. Bershtein VA and Egorov VM (1990) DSC in Physico-Chemical Inverstigations of Polmyers [in Russian], Khimiya Moscow
69. Prest WM and Roberts J (1981) Ann N Y Acad Sci *371:* 67

Subject Index

amorphous c-texture 142
amorphous polymers, heat capacity below 1 K 21
amorphous texture 223, 240
—, PET 154
amorphous regions, intrafibrillar 150
anisotropy coefficient 60
annealing, drawn crystalline polymers 153
—, high pressure 154

ballistic mode 289
biological membranes, thermomechanical approach 202
block copolymers, mesophase 198
—, microphase-separated 158
—, single crystal morphology 158
brittle fracture 249, 250
brittle-plastic fracture 251
"bundle-of-tubes" model 81

calorimetry, adiabatic 107
—, deformation 127, 175, 284
—, differential scanning 107
chain slippage and breakage 191
chains, limited extensibility 183, 185
characteristic parameter, Hookean solid 131
characteristic ratio 129
cold drawing 141, 190
—, local heating 214
—, temperature effects 212
—, thermodynamic characteristics 226, 227
cold working of metals and alloys, latent energy 232
compression, plastic deformation 231
conformational elasticity 147
conformational energy 177
conformational transitions 234
constrained junction fluctuation model 203
continuum models 7
crack growth, local thermal processes 254
crazing 235
critical stress 224

crystalline polymers, negative thermal expansion 94
—, two-phase model 36
crystallization, orientational 223, 229
—, strain-induced 183
cyclic deformation, self-heating 268

de Senarmont method 120
Debye function 5
Debye model 4
Debye temperature 5
deformation, energy balance 250
—, latent energy 237
—, micromechanics 262
degradation, thermal effect 260
degradation of macromolecules, energetic effects 263
dilatation, of an elastomer 182
dilatometry, linear 121
dislocations 233
dissolution calorimetry 292
drawn crystalline polymers 143
—, negative thermal expansion 96
dynamic transition, disorder-order type 224

Einstein function 3
Einstein model 3
Einstein temperature 3
elastic inversion, heat 179
elasticity modulus, amorphous regions 146
elastomeric blends, thermomechanical properties 201
energetic components, 164
energy stored 266
entropic components, 164
equation of state 169, 172
—, amorphous polymers 78
—, crystalline polymers 77
—, thermomechanical 128, 130, 132
—, torsion 168
extension-contraction cycle 264

fibrillization 235
filled elastomers, energy contributions 193

—, entropy contributions 193
filled rubbers 192
formation, free radicals 261
—, new groups 261
fracture, thermal behavior 249
fracture zone, temperature profile 257
free volume 29

gas calorimeters 284
gauche-isomers 29
glassy polymers, densified 241
—, plastically deformed 236
graft copolymers, energy contribution
    194
Gruneisen parameter 75
—, experimental methods for determin-
    ing 80
—, lattice γ_L 79
—, temperature dependence 80
—, thermodynamic γ_T 79
Guth–Smallwood equation 192

hard elastic, thermomechanical behavior
    244
heat capacity, conformational 31
—, conformational contribution 29
—, contribution of holes 29
—, glass transition 28
—, interchain 80
—, temperature dependence 6
—, vibrational 33
—, vibrational contribution 29
heat-conducted calorimeters 288
helix-like macromolecules, low tempera-
    ture heat capacity 15
Hess-Statton-Hearl model 148
homologous temperature 233
Hosemann-Bonart model 149
hot zone, temperature profile 257, 280

intensity factor 251
interchain effects 179
internal pressure 75
internal stresses 234, 243
intrachain energy effects 175
inversion, heat 166
—, internal energy 145, 166, 179
IR radiation 274
IR-microscope 279
isoentropic measurements 272
isomeric state theory 177
isometric measurements 282
isothermal bulk modulus 130
isothermal compressibility 166
isothermal measurements 284

Kelvin (Thompson) Equation 134
Kelvin effect 129

lattice vibrations 43
Lifshitz model 9, 15

linear thermal expansion, isoenergetic
    chains 167
linear thermal expansion coefficients, for
    crystalline polymers 139
—, glassy polymers 139
—, metallic materials 139
liquid crystalline networks 186
low temperature conductivity, influence
    of the crystallinity 53
low temperature thermal conductivity of
    amorphous solids 47

material, ideal elastic energy 130
Maxwell model 51
Maxwell model, modified 62
membrane effect 85
metastable linear defects 238
microcracks 246
microexplosions 256
microfibrils 262
model of statistical intercrystalline
    bridges 63
Mooney-Rivlin equation 171
multiblock copolymers, thermomechanic
    characteristics 196

neck formation 190, 214
neck, self-oscillated 221
—, uniform 214
—, unstable 222
necking 214
—, temperature rise 276
necking zone, softening 214
negative thermal expansivity, helical
    chains 92
—, planar zig-zag 87
normal mode spectrum 11

overstressed macromolecules, rupture 256
overstressed tie-molecules 262

PBT 155
PDMS networks, bimodal 185
PE, dispersion curves 12
—, glass transition temperature 38
—, Gruneisen parameters 81
—, temperature dependence of C_p 18
—, thermal conductivity 62
—, vibrational modes 12
—, volume thermal expansion 89
Peterlin-Prevorsek model 100, 150
phonon scattering 43
phonon velocity 43
physical aging 236
plastic deformation 146, 251, 264
—, "jump-like" behavior 224
—, ideal 225
—, temperature effects 212
—, thermodynamic characteristics 227
Poisson ratio, crystalline polymers 140
—, drawn crystalline polymers 150
—, HB microfibril 149

—, undrawn 140
polyesters, thermomechanical behavior 195
polymer glasses, anomalous heat capacity 16
polymeric crystals, thermal expansivity 84
polymorphism in deformation 155
primitive path models 203

quasi-brittle fracture 251

random copolymers, thermoelasticity 200
re-drawing 243
recrystallization, during drawing 218
reformation, rigid phase in SBS 192
reinforcement, thermomechanical measurements 192
β-relaxation 35
γ-relaxation 35
rotational isomers 29
rubber elasticity, new models 203
rupture of macromolecules 235
rupture of polymers, temperature rise 278

SBS thermoelastoplastics, glassy domains 197
SBS, single crystal 191
secondary relaxations 34
segmented polyurethanes, thermomechanical behavior 195
shear bands 239
Shen 169
skeleton approximation 14
sliplink models 203
statistical crystal bridge model 156
Stockmayer-Hecht model 14, 15, 89
stored energy 226, 231
—, mechanisms 233
—, nature 233
stress relaxation, heat 140, 144
stress softening, of filled elastomers 189
stress-strain isotherms 185
stressed chains, energetics of scission 260
stressed macromolecules, degradation 258
—, energetics 259
—, ruptures 251
—, thermodynamics 235
stretched elastomers, linear thermal expansion 95
—, negative expansion 96
structural models, thermomechanical behavior 147
sublass secondary relations 35
superdrawn polymers, thermomechanical properties 156
surface energy 249
system, ideal entropy elastic 130

Tarasov model 7
T_g-transition 36

temperature rise, at the tip of propagating cracks 252
temperature rises, local 215, 221
term conductivity, van der Waals bonds 57
theory, configurational entropy 33
thermal conductivity, aggregate model 64
—, amorphous solids 44
—, anisotropy 59
—, degree of crystallinity 49
—, dependence upon molecular weight 57
—, drawn polymers 61
—, DSC technique 117
—, molten state 57, 58
—, percolation 69
—, percolation approach 69
—, polymer composites 68
—, pressure dependence 66
—, radiation effects 65
—, steady state measurements 111
—, temperature dependence 45
—, two phase system 50
—, unsteady-state methods 114
thermal contraction 86
thermal diffusivity, temperature dependence 70
—, unsteady state methods 111
thermal expansion, anisotropy 84, 98
—, linear 84, 85
—, negative 85, 86
—, polymer composites 101
—, polymeric crystals 90
thermal expansivity, anharmonicity 75
—, layer structures 85
thermal microexplosions 261, 267
thermal pressure 75
thermal resistance, interphase 55
thermodynamics of deformation, solids 129
thermoelastic coefficients 136
thermoelastic effect 129, 134, 273
thermoelasticity, liquid crystalline networks 174
thermoelastoplastics 189
—, energy contribution 193
—, limited chain extensibility 194
—, loading-unloading cycle 190
thermomechanical effects 275
thermomechanical inversion, internal energy 133
—, internal energy in Hookean solids 131
thermomechanics, crystalline polymers 137
—, drawn amorphous polymers 141
—, elastic materials 128
—, Gaussian networks 163
—, non-Gaussian networks 170
—, undrawn glassy polymers 137
thixotropy effects 144
tie-molecules 151
Tobolsky 169

trans-isomers 29
Tschoegl et al. 172
two-dimensional compressibility 81
two-phase model 95

unperturbed dimensions of chains, temperature coefficient 165, 177

vacancies 233
Valanis-Landel equation 171
van der Waals model 172

vibrational entropy 134
vibrational modes, anharmonicities 34
vibrational spectra 3, 4
vibrations, acoustic 6
—, lattice 6
—, skeleton 6
volume dilatation 166
—, strain-induced 169, 182

waves, longitudinal and transverse 45

Advances in Polymer Science

Edited by an international board of experts

Volume 99

L. A. Errede, 3M Corporate Research Laboratories, St.Paul, MN

Molecular Interpretations of Sorption in Polymers Part 1

1991. VIII, 115 pp. 64 figs. 20 tabs. Hardcover ISBN 3-540-53497-0

This book reviews recent advances in polymer swelling resulting from the use of novel microporous composite films. It offers a new approach to understanding sorption processes in polymer-liquid systems based on the molecular structures of the sorbed molecules and the repeat unit of the sorbent polymer. It is shown how the adsorption parameters obtained in these studies relate meaningfully with the Flory-Huggins interaction parameters. This implies that these adsorption parameters have relevance not only for swelling and drying of polymers, but also for other phenomena in which molecular sorption plays an important role, such as in chromatography and in membrane permeation.

The series **Advances in Polymer Science** will no longer be distributed in Eastern and Central European countries (WP) by Akademie-Verlag. Orders should be sent directly to Springer-Verlag.

Springer-Verlag

Berlin
Heidelberg
New York
London
Paris
Tokyo
Hong Kong
Barcelona
Budapest

N. W. Tschoegl, Pasadena, CA, USA

The Phenomenological Theory of Linear Viscoelastic Behavior

An Introduction

Cover illustrations: C. A. Tschoegl

1989. XXV, 769 pp. 227 figs. 25 tabs.
Hardcover ISBN 3-540-19173-9

Contents: Introductory Concepts. – Linear Viscoelastic
Response. – Representation of Linear Viscoelastic Behavior
by Series-Parallel Models. – Representation of Linear
Viscoelastic Behavior by Spectral Response Functions. –
Representation of Linear Viscoelastic Behavior by Ladder
Models. – Representation of Linear Viscoelastic Behavior
by Mathematical Models. – Response to Non-Standard
Excitations. – Interconversion of the Linear Viscoelastic
Functions. – Energy Storage and Dissipation in a Linear
Viscoelastic Material. – The Model-
ling of Multimodal Distributions of
Respondance Times. – Linear Visco-
elastic Behavior in Different Modes
of Deformation. – Appendix: Trans-
formation Calculus. – Solutions to
Problems. – Epilogue. – Notes on
Quotation. – List of Symbols. –
Author Index. – Subject Index.

Springer-Verlag
Berlin
Heidelberg
New York
London
Paris
Tokyo
Hong Kong
Barcelona
Budapest